MW01436118

Intelligent Vision Systems for Industry

Springer
*London
Berlin
Heidelberg
New York
Barcelona
Budapest
Hong Kong
Milan
Paris
Santa Clara
Singapore
Tokyo*

Bruce G. Batchelor and Paul F. Whelan

Intelligent Vision Systems for Industry

Springer

Professor Bruce G. Batchelor
Department of Computer Science
University of Wales, P.O. Box 916, Cardiff CF2 3XF, UK

Dr Paul F. Whelan
School of Electronic Engineering
Dublin City University, Glasnevin, Dublin 9, Ireland

ISBN 3-540-19969-1 Springer-Verlag Berlin Heidelberg New York

British Library Cataloguing in Publication Data
Batchelor, Bruce G. (Bruce Godfrey), 1943-
  Intelligent vision systems for industry
  1.Computer vision - Industrial applications
  I.Title II.Whelan, Paul
  670.4'272'637
ISBN 3540199691

Library of Congress Cataloging-in-Publication Data
Batchelor, Bruce G.
  Intelligent vision systems for industry / B.G. Batchelor and P.F. Whelan.
    p. cm.
  Includes bibliographical references and index.
  ISBN 3-540-19969-1 (hardback : alk. paper)
  1. Computer vision. 2. Image processing. I. Whelan, Paul F., 1963- . II. Title.
TA 634.B37 1997
670.42'7—dc21   96-37943

Apart from any fair dealing for the purposes of research or private study, or criticism or review, as permitted under the Copyright, Designs and Patents Act 1988, this publication may only be reproduced, stored or transmitted, in any form or by any means, with the prior permission in writing of the publishers, or in the case of reprographic reproduction in accordance with the terms of licences issued by the Copyright Licensing Agency. Enquiries concerning reproduction outside those terms should be sent to the publishers.

© Springer-Verlag London Limited 1997
Printed in Great Britain

The use of registered names, trademarks etc. in this publication does not imply, even in the absence of a specific statement, that such names are exempt from the relevant laws and regulations and therefore free for the general use.

The publisher makes no representation, express or implied, with regard to the accuracy of the information contained in this book and cannot accept any legal responsibility or liability for any errors or omissions that may be made.

Typesetting: Camera ready by author
Printed and bound by Cambridge University Press, Cambridge, England
69/3830-543210 Printed on acid-free paper

To Caroline, Brian and Phyllis.

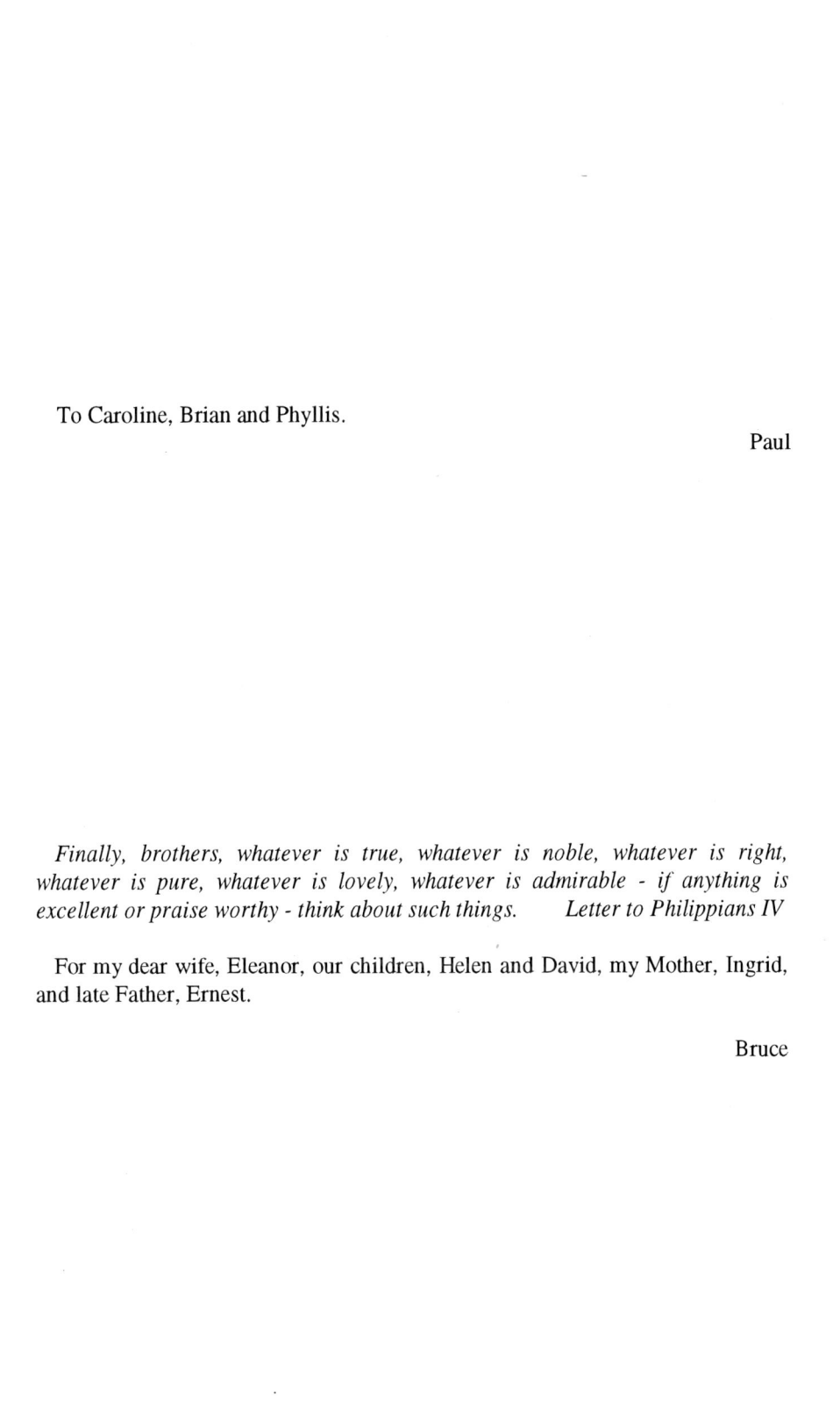

Paul

Finally, brothers, whatever is true, whatever is noble, whatever is right, whatever is pure, whatever is lovely, whatever is admirable - if anything is excellent or praise worthy - think about such things.   *Letter to Philippians IV*

For my dear wife, Eleanor, our children, Helen and David, my Mother, Ingrid, and late Father, Ernest.

Bruce

Preface

During the period 1970 - 1990, Japan taught Europe and America the importance of quality in manufactured goods. The West learned the hard way: markets were quickly lost to companies whose names were hitherto unknown. Many long established and well respected Western companies were unable to meet the challenge and consequently failed to survive. Those that did were often faced with difficult years, as their share of the market shrank. Most companies in Europe and America have largely come to terms with this and now realise that quality has a vital role in establishing and maintaining customer loyalty. In the present climate of opinion, any technology which improves or simply guarantees product quality is welcome.

Machine vision is a relatively new technology, which has much to offer manufacturing industry in improving product quality and safety, as well as enhancing process efficiency and operational safety. Machine vision owes its rising popularity to one major factor: optical sensing is inherently clean, safe (because it a non-contacting technology) and very versatile. It is possible to do certain things using vision (both human and machine) that no other known sensing method can achieve - imagine trying to sense stains, rust or surface corrosion by any other means.

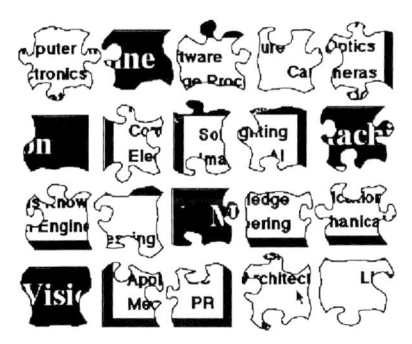

Designing a machine vision system is like assembling a jigsaw.

Among other component technologies machine vision involves the digitisation, manipulation and analysis of images, usually within a computer, a subject which is also covered by the terms *image processing* and *computer vision*. However, we must emphasise that *machine vision, computer vision* and *image processing* are not synonymous. None is a subset of either of the others. *Computer vision* is a branch of *Computer Science*, while machine vision is an area of specialisation within *Systems Engineering*. Notice, in particular, the use of the words "Science" and "Engineering" here. Machine vision does not necessarily imply the use of a computer; specialised image processing hardware is often used to obtain higher processing speeds than a conventional computer can achieve.

Machine vision system for industry first received serious attention in the mid-1970s, although the proposal that a video system be used for industrial inspection was first made in the 1930s. Throughout the early 1980s, the subject developed slowly, with a steady contribution being made by the academic research community, but with only limited industrial interest being shown. It seemed in the mid-1980s that there would be a major boost to progress, with serious interest being shown in vision systems by the major American automobile manufacturers. Then, came a period of serious disillusionment in the USA, with a large number of small vision companies failing to survive. In the late 1980s and early 1990s, interest has grown markedly, due largely to significant progress being made in making fast, dedicated image digitisation and processing hardware. In the mid-1990s, the role of the general purpose processor is being revised, with the modern RISC processors offering high processing speed on a standard computing platform. Throughout this period, academic workers have been steadily proving feasibility in a very wide range of products, representing all of the major branches of manufacturing industry.

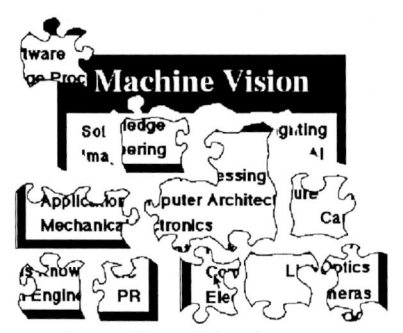

Soon after starting work, machine vision is seen as a confusing jumble of disconnected ideas.

Industrial image processing systems, which necessarily form part of a vision system, have developed very considerably in the last decade. In addition, there have been major advances in other component technologies: image sensors, specialised lighting units, lenses and advisor (CAD) programs, which guide a vision engineer through the initial stages of the design process. However, *systems integration* remains the key factor for the successful design and operation of a machine vision system.

Having separated the subjects of machine vision and image processing, our first task in this book is to introduce the reader to the basic concepts of image processing, as they apply to our subject. (Chapter 2) There are numerous techniques for manipulating images that are either not used, at all, or are used very infrequently in machine vision. Wherever there are problems of computational speed, machine vision systems engineers will either seek another solution, or avoid the problem entirely. Standard image processing techniques are able to achieve some remarkable results but they could not be described as being intelligent. By adding a certain level of intelligence, through the integration of image processing software and the AI language Prolog, we are able to do certain things that would otherwise be impossible. For example, analysing the image of a standard ("analogue") clock, in order to tell the time is one such task that could not be solved using "traditional" image processing methods working on their own. We shall, therefore, devote much of the discussion in this book to explaining how intelligence can be provided to image processing, or alternatively, how Artificial Intelligence can be given "eyes". All of this is done with one goal in mind: to improve the prospects for installing machine vision systems in factories.

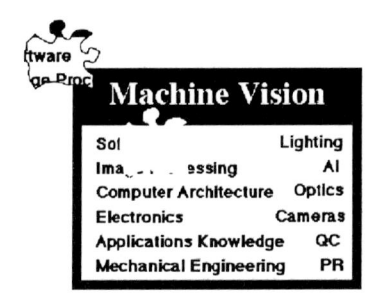

Machine Vision

Sof	Lighting
Ima... essing	AI
Computer Architecture	Optics
Electronics	Cameras
Applications Knowledge	QC
Mechanical Engineering	PR

Eventually, the pieces fit together. However, if one piece is missing, the result is imperfect; system integration is incomplete.

There is a serious bottleneck in the design of machine vision systems: a high level of skilled man-power is needed to achieve an effective design. To illustrate the problem, consider the case of just one organisation, which has over 60000 products. That company operates a policy which tries to maintain at least 25% of its sales on products that are not more than 5 years old. Simple arithmetic shows that over 10 new product are being introduced by that one company alone, every working day. If we were to use a machine vision system on only 1% of those new product lines, we would need to design, build, install and test a new system once every 2 weeks. At the moment, the design process typically takes several months and there are simply not enough machine vision engineers to provide that kind of level of support, even for that one company. We desperately need more well-educated machine vision systems engineers. We also need improved design tools. By claiming that machine vision is a flexible technology, without having the man-power available to fulfil that boast is simply foolish. Such "overselling" of the virtues of machine vision technology was largely responsible for the collapse in credibility and confidence in the mid-1980s, to which we referred earlier. We need both improved educational material and better engineering tools, if we are to meet the challenge that this subject imposes upon us. (Chapter 3) Working in such a flexible and potentially beneficial technology carries responsibilities, because it is upon our shoulders that its future development and exploitation lies.

The user interface is all important, since this will either make a system acceptable or damn it to certain failure. For this reason, we shall discuss the prospects for using multi-media interfaces, including, hypertext, speech synthesis, speech recognition and natural language understanding. In Chapter 4, we also discuss the use of multi-camera and multi-processor systems, since it is clear that considerable advantage can be obtained from the use of systems that are able to communicate information about what they see to each other.

Machine Vision	
Software	Lighting
Image Processing	AI
Computer Architecture	Optics
Electronics	Cameras
Applications Knowledge	QC
Mechanical Engineering	PR

System integration is complete; all elements are now in place. The result is perfect (machine) vision.

Industrial machine vision systems would be virtually useless if it were not possible to control external devices, such as lamps, cameras, lenses, robots, etc. A good deal of attention will therefore be paid to this topic. (Chapter 5) We devote a whole chapter (Chapter 6) to the task of recognising coloured objects. The approach we take here is one which has not always found favour with Colour Scientists - but it works!

We conclude by discussing several case studies, which may seem to concentrate on unimportant tasks, such as recognising playing cards, telling the time, etc. However, all of the applications that we discuss in Chapter 7 reflect industrially important tasks, in a way which allows us to write freely about the technical issues, without violating commercial confidence.

It is customary in many areas of public writing to use so called gender-neutral phrases, such as *"he / she"*, *"his / her"* , *"s/he"* etc. We regard these as being both clumsy and counter-productive. In this book, we use the words, *"he"* and *"him"* in the traditional way, to include both sexes, without claiming precedence for either. This is done to improve clarity and to avoid placing women after men, as *"he / she"* does.

While many of the ideas the ideas outlined in this book can be implemented on a range of computers, an integrated software package, called *PIP* (*Prolog Image Processing*), has been designed specifically for this and runs on Macintosh™ computers. Readers who wish to gain access to PIP should contact Bruce Batchelor or Andrew Jones at the University of Wales, Cardiff. An interactive image processing package for the Windows® environment, without Prolog, has also been developed. This is called *MvT* (*Machine Vision Tutorial*) and is available from Paul Whelan at Dublin City University. For current information on our research and the status of the *PIP* and *MvT* software packages, please consult our WWW sites.

Bruce G. Batchelor
Department of Computer Science
PO Box 916, University of Wales
Cardiff, CF2 3XF, Wales
UK

bruce@cs.cf.ac.uk
http://www.cs.cf.ac.uk/

Paul F. Whelan
School of Electronic Engineering
Dublin City University
Glasnevin, Dublin 9
Ireland

whelanp@eeng.dcu.ie
http://www.eeng.dcu.ie/~whelanp/home

Acknowledgements

It is our pleasure to acknowledge and thank all of the many people who have helped us to formulate our ideas through many fruitful discussions and who have given us so much encouragement. These include Frederick Waltz, Mike Snyder, John Chan, Nicky Johns, Derek Molloy, Don Braggins, Robert Churchhouse and Nick Fiddian.

We also wish to thank the following for their specific help and encouragement:

- Clive Spenser, Logic Programming Associates Ltd., London, England, UK for kindly supplying several copies of MacProlog for use in our software development program.
- Andrew Jones and Ralf Hack for their contribution to Appendix D.
- Andrew Jones, Ralf Hack, Steve Palmer and Eric Griffiths for their contribution to the development of the PIP software package.
- Michael Daley for developing the MMB interfacing hardware.
- Stephen Cooper, formally of Uppsala University, for his design for the interface between MacProlog and Hypercard.
- Tommy McGowan for his work in developing the Prolog speech synthesis interface.
- Ken McClannon for his work on machine vision in process control.
- John Miller and his colleagues for porting the *Lighting Advisor* to the DOS®/Windows® platforms.
- Mark Graves for his *Darkroom* program which acted as the inspiration for the Machine Vision Tutorial (MvT) software package.
- Prof. Charles McCorkell, Head of the School of Electronic Engineering, Dublin City University, for his support of the *Vision Systems Laboratory* at Dublin City University and for enabling Bruce Batchelor to visit Dublin on a regular basis.
- The Commission of the European Union (Agro-Industry Research Programme) for their financial support, which helped to develop the Prolog Image Processing (PIP) software (Grant no AIR2-CT93-1054).
- Eolas (Forbairt) and the British Council, whose financial support enabled us to establish the initial link between the University of Wales and Dublin City University.

Many of the ideas that are outlined in this book were formed during our time working with industry. Therefore we would like to acknowledge all our industrial colleagues and thank them for their permission to use some of the material cited in this book. Special thanks are due to Nicholas Pinfield of Springer-Verlag for his commitment to this book.

Apple, Macintosh and the Apple logo are registered trademarks of Apple Computer, Inc.
3M and the 3M logo are registered trademarks of the 3M Company.
VISA and the VISA logo are registered trademarks of VISA International.
DOS and Windows are registered trademarks of Microsoft Corporation.

Hypercard is a product of Apple Computer, Inc.
MacProlog is a product of Logic Programming Associates Ltd.
Intelligent Camera is a product of Image Inspection Ltd.
VCS is a product of Vision Dynamics Ltd.

Table of Contents

1 Basic Concepts 1
  1.1 Industrial Vision Systems...1
      1.1.1 Justification..3
      1.1.2 Limitations of Present Systems 4
      1.1.3 Flexible Manufacturing Systems 6
      1.1.4 Process Control...7
  1.2 Systems Engineering 9
      1.2.1 Importance of Context..9
      1.2.2 Industrial Examples 10
  1.3 Intelligent Vision..13
      1.3.1 Heuristics and Algorithms..15
      1.3.2 Artificial Intelligence (AI) Languages 17
  1.4 Book Outline 17

2 Basic Machine Vision Techniques..19
  2.1 Representations of Images..19
  2.2 Elementary Image Processing Functions 21
      2.2.1 Monadic, Point-by-point Operators..22
      2.2.2 Dyadic, Point-by-point Operators 24
      2.2.3 Local Operators 25
      2.2.4 Linear Local Operators..25
      2.2.5 Non-linear Local Operators...28
      2.2.6 N-tuple Operators 32
      2.2.7 Edge Effects...32
      2.2.8 Intensity Histogram 33
  2.3 Binary Images..35
      2.3.1 Measurements on Binary Images 41
      2.3.2 Shape Descriptors 43
  2.4 Binary Mathematical Morphology 43
      2.4.1 Opening and Closing Operations...46
      2.4.2 Structuring Element Decomposition 47
  2.5 Grey Scale Morphology 49
  2.6 Global Image Transforms..51
      2.6.1 Hough Transform 51
      2.6.2 Two-dimensional Discrete Fourier Transform.........................53
  2.7 Texture Analysis..56
      2.7.1 Statistical Approaches 56
      2.7.2 Co-occurrence Matrix Approach 57
      2.7.3 Structural Approaches 59
      2.7.4 Morphological Texture Analysis 60

2.8	Implementation Considerations	60
2.8.1	Morphological System Implementation	61
2.9	Commercial Devices	61
2.9.1	Plug-in Boards: Frame-stores	62
2.9.2	Plug-in Boards: Dedicated Function	63
2.9.3	Self-contained Systems	63
2.9.4	Turn-key Systems	64
2.9.5	Software	64
2.10	Further Remarks	64

3 Intelligent Image Processing 66

3.1	Interactive Image Processing	66
3.1.1	Modus Operandi	67
3.1.2	Prototyping Inspection Systems	67
3.1.3	Building Simple Programs	69
3.1.4	Interaction and Prolog	70
3.2	Introducing Prolog+	71
3.3	Review of Prolog	73
3.3.1	Sample Program	76
3.3.2	Sample Queries	78
3.4	The Nature of Prolog+	79
3.5	Prolog+ Programs	81
3.5.1	Recognising Bakewell Tarts	81
3.5.2	Recognising Printed Letters	83
3.5.3	Identifying Table Cutlery	84
3.5.4	Analysing all Visible Objects	87
3.5.5	Recognising a Table Place Setting	87
3.6	Abstract Concepts in Prolog+	90
3.6.1	Describing a Simple Package	90
3.6.2	Abstract Spatial Relationships	91
3.6.3	Geometric Figures	93
3.7	Implementation of Prolog+	96
3.7.1	The # Operator	96
3.8	Comments	99

4 Enhanced Intelligent Systems 101

4.1	Prolog+ Environment: A Tool-box for Machine Vision	102
4.1.1	Defining New Predicate Names	102
4.1.2	Default Values for Arguments	103
4.1.3	Useful Operators	103
4.1.4	Program Library	106
4.1.5	Auto-start	106
4.1.6	Interactive Mode	108
4.1.7	User Extendible Pull-down Menus	108
	Mechanism for Extending Menus	109

4.1.8 Command Keys	113
4.1.9 Graphical Display of a Robot Work Cell	115
4.1.10 Speech Synthesis and Recorded Speech	116
4.1.11 On-line HELP	117
4.1.12 Cursor	118
4.1.13 Automatic Script Generation and Optimisation	120
4.1.14 Linking to Other Programs	120
Hypercard Controller for a Flexible Inspection Cell	122
4.2 Understanding Simple Spoken Instructions	124
4.2.1 Speech Recognition	124
4.2.2 Natural Language Understanding	127
4.2.3 Automatically Building a Pull-down Menu	127
4.2.4 Understanding NL Commands for an (X,Y,θ)-table	129
4.2.5 Sample Sentences	132
4.2.6 Interpreting the Parser Output	132
4.2.7 Review	133
4.3 Aids for Designing Vision Systems	134
4.3.1 Lighting Advisor	135
Stack Structure	136
Search Mechanisms	136
Remarks About the Lighting Advisor	141
4.3.2 Other Design Aids for Machine Vision	143
4.4 Multi-camera Systems	144
4.4.1 Multiplexed-video Systems	144
4.4.2 Networked Vision Systems	147
4.4.3 Master-Slave System Organisation	150
4.4.4 Remote Queries	152
Interactive Operation of the Remote Process	153
4.4.5 Blackboard	154
Master and Slave Program Elements	155
4.4.6 Controlling the Master-Slave System	155
Starting the System	156
Stopping a Slave	156
Passing a Message to the Slave	156
Receiving Data from a Slave	157
Slave Program	157
Blackboard (Snapshot of Database, Changing Constantly)	157
4.4.7 Crash Recovery	158
Programming the Slave from the Master	158
4.5 Comments	158
5 Controlling External Devices	**160**
5.1 Devices and Signals	160
5.2 Protocols and Signals	161
5.2.1 Interfacing to Commercial Systems	162

XVIII

5.3 Programmable Logic Controller 166
5.4 General Purpose Interface Unit 169
    5.4.1 Motivation for the Design 171
    5.4.2 Hardware Organisation 172
    5.4.3 Programs 173
    5.4.4 Digression on Lighting 173
    5.4.5 Languages for Robotics 176
5.5 Flexible Inspection Cell, Design Issues 177
    5.5.1 Lighting Arrangement 177
    5.5.2 Mechanical Handling 178
    5.5.3 Cameras and Lenses 179
    5.5.4 MMB-Host Interface Protocol 180
    5.5.5 Additional Remarks 181
    5.5.6 HyperCard Control Software for the FIC 183
5.6 Prolog+ Predicates for Device Control 183
5.7 System Calibration 185
    5.7.1 FIC Calibration Procedure (Overhead Camera) 186
    5.7.2 Calibration, SCARA and Gantry Robots (Overhead Camera) 190
    5.7.3 Calibration Procedure (Overhead Narrow-view Camera) 191
    5.7.4 Calibration Procedure (Side Camera) 193
5.8 Picking up a Randomly Placed Object (Overhead Camera) 194
    5.8.1 Program 197
5.9 Grippers 198
    5.9.1 Suction Gripper 198
    5.9.2 Magnetic Gripper 198
    5.9.3 Multi-Finger Gripper 199
    5.9.4 Further Remarks 200
5.10 Summary 201

6 Colour Image Recognition **206**
6.1 Introduction 206
6.2 Applications of Coarse Colour Discrimination 207
6.3 Why is a Banana Yellow? 209
6.4 Machines for Colour Discrimination 213
    6.4.1 Optical Filters 213
    6.4.2 Colour Cameras 215
    6.4.3 Light Sources for Colour Vision 216
    6.4.4 Colour Standards 218
6.5 Ways of Thinking about Colour 219
    6.5.1 Opponent Process Representation of Colour 220
    6.5.2 YIQ Colour Representation 220
    6.5.3 HSI, Hue Saturation and Intensity 221
    6.5.4 RGB Colour Space: Colour Triangle 221
    6.5.5 1-Dimensional Histograms of RGB Colour Separations 224
    6.5.6 2-Dimensional Scattergrams 224

6.5.7 Colour Scattergrams	226
6.6 Programmable Colour Filter (PCF)	227
6.6.1 Implementation of the PCF	228
6.6.2 Programming the PCF	229
6.6.3 Recognising a Single Colour	235
6.6.4 Noise Effects	235
6.6.5 Recognising Multiple Colours	236
6.6.6 Pseudo-Colour Display for the PCF	237
6.6.7 Recent Teaching of the PCF Dominates	238
6.6.8 Prolog+ Software for Operating the PCF	239
Plot Colour Scattergram	241
Draw Colour Triangle Outline	242
Clear LUT	242
Store Current LUT	242
Reload Stored PCF	242
Reverting to Monochrome Operation	242
6.6.9 Programming the PCF using the Colour Scattergram	242
6.6.10 Programming the PCF by Image Processing	243
6.6.11 "Hue" PCF	244
6.6.12 Analysing Output of the Hue PCF	246
6.6.13 "Segmented" PCF	248
6.6.14 Measuring Colour Similarity and Saturation	248
6.6.15 Detecting Local Colour Changes	249
6.6.16 Colour Generalisation	250
6.7 Colour Recognition in Prolog+ Programs	252
6.7.1 Counting Coloured Objects	252
6.7.2 Recognising a Polychromatic Logo, Program 1	254
6.7.3 Recognising a Polychromatic Logo, Program 2	256
6.7.4 Recognising a Polychromatic Logo, Program 3	257
6.7.5 Multiple Exemplar Approach to Recognition	258
6.7.6 Learning Proportions of Colours in a Scene	260
6.7.7 Superior Program for Learning Colour Proportions	262
6.7.8 Teaching the PCF by Showing	263
6.7.9 Template Matching of Colour Images	266
6.7.10 Using Colour for Object Orientation	269
6.7.11 Approximating an Image by a Set of Overlapping Discs	271
6.7.12 Interpreting Resistor and Capacitor Colour Codes	273
6.8 Discussion and Conclusions	275
7 Applications of Intelligent Vision	**295**
7.1 Recognition of Printed Patterns	295
7.1.1 Non-picture Playing Cards	295
7.1.2 "Stars"	296
7.1.3 "Smiley Faces"	297
7.1.4 Alphanumeric Characters	298

Program	299
Comments	301
Logical and Analogue Shape Measurements	302
7.2 Manipulation of Planar Objects	303
7.2.1 Assumptions	303
7.2.2 Significance	304
7.2.3 Simple Shape Measurements	304
7.2.4 Learning and Recognition	306
7.2.5 Program Listing	308
7.2.6 Sample Output of Recognition Phase	310
7.3 Packing and Depletion	311
7.3.1 Geometric Packer Implementation	312
7.3.2 Heuristic Packing Techniques	313
Blob Packing	314
Polygon Packing	316
7.3.3 Performance Measures	319
Predicates	319
7.3.4 Robot Gripper Considerations	321
7.3.5 Packing Scenes with Defective Regions	322
7.3.6 Discussion	323
7.4 Handedness of Mirror-Image Components	323
7.4.1 Handedness and Chirality	323
Relating Chirality and Handedness	324
7.4.2 Concavity Trees	326
Formal Definition	328
Generating Concavity Trees	329
Sample Concavity Trees	331
Canonical Form of Concavity Trees	334
Program to find Chirality	336
7.4.3 Properties of Concavity Trees	336
Instability	338
7.4.4 Simpler Tests for Chirality	339
Second Program	340
Third Program	341
Fourth Program	341
Fifth Program	342
7.5 Telling the Time	343
7.5.1 Significance	343
7.5.2 Simplifying Assumptions	344
7.5.3 Lighting	344
7.5.4 First Program	345
7.5.5 Other Methods	347
7.5.6 Concluding Remarks	348
7.6 Food and Agricultural Products	349
7.6.1 Objective	349

  7.6.2 Industrial Relevance 349
  7.6.3 Product Shape, Two-dimensions 351
    Image Acquisition 352
    Rectangular and Circular Biscuits 352
    Slices of Bread..355
    Locating the Base and Determining Orientation 356
    Locating Straight Sides...357
    Measuring Overspill 358
    Radius of Curvature of Top Edge 358
  7.6.4 Analysing the 3D Structure of an Uncut Loaf 360

8 Concluding Remarks..380

References 383

A Proverbs, Opinions and Folklore 390

B Factors to be Considered when Designing a Vision System 401

C General Reference Material 404

D PIP - Software Implementation of Prolog+..414

E Prolog+ and PIP Commands 425

Glossary of Terms 434

Index of Predicates, Operators and Grammar Rules 447

Index 451

1

Basic Concepts

As a mature research topic, machine vision dates back to the mid-1960s. Early work at a range of institutions, including the National Physical Laboratory (UK), SIRA (UK), SRI, MIT and Edinburgh University, demonstrated the potential of machine vision in inspection, robotic control and automated assembly. Machine vision is an umbrella term used to describe many different types of vision systems, but in general, machine vision systems are used in the automated processing, analysis and understanding of images in an industrial environment. A more formal definition is given as follows:

> *"The use of devices for optical, non-contact sensing to automatically receive and interpret an image of a real scene in order to obtain information and/or control machines or processes."* [AVA-85]

Machine vision systems should not necessarily be modelled on, or attempt to emulate human vision [HOC-87]. Whereas the analysis of human vision is useful to those working in perception psychology and computer vision, it is not as relevant to vision engineers trying to solve industrial problems. This does not mean that researchers should abandon the goal of trying to develop human-like vision systems. As well as the obvious results of such research, the pursuit of such goals may result in some useful techniques that can be applied in a more practical context. Human analogies, while useful stimulants for ideas, should not be followed dogmatically [LEE-89]. The danger in relying on such human driven approaches to the development of industrial vision systems is that simpler, and perhaps more elegant, solutions may be overlooked.

1.1 Industrial Vision Systems

The design of industrial vision systems, see Figure 1.1, requires a broad spectrum of techniques and disciplines [BAT-85]. These include electronic engineering (hardware and software design), engineering mathematics, physics (optics and lighting) and mechanical engineering (since industrial vision systems deal with a mainly mechanical world). Detailed descriptions of the techniques and algorithms involved in the analysis, processing and interpretation of digital images can be found in a growing number of text books that cover the field of machine vision (see Appendix C). A summary of the techniques and issues involved in the design of industrial vision systems can be found in a collection of papers on industrial machine vision systems collated by Batchelor and Whelan [BAT-94].

However, many industrial vision systems continue to be designed from a purely software engineering perspective, without consideration for any of the other system disciplines. While it is acknowledged that the software engineering task in machine vision is a critical one, the other system elements are neglected at our peril. No single discipline should be emphasised at the expense of the others. Lately, a number of researchers [HAR-92, PAV-92] have argued for the design of vision systems to be firmly placed back into a systems engineering framework. This arises from the belief that an inadequate amount of vision research deals with the genuine design and systems problems involved in the implementation of industrial vision systems. [SIM-81]

One of the reasons for the current growth of machine vision systems in manufacturing is the falling cost of computing power. This has led to a spread in the technology and has enabled the development of cheaper machine vision systems. This, in turn, has enabled medium-sized manufacturing companies to consider the option of using machine vision to implement their inspection tasks. To a lesser extent, the availability of a well educated work-force, a small proportion of which has an awareness of machine vision, has also aided the growth, and acceptance, of industrial vision systems.

The main reason, however, for this growth is strategic. That is the realisation within many industries that machine vision is an integral component of a long term automation development process, especially when one considers the importance of quality in manufacturing. This, combined with the legal liabilities involved in the production and sale of defective products, highlights the strategic case for the use of machine vision in automated inspection. A similar argument applies to the application of vision to robotics and automated assembly.

The main application areas for industrial vision systems occur in automated inspection and measurement and, to a lesser extent, robotic vision. Automated visual inspection and measurement systems have, in the past, tended to develop faster. In fact, quality control related applications such as inspection, gauging and recognition, currently account for well over half of the machine vision market. This has been mainly due to the lower cost and the ease of retrofitting such inspection systems onto existing production lines, compared to the large capital investment involved in developing a completely new robotic work cell and the extra uncertainty and risks involved in integrating two new and complex technologies.

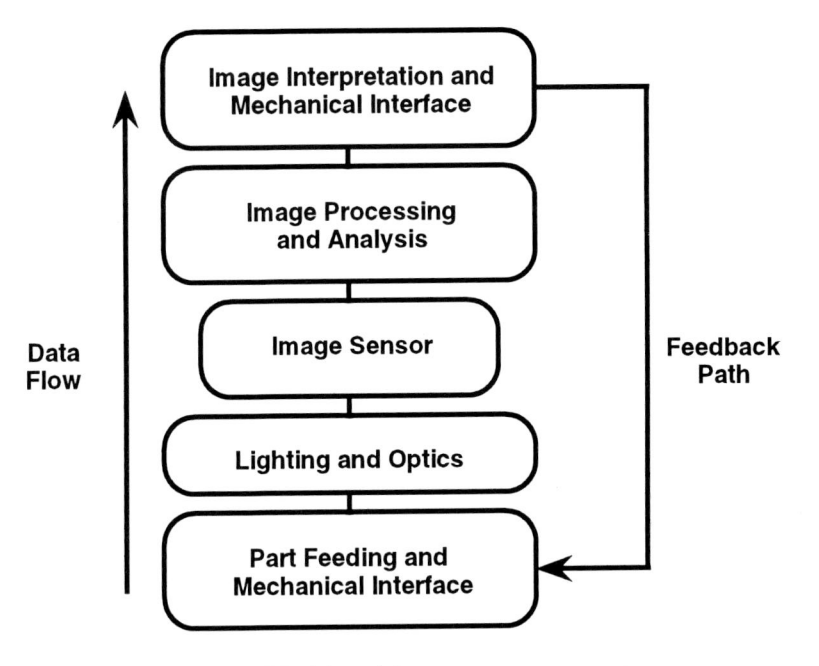

```
                    ┌─────────────────────────┐
                    │ Image Interpretation and│──┐
                    │   Mechanical Interface  │  │
                    └─────────────────────────┘  │
                    ┌─────────────────────────┐  │
                    │    Image Processing     │  │
                    │      and Analysis       │  │
                    └─────────────────────────┘  │
   Data             ┌─────────────────────────┐  │  Feedback
   Flow             │      Image Sensor       │  │    Path
                    └─────────────────────────┘  │
                    ┌─────────────────────────┐  │
                    │   Lighting and Optics   │  │
                    └─────────────────────────┘  │
                    ┌─────────────────────────┐  │
                    │    Part Feeding and     │◄─┘
                    │   Mechanical Interface  │
                    └─────────────────────────┘
```

Figure 1.1 Machine vision system components.

1.1.1 Justification

Machine vision is maturing as a technology as more engineers are entering the field and more companies are availing of its benefits. Many others, however, are hesitant or unwilling to commit themselves to using vision, because they fear the capital, development, installation and maintenance costs involved. These reservations are understandable if they try to justify the investment in terms of the primary financial measurements: return on investment, return on capital employed and pay-back periods. There are, however, many tangible benefits that can be used to justify the investment, such as improved product quality and safety, increased productivity, improved operational safety and reduced waste. The subsequent reduced warranty and reshipment costs, increased accuracy and repeatability, and lower error rate compared to manual inspection are all significant benefits.

For the majority of machine vision applications the cost of the vision system is small, relative to the total cost (and overall technology content) of automating a new production line [KRU-81]. It is vital that the installation of a machine vision system does not hinder the overall operation of the production line. Introducing a machine vision system into a production process, without fully considering all the implications will result in false expectations of the system's capabilities [HOL-84]. (See Appendix A, for a light-hearted look at the opinions and folklore that surround machine vision, and Appendix B, for a list of some of the factors involved in the design and installation of a vision system.) Some of the key questions that must be considered by a manufacturer prior to the commissioning of a new vision system, are given below (see [HOL-84] for a more detailed discussion of these issues).

- Will the inclusion of the machine vision system affect the production speed ?
- Will the manufacturing process have to be modified to accommodate the introduction of the vision system ?
- Will the production line have to be retrofitted with the automated vision system, or does the vision integrator have total control over the inspection environment ?
- Will the vision system require custom plant, process and/or environment changes ?
- As the production demands change, can the vision system be easily reconfigured ?
- How often will the vision system need to be serviced and can this be accommodated by the overall line service schedule ?

Machine vision inspection systems now appear in every major industrial sector, areas such as electronics, automotive, medical, food, and manufacturing industries. (See [BAT-94, CHI-82, CHI-88 WAL-88] for a more complete discussion of machine vision applications.) Such applications still tend to use automated visual inspection as open-loop systems. That is they allow manufacturers to inspect every single product, without suffering a loss in product throughput, but without having a direct affect on the processing of the product itself. As manufacturing technology becomes more complex, there is a growing requirement to integrate the inspection process more closely with the overall manufacturing process [MCC-92, MCC-93]. This moves the application of automated inspection from a quality control to a process control role, that is, from defect detection to defect prevention.

1.1.2 Limitations of Present Systems

Unfortunately, machine vision has had a rather chequered background. In the past, customers have had unrealistic expectations, often fuelled by the vision industry. Over the last two decades, some vision integrators have unsuccessfully stretched the use of vision systems, to the extent that certain industries have had their 'fingers burnt' after receiving false and unrealistic promises and disappointing results. However, it must be emphasised that these examples remain in the minority. Difficulties were often compounded by the fact that many end users did not know how to assess the performance of vision systems [RUM-89].

It may seem obvious to say that one of the key steps in any automated inspection application is to know exactly what you are inspecting [FRE-88]. Unfortunately, vision systems will often be applied to products and tasks that are outside its original specification, without any appreciation of the different visual characteristics of the new application. Therefore, it is important for vision system designers to outline the strengths, and more importantly, the weaknesses of their vision systems from the beginning of the design process.

While the use of machine vision systems in industry has grown in the last few years, and continues to be seen as an area of automation with enormous potential, it still has a long way to go before it is universally accepted as a standard automation tool. Pavlidis [PAV-92] has identified some of the reasons for this slow growth and these are summarised below:

- Absence of systematic testing and experimentation. This suggests that machine vision lacks one of the key features of engineering.
- Researchers are not facing up to how difficult the problem is.
- No accepted sub-goals. There is a tendency to adopt 'all or nothing' research strategies.

Machine vision systems are not perfect tools and researchers and engineers must be aware of the realities of a given application, as well as the ultimate aim of the inspection and/or assembly task. For example, the application of vision to automated assembly can be impressive to watch, but often deceptive. If one of the pieces to be assembled is rotated or moved slightly, then the system may not be able to cope with this change in its working environment [LEE-89]. However, if the constraints of the system, such as its inability to cope with such environmental changes, are made clear, then the system can serve a useful purpose.

Haralick [HAR-92] emphasises the importance of characterising the performance of vision systems and procedures. He makes the point that, whether it is called an adaptive, intelligent or a self-learning system, all such systems are making estimates. Therefore, there is a need to measure such estimates by the application of rigorous engineering performance criteria. He calls for a more rigorous approach when discussing system errors and for a systems engineering framework that will meet the realities of the manufacturing process.

There is also a need to educate the customer about vision in a broader context, rather than just concentrating on their immediate application needs. This education process should be continuous, beginning at the feasibility study stage, right up to the final installation, and not just a token gesture undertaken towards the end of the installation phase. A customer who has a reasonable knowledge of the vision application will be more open to suggesting changes in the process. This will be true, especially if the system integrator can show that there is a possibility of reducing the complexity of the image analysis (and systems cost), thus leading to a better engineered solution.

Education is vitally important but this is not the total solution, since there is also a need for the development of more flexible vision systems that can handle a larger class of objects, under less constrained manufacturing conditions. Vision engineers should also begin providing standard solutions to automation problems and not selling machine vision technology for its own sake. This requires an understanding of the problem, at a systems level. Any tendency for vision engineers to shy away from the systems problems will reduce the likelihood of a successful application implementation.

1.1.3 Flexible Manufacturing Systems

When considering *all* of manufacturing industry, the presence of 'smart' automation is minimal in extent at the moment, although there are high local concentrations. It requires new flexible techniques that combine the strengths of the work that has been done in the development of industrial material handling and automated inspection systems, combined with the growing research into assembly strategies. Such systems would avoid the need for substantial retooling between product changes, and would enable manufacturing systems to cope with an increasing number of product variants [RUM-89].

Such systems would also have the flexibility to respond to changes in the production line, manufacture or assembly procedures [HOS-90]. Depending on the design of a product, additive, multiple-insertion or combinational assembly processes are used. With multiple-insertion, the inspection process can be carried out at the end of the manufacturing cycle. However, with additive and combinational assembly processes, inspection must be carried out on each part, as it is inserted. Therefore, visually controlled assembly systems also have the added bonus of some form of gross inspection of the product under assembly, even if this is only finding the nature and orientation of the parts to be assembled [WAL-88].

The majority of industrial assembly systems are either manually operated, or use semi-automation to some degree. However, these systems can be unreliable. Reasons for such unreliability include the lack of any automated visual feedback and/or discrepancies of the human operators. Therefore, such systems tend to be expensive to operate. This is especially the case in Western Europe and the US, where it is difficult for manufacturers to match the labour costs involved in manual assembly, when compared to the Far East and the former Eastern Bloc countries. The use of robots in materials handling eliminates the need to have human beings performing monotonous, exhausting or hazardous work. This is an increasingly important factor, since it is generally becoming socially unacceptable[1] for people to perform boring, repetitive, 'robot-like' jobs. Hence,

---

[1] This is not always the case. There are important social issues at stake here [BAT-95]. However, a discussion of these concerns is beyond the scope of this book.

the need for automated systems is not necessarily about the *displacement* of labour [NEM-95], but is concerned instead with the growing expectations of an increasingly educated labour force and economic realities of the industrialised world.

Although the application of robotics and vision to parts assembly has great potential [OWE-85, HAR-87] and will strongly influence the competitiveness of the European Community, it is currently lacking in European industry [DEL-92]. This has been recognised by the European Community through its funding of major projects such as ESPRIT, BRITE and more specifically the EUREKA projects that fall under the umbrella term FAMOS (a German acronym for flexible automated assembly systems). The FAMOS-EUREKA projects have targeted one of the weakest points in Europe's manufacturing industries, with the objective of reversing the decline of more than two decades. This is especially relevant to the manufacture of products such as cameras, motorcycles and domestic appliances. Its aim is to create automated assembly systems which are flexible enough to enable manufacturers to change product lines when needed and to produce small batches of products efficiently. These projects include participants from a wide range of European industries and universities [EUR-89].

In the past, automated assembly systems have been developed mainly for handling high volume production (greater than 100,000 parts per annum), with a low number of variants (between 1 and 3 different types). However, current production assembly demands include:

- A high degree of flexibility.
- Wider range of applications with greater numbers of different versions and models.
- Small batch runs and shorter production times. 75% of applications are in small to medium batches ($\leq$ 50 items).
- Integrated quality control.
- Long unmanned operation periods with unmanned transfer vehicles.
- Ease of integration into the current production line.
- Ability to handle customised products.

In reporting on a review of the key features of automated assembly systems based on 22 German companies, Delchambre [DEL-92] highlights the fact that 98% of products are made of fewer than 25 parts, and that 90% of parts weigh less than 1Kg.

1.1.4 Process Control

In the modern manufacturing environment, economy of *scope* is becoming as important as economy of *scale*. Companies must be able to produce a variety of products using a flexible manufacturing system, while maintaining a very high level of quality. There is a need to extend the role of machine vision beyond that of inspection, to become the key controlling element in a closed loop process. Such integration will allow flexible control of the production line, using defect information to locate fault sources and allowing automatic feedback for adjustment and correction, as well as monitoring the overall efficiency.

Product lifetimes are being reduced all the time. This, coupled with an ever increasing demand for higher quality, is forcing manufacturers to produce a larger variety of products, to higher standards, with a much shorter lead time, from initial design to a commercial product reaching the market place. Satisfying these new manufacturing conditions necessitates the use of flexible automation, with better process control and a far higher level of integration. There are considerable technological challenges that must be overcome to attain these goals. Machine vision should be able to make a significant contribution to their achievement.

There are enormous benefits to be attained from integrating vision into the process control of *Discrete Event Dynamic Systems*, *(DEDS)*, producing discrete parts or products. This type of integration will make the automated process far more flexible, making it easier to accommodate product design changes and the introduction of new products, thereby reducing the cost of short manufacturing runs. It should allow the control system in a multi-product plant to handle a large mix of products, by using appropriate processing and analysis software for each product. The vision system will achieve quality assurance through process feedback, giving better built-in quality. It will aid in the fine tuning of the process thus reducing variance. The vision system can be used to monitor the effects of process changes, the introduction of new machines, maintenance, and process improvements. It should reduce the response time for the correction of fault sources in comparison to the manual equivalent because the integrated system can collect and analyse large amounts of data very quickly.

Such systems should allow defect prevention by monitoring trends and generating the appropriate feedback signals for automatic correction of the process. However, in some situations, the process monitoring system will merely alert the factory personnel, so that they can schedule preventative maintenance, before defective products are produced. Adjusting machines before they have any serious problems should increase uptime, which is very important in any plant, but particularly if a "Just-In-Time" system of manufacturing is employed. In-process monitoring will also facilitate automatic and dynamic construction of inventory, allowing reduced buffer storage, product routing (thus improving machine utilisation), and general production scheduling. It will free quality assurance personnel from time-consuming data collection, processing and interpretation of results, allowing them to concentrate on process improvements and manual trouble-shooting.

Vision provides a wealth of information about a process, in comparison with the majority of other sensors, which for the most part only provide binary (on/off), information and has limited use in generating control signals. Any control system has to be well informed to make good control decisions! The parts being manufactured are like 'windows' into the process. A control system can use machine vision to look through these windows. An intelligent vision-based controller using *a priori* process knowledge, could locate the cause of the problem. It may even be able to fix it automatically. Mechanical breakdowns and component failures would have to reported to an operator, while parts are re-routed away from the faulty machines.

1.2 Systems Engineering

Machine intelligence is not an exercise in philosophy but an engineering project. [MIC-86]

The aim of this section is to define the current state of machine vision, as seen from a systems engineering perspective. An essential part of this is a discussion of some of the research areas that must be studied, in order to advance the development of industrial vision systems. The views presented in this section have arisen from the authors involvement with machine vision systems engineering, in both industry and academia.

1.2.1 Importance of Context

During the development of machine vision systems over the last 30 years there have been two main approaches. One approach that researchers took was the development of general purpose vision systems. (Section 2.9 discusses commercial vision systems.) These systems mainly concentrated on the software aspect of the vision task, and due to the generality of such systems, vision integrators were faced with a wide and varied range of image processing, and to a lesser extent, image analysis techniques. The main challenge facing system designers is to reduce the complexity of the system, to enable it to carry out the required inspection functions, under the tight budgetary and operating conditions required by industry. The success of such systems in the manufacturing environment have been limited, since they require a significant amount of work and reprogramming to get them to perform a practical vision task.

The second approach is based on generating turn-key vision systems which provide total solutions to a given industrial task. These systems have the advantage of being tuned to a specific application. They tackle the problem rather than trying to fit the task to a collection of software procedures which are looking for an application. However, the second approach will only work effectively if the designer takes into account the context of the industrial application.

So, what is meant by the *context* of a machine vision system? The Collins English dictionary definition of "context" is given as *"conditions and circumstances of an event"*. For example, one can recognise and understand abstract words in the context of a sentence structure with less difficulty when compared to viewing/hearing such words in isolation [DRE-86]. This highlights the strength and importance of context in trying to make sense of the world around us. Likewise, in the successful development of machine vision systems, whether inspection or robotic vision, it is necessary to view the problem in its entirety. All possible considerations, electronic, optical and mechanical must be considered. This is not an easy task, and many vision system designers feel uncomfortable dealing with system issues, which are often outside their own area of expertise.

The complexity of a machine vision application is largely a reflection of the complexity of the environment in which it finds itself [SIM-81]. Therefore, a successful vision application requires a total systems approach and requires a range of engineering and practical skills to deal with the complex industrial environment. When faced with a specific application requirement, it is always well worthwhile analysing the problem from a systems engineering perspective. By adopting a systems approach, the maximum use is made of problem-specific "contextual" information, derived, for example, from the nature of the product being handled, the process used to manufacture it and the special features of the manufacturing environment. Doing this, it is often found that the complexity of the application can be reduced.

Researchers and engineers must also be open to the idea that vision may not be the appropriate or ideal approach for the task at hand. Some tasks that the end user may see as being a suitable application for machine vision, may in fact be better served by using other engineering approaches, such as the use of mechanical sensors, optical and electronic transducers. Some of the unsuccessful vision applications of the past have been caused by applying vision technology in inappropriate ways. For machine vision to become generally accepted by the manufacturing community, it must concentrate on tackling the industrial problems, rather than try to employ a given technology for its own sake. There are considerable benefits in adhering to the Japanese philosophy of restricting the tasks to suit the capabilities of the equipment.

1.2.2 Industrial Examples

The two case studies discussed in this section illustrate the complexities of designing and building an industrial vision system and emphasise how detailed knowledge of the application context can simplify the vision system design. The purpose of including these case studies here is to explain the development of industrial vision systems while concentrating on the systems engineering approach to the vision problem, rather than the image analysis and processing routines.

The first case study outlines the automated on-line inspection of plastic screw-on bottle tops. At a certain stage during the manufacture, the preformed plastic bottle tops are passed through a sealing machine, which inserts a grey plastic seal into the bottle top. The product loading station then places the bottle tops, in single file, onto the large product placement star-wheel shown in Figure 1.2. This transports the bottle tops beneath the camera and lighting inspection head. The image of the bottle top is then analysed and an *accept/reject* decision is made.

During the analysis stage, the bottle tops are moved into the product unloading station. By the time, the bottle tops arrive there, the product has already been classified and the unloading station removes the product from the starwheel. It is then placed on one of two conveyors depending on whether an accept or reject decision has been made.

Due to tight budgetary constraints and the computational overhead involved in colour processing, the use of a colour camera was not a feasible option. This difficulty was overcome by placing a motor-controlled colour filter carousel between the inspection head and the product placement starwheel (Figure 1.2). The carousel places a colour filter in the inspection system's optical path. The selection of the colour filter depends on the colour of the bottle top to be inspected. The choice of the filter is under menu control and is selected to achieve maximum contrast between the bottle top and its grey seal. Although changing of the colour filter is slow compared to the inspection speed of the bottle tops, this is not a problem, since the bottle top's colour only change between batches, and not within a given batch cycle. This leaves ample time for the vision system to change the colour filter automatically, based on the menu settings chosen by the line operator.

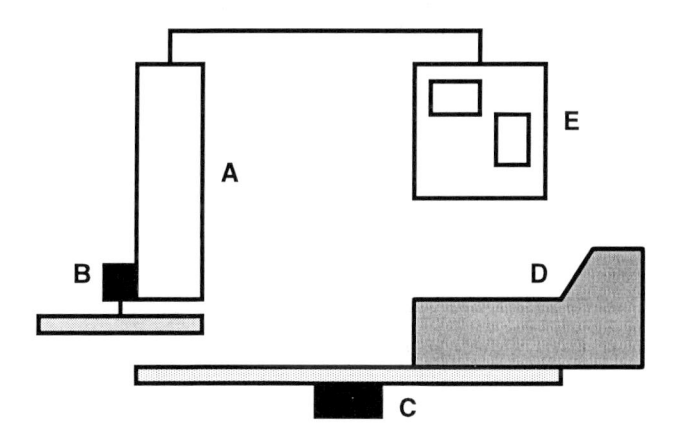

Figure 1.2 Plastic bottle top handling and inspection system. *A.* Inspection head - camera and lighting unit, *B.* Colour filter carousel, *C.* Product placement star-wheel, *D.* Product loading station (This feeds the bottle tops from the sealing machine and places them on the star-wheel for inspection) *E.* Image analysis system.

It is often surprising the extent to which a product's design can be constrained to suit the limitations of the vision system, without adversely affecting the product's functionality, aesthetics or the ability to manufacture the product [RED-91]. Although it can be argued that this imposes intolerable constraints on the product design, these restrictions need not be any more rigid than those imposed by good design for 'manufacturability'. For example, in the second case study the vision system designers were faced with the task of checking for colour mis-registration on high quality printed cartons. In this case, the product was slightly modified, to simplify the image analysis task.

The manual method of inspecting for colour mis-registration requires the examination of the printed sheets, after they have been cut into individual cartons, folded and passed through the gluing stage. Gross registration errors are obvious to the inspector after an initial glance at the carton, whereas slight registration errors are found by viewing the printer's registration mark. (This mark is printed on a part of the carton that is hidden from consumer, once the carton is assembled. See Figure 1.3.) Due to the highly automated nature of the printing process, there are few gross registration errors. In practice, the majority of errors are due to slight slippages in the printing process. These slight registration errors are difficult to find and classify manually.

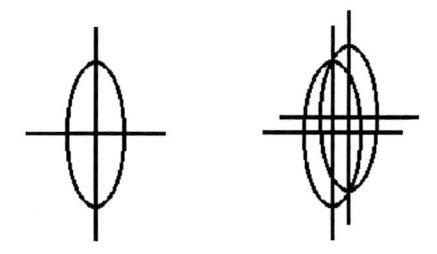

Figure 1.3 Registration symbol for manual inspection. The figure on the left indicates correct registration. The figure on the right indicates incorrect registration of two overlaid registration marks.

Figure 1.3 shows an example of a manual registration mark, initially printed in black on a white background As each new colour is applied to the carton, a new registration mark, of the same design but in the new colour, is overlaid on the original printed mark. Therefore, if all the colours are registered correctly, they produce a single well defined registration mark. However, if any type of mis-registration occurs, the registration mark for that colour appears shifted with respect to the black reference mark.

The inspection of the original design for the registration mark (Figure 1.3) was difficult for the machine vision system to handle. The registration mark is not only difficult to describe, but if mis-registration occurs the image becomes more complex and hence more difficult for a machine vision system to analyse[2].

In this instance, the product modification simply involved the redesign of the registration mark (Figure 1.4). This new registration mark consists of an outer black circle which contains a number of solid coloured disks, one for each of the subtractive primaries (magenta, yellow and cyan), and a fourth solid disk,

---

[2] Humans and machine vision systems often use different recognition criteria. Therefore, the two approaches should not be confused.

representing the extra solid colour to be printed (green in this application). This is printed on a white background. The black ring is laid down first and acts as the registration reference colour. As each colour is applied by the printing process, a solid disk of that colour is also printed inside the black reference ring. The offset of each of these disks, measured from the centre of the black ring, gives a measure of the position for that colour imprint with reference to black.

The ability to modify the product to suit the vision system's capabilities and strengths, highlights the benefits of holding detailed discussions with the end user during the feasibility study. If the end user is involved from the beginning of the design process, the designer may be fortunate to find that the customer is willing to consider changes in product presentation which will simplify the vision task. This is more likely, of course, if it can be shown that system costs can be reduced by doing so. The use of this custom registration mark, developed in conjunction with the end user, transformed a potentially difficult and expensive vision task into a much simpler one.

Complete knowledge of the application context cannot always be achieved. Therefore, there is a need for vision systems to contain procedures that can deal reliably with missing or ambiguous information. Also, in many applications, only partial control over the working environment can be realistically achieved. There will always be some application specific obstacles that cannot be removed by the use of the systems engineering approach to the task [LEE-89]. The trade-off of potential usage (i.e. generality of the application and flexibility) versus simplicity is an important decision to be made during the design of a machine vision system.

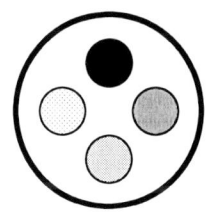

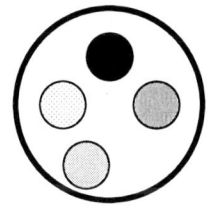

Figure 1.4 Modified colour registration symbol. The figure on the left indicates correct registration. The figure on the right indicates incorrect registration of the lower inner disk.

1.3 Intelligent Vision

There is more to (machine) vision than meets the eye.

As mentioned earlier, the majority of industrial vision applications are concerned with the inspection and/or automated assembly of simple, well defined, mass produced goods. Nevertheless this only forms a small proportion of the overall manufacturing industry; the majority of manufactured goods are made in batches of 50 or less [BAT-91]. Consequently, there is a need to make vision systems more flexible to cope with the different demands of small batch manufacture, particularly the ability to have a fast application turnaround.

This points towards the need to develop a new generation of 'intelligent' (or adaptive[3]) industrial vision systems. Intelligence is needed

- to interpret the description of the object to be recognised
- to interpret a complex visual scene
- to plan actions following the recognition process.

It is clear from even simple situations that intelligence and vision are intrinsically dependent upon each other. Intelligence needs vision to supply it with sensory data. Vision needs intelligence to resolve ambiguities in visual scenes and to make high-level judgements about what a complex scene contains.

To ensure that this new generation of vision systems is flexible, it really is necessary to use techniques that can cope with less constrained manufacturing environments, through the use of heuristics in conjunction with algorithmic procedures.

There is also a need to develop robotic vision systems which have a more adaptive visual feedback capability, such as the ability to manipulate arbitrary shapes under visual control [WHE-93]. The development of such adaptive visually controlled work cells will accelerate the growth of robotic vision systems in industry.

The development of generic tools to deal with visual cues, such as shape, size, colour and texture, must still have a high priority. Indeed, this continues to be one of the key research challenges for the future. These generic descriptors will aid in the development of machine vision applications, but when faced with a specific application the problem should be viewed within a systems engineering framework. The use of the 'contextual' information should be maximised to simplify the task. For example, in a visually controlled 'pick and place' machine, there is often no inherent reason why each item cannot be presented to the machine in a predetermined place and orientation. Therefore by mechanically restricting the orientation and positioning of the device under inspection, the

---

[3] The term 'intelligent' can be interpreted in different ways, but it is often taken to imply the imparting of human intelligence to a machine. This is not what we are necessarily interested in as machine vision designers, but rather the development of vision systems that will have the capability of adapting to the changing world around it. This may use artificial intelligence techniques but will not necessarily depend on them. Some authors prefer to use the term 'adaptive' rather than 'intelligent', however, the use of the term 'artificial intelligence' is now so ingrained in engineering and science communities, for both good and bad reasons, that it is not possible to dispense with it entirely. Therefore, it is advisable to qualify the use of such a term.

visual inspection task can be simplified. This type of demand may not always be unreasonable and should always be pursued [DRE-86].

To advance from the current generation of machine vision systems to a new, more flexible family requires addressing a number of key issues:

- Development of adaptive (intelligent) machine vision systems.
- Application of a systems engineering approach to industrial vision tasks.
- Maximise the use of contextual information available from the product, process and application environment.
- The production of standard solutions to industrial problems.
- Tackling of sub-goals.
- Widening the application base.
- The use of vision in a process and quality control role.
- Performance characterisation tools.
- Ability to deal with unclear or missing information.
- Systematic testing and repeatable experimental results.
- Generic tools to deal with common analysis features such as shape, size, colour and texture.
- Investigation of algorithmic and heuristic procedures.
- Flexible, user friendly interfaces.
- Broader education of the systems issues.

Machine vision can only progress and become fully accepted in manufacturing industry, if it employs advances in vision research in a sensible way [PAV-92].

1.3.1 Heuristics and Algorithms

While many of the current industrial applications of machine vision rely on implementing algorithmic procedures, the next generation of systems will make use of both algorithmic and heuristic approaches. The proper combination of these approaches will allow a more flexible approach to problem solving in the industrial domain.

The heuristic approach to problem solving, is regarded by some researchers as a soft option, since it is perceived as relying on common sense rather than mathematical rigour. However, if often happens that the development of solutions based on heuristic techniques is a sensible option, and often the only one !

On a practical level, many important and varied industrial vision problems are full of peculiarities, that are difficult or even impossible to state mathematically. Hence there is a need for alternative approaches. This argument does not imply that heuristics are better, or worse, than a given algorithmic approach, but rather that the proper use of heuristic methods offers a powerful alternative and should always be considered when faced with difficult system design issues [PEA-84, TAY-88].

A heuristic method, as defined by Silver, Vidal and DeWerra [SIL-80], is a *"procedure for the solving of well defined mathematical problems by an intuitive approach in which the structure of the problem can be interpreted and exploited intelligently to obtain a reasonable solution"*, and is not guaranteed to give an optimal solution. They also point out the main motivations for using heuristics, and although their paper is concerned with the use of such techniques in an operational research environment, these reasons have been generalised and are listed below.

- The problem is such that an analytic or iterative solution is unknown.
- An exact analytic or iterative solution may exist, but the implementation may be computationally expensive and therefore impractical.
- A heuristic method, may be simpler for the design engineer to understand.
- For a well-defined problem that can be solved optimally, a heuristic method can be used for learning purposes.
- A heuristic method may be used as part of an iterative procedure that *guarantees* the finding of an optimal solution.
- Heuristics can be used to give a good starting solution in implicit enumeration approaches to a problem. This can help to reduce the computational effort needed to search for an optimal solution.

One of the qualities that a good heuristic procedure should possess includes an average performance close to that of the 'optimal' solution (i.e. that is the closeness of the solution to optimal, rather than the time taken to compute the answer). Of course, such a performance measurement may not be possible in many applications, since one of the major reasons for using heuristics in the first place is that it may be impossible to find an optimal solution. Therefore, the use of heuristics requires quantitative performance measures to decide if the procedure is "good enough" (satisfactory). Other key performance considerations include fast heuristic execution, a small probability of worst-case performance occurring and that the solution should be simply designed and implemented easily and cheaply.

There are problems relying on a purely heuristic approach: such approaches tend to be memory intensive. Moreover uncontrolled searches, if allowed, are time consuming. Therefore, heuristic procedures are most often used in applications where "intelligence" is more important than speed. For the benefit of the heuristic approach to be maximised, it is important for the designer to have an appreciation of the theoretical problem under consideration and the systems issues contained in it. The use of heuristics is no excuse for a reduction in engineering and scientific rigour.

1.3.2 Artificial Intelligence (AI) Languages

Artificial intelligence languages are currently found in a wide range of expert systems that aid knowledge representation, retrieval and presentation. The main families of languages used for AI programming include [ROB-89]:

- Functional application languages (e.g. Lisp)
- Logic programming languages (e.g. Prolog, Parlog)
- Object oriented languages (e.g. Prolog++, Smalltalk).

There is no single dominant language for AI applications. As with any programming task, the designer should chose the language that will allow him to carry out the task with the minimum of effort. In reality, many programmers have favourite languages, ones which they feel most comfortable using. The reasons for language choice can also be geographical or cultural. For example, Lisp is dominant in the US, while Prolog (which was developed in Europe) is commonly used in Europe and Japan. In the case of the work outlined in this book the authors have used Prolog. (The reasons for choosing this language will be outlined in Chapter 3.) However, this does not exclude the implementation of the ideas outlined in this book in other languages.

The ideas behind the integration of AI languages and industrial vision applications are more recent. AI based industrial vision applications include:

- Automated packing of arbitrary shapes [WHE-93].
- The use of expert systems in process control [MCC-93].
- The use of expert systems in cake inspection [BAT-91].
- Inspection of food [BAT-91].

1.4 Book Outline

Machine vision research and engineering has been discussed in a wide range of books and other forums. As a result, it often difficult for an author to find a fresh approach to the subject. The aim of this book is to look at the development of tools, techniques and systems that will enable vision engineers to design the next generation of industrial vision systems. Such systems will have to be more adaptive than their predecessors to their environment and for this reason have the appearance of intelligence.

Chapter 2 lays the framework for this discussion, by reviewing the current state of machine vision engineering and pays particular attention to basic machine vision techniques. This chapter is aimed at readers with minimal prior experience of machine vision. More experienced readers will find most of this material familiar, although they may wish to use this chapter as a reference to the Prolog+ commands used in the remainder of the book. Chapter 3 introduces the reader to intelligent image processing. This discussion will include an introduction to interactive image processing and the Prolog+ vision language used by the authors in the development of intelligent vision systems. A number of Prolog+ programs are include to illustrate the power of this approach to image processing and analysis.

Chapter 4 discusses intelligent systems that have been enhanced by expanding the basic Prolog+ concepts introduced in the previous chapter. As machine vision applications become more complex, the knowledge-based functions will also need to be automated. The use of expert systems to aid in the design of a vision systems optical arrangement, lighting configuration and even camera selection will become commonplace. The ideas behind this knowledge automation are also outlined in Chapter 4, which also deals with understanding simple spoken expressions and the integration of intelligent multi-camera systems within the Prolog+ environment.

Since machine vision systems interact with a (mainly) mechanical world, the need for intelligent control of external devices is a key factor in the overall design of the vision system. Chapter 5 introduces a general purpose interface unit, developed for use with a flexible inspection cell in conjunction with Prolog+. Vision system calibration and a range of general system issues are also discussed in this chapter. Chapter 6 introduces the issues involved in colour image processing and analysis. It outlines a number of approaches to the colour imaging task. Chapter 7 puts the ideas outlined in the previous chapters into practice. A number of applications of intelligent vision systems to a range of industrial problems including food inspection and automated packing systems are covered.

There are five appendices. Appendix A presents some of the proverbs, opinions and folklore that surround machine vision. While this section is offered in a light-hearted manner, it encapsulates some important lessons that we have learned but which are unfortunately not universally acknowledged or understood. Appendix B outlines some of the important factors that must be considered when designing a vision system. Appendix C contains a compilation of general reference material, useful for machine vision designers. This includes machine and computer vision texts, conference proceedings, special issues of relevant journals, survey and review papers, lists of periodicals, journals and magazines relating to machine vision and references to a wealth of on-line Internet resources. Appendix D outlines the issues relating to a general purpose software implementation of Prolog+, while Appendix E summarises the Prolog+ commands used throughout this book. Finally, a glossary of machine vision terms is included.

2

Basic Machine Vision Techniques

The purpose of this chapter is to outline some of the basic techniques used in the development of industrial machine vision systems. These are discussed in sufficient detail to understand the key ideas outlined elsewhere in this book. For a more detailed explanation of image processing and image analysis techniques, the reader should refer to the general reference material in Appendix C. In the following discussion we shall frequently indicate the equivalent Prolog+ operators for the vision techniques described. (A more detailed discussion of Prolog+ operators can be found in Chapter 3 and Appendix E.) Prolog+ commands appear in square brackets. In certain cases, sequences of Prolog+ commands are needed to perform an operation and these are similarly listed.

2.1 Representations of Images

We shall first consider the representation of *Monochrome* (grey-scale) images. Let i and j denote two integers where $1 \leq i \leq m$ and $1 \leq j \leq n$. In addition, let f(i,j) denote an integer function such that $0 \leq f(i,j) \leq W$. (W denotes the white level in a grey-scale image.) An array F will be called a *digital image*.

$$F = \begin{vmatrix} f(1,1), & f(1,2), & \ldots & f(1,n) \\ f(2,1), & f(2,2), & \ldots & f(2,n) \\ . & . & \ldots & . \\ . & . & \ldots & . \\ f(m,1), & f(m,2), & \ldots & f(m,n) \end{vmatrix}$$

An address (i,j) defines a position in F, called a *pixel, pel* or *picture element*. The elements of F denote the intensities within a number of small rectangular regions within a real (i.e. optical) image. (See Figure 2.1) Strictly speaking, f(i,j) measures the intensity at a single point but if the corresponding rectangular region is small enough, the approximation will be accurate enough for most purposes. The array F contains a total of *m.n* elements and this product is called the *spatial resolution* of F. We may *arbitrarily* assign intensities according to the following scheme:

f(i,j) = 0	black
0 < f(i,j) ≤ 0.33W	dark grey
0.33W < f(i,j) ≤ 0.67W	mid-grey
0.67W < f(i,j) < W	light grey
f(i,j) = W	white

Let us consider how much data is required to represent a grey-scale image in this form. Each pixel requires the storage of $log_2(1+W)$ bits. This assumes that (1+W) is an integer power of two. If it is not, then $log_2(1+W)$ must be rounded up to the next integer. This can be represented using the ceiling function, < ... >. Thus, a grey-scale image requires the storage of $< log_2(1+W) >$ bits. Since there are $m.n$ pixels, the total data storage for the entire digital image F is equal to $m.n.< log_2(1+W) >$ bits. If m = n ≥ 128, and W ≥ 64, we can obtain a good image of a human face. Many of the industrial image processing systems in use nowadays manipulate images in which m = n = 512 and W = 255. This leads to a storage requirement of 256 Kbytes/image. A *binary image* is one in which only two intensity levels, black (0) and white (1), are permitted. This requires the storage of $m.n$ bits/image.

An impression of colour can be conveyed to the eye by superimposing *four* separate imprints. (Cyan, magenta, yellow and black inks are often used in printing.) Ciné film operates in a similar way, except that when different colours of light, rather than ink, are added together, *three* components (red, green and blue) suffice. Television operates in a similar way to film; the signal from a colour television camera may be represented using three components: $R = \{r(i,j)\}$; $G = \{g(i,j)\}$; $B = \{b(i,j)\}$, where R, G and B are defined in a similar way to F. The vector $\{r(i,j), g(i,j), b(i,j)\}$ defines the intensity and colour at the point (i,j) in the colour image. (Colour image analysis is discussed in more detail in Chapter 6.) Multispectral images can also be represented using several monochrome images. The total amount of data required to code a colour image with r components is equal to $m.n.r.< log_2(1+W) >$ bits, where W is simply the maximum signal level on each of the channels.

Ciné film and television will be referred to, in order to explain how moving scenes may be represented in digital form. A ciné film is, in effect, a time-sampled representation of the original moving scene. Each frame in the film is a standard colour, or monochrome image, and can be coded as such. Thus, a monochrome ciné film may be represented digitally as a sequence of two-dimensional arrays [F_1, F_2, F_3, F_4,....]. Each F_i is an $m.n$ array of integers as we defined above, when discussing the coding of grey-scale images. If the film is in colour, then each of the F_i has three components. In the general case, when we have a sequence of r-component colour images to code, we require $m.n.p.r.< log_2(1+W) >$ bits/image sequence, where the spatial resolution is $m.n$ pixels, each spectral channel permits *(1+W)* intensity levels, there are *r* spectral channels and *p* is the total number of "stills" in the image sequence.

We have considered only those image representations which are relevant to the understanding of simple image processing and analysis functions. Many alternative methods of coding images are possible but these are not relevant to this discussion. (See the general reference material in Appendix C for more information on this subject.)

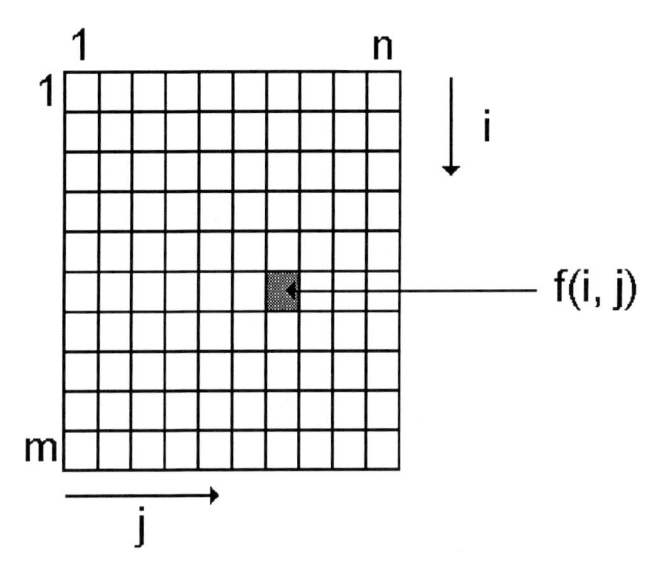

Figure 2.1 A digital image consisting of an array of *m.n* pixels. The pixel in the ith row and the jth column has an intensity equal to *f(i,j)*.

2.2 Elementary Image Processing Functions

The following notation will be used throughout this section, in which we shall concentrate upon grey-scale images, unless otherwise stated.

- *i* and *j* are row and column address variables and lie within the ranges: $1 \le i \le m$ and $1 \le j \le n$. (Figure 2.1)
- $A = \{a(i,j)\}$, $B = \{b(i,j)\}$ and $C = \{c(i,j)\}$.
- *W* denotes the white level.
- g(X) is a function of a single independent variable X.
- h(X,Y) is a function of two independent variables, X and Y.
- The assignment operator '←' will be used to define an operation that is performed upon one data element. In order to indicate that an operation is to be performed upon all pixels within an image, the assignment operator '⇐' will be used.
- k, k1, k2, k3 are constants.

- N(i,j) is that set of pixels arranged around the pixel (i,j) in the following way:

(i-1, j-1)	(i-1, j)	(i-1, j+1)
(i, j-1)	(i, j)	(i, j+1)
(i+1, j-1)	(i+1, j)	(i+1, j+1)

Notice that N(i,j) forms a 3x3 set of pixels and is referred to as the *3x3 neighbourhood* of (i,j). In order to simplify some of the definitions, we shall refer to the intensities of these pixels using the following notation:

A	B	C
D	E	F
G	H	I

Ambiguities over the dual use of A, B and C should not be troublesome, as the context will make it clear which meaning is intended. The points {(i-1, j-1), (i-1, j), (i-1, j+1), (i, j-1), (i, j+1), (i+1, j-1), (i+1, j), (i+1, j+1)} are called the *8-neighbours* of (i, j) and are also said to be *8-connected* to (i, j). The points {(i-1, j), (i, j-1), (i, j+1), (i+1, j)} are called the *4-neighbours* of (i, j) and are said to be *4-connected* to (i, j).

2.2.1 Monadic, Point-by-point Operators.

These operators have a characteristic equation of the form:

$c(i,j) \Leftarrow g(a(i,j))$ or $E \Leftarrow g(E)$

Such an operation is performed for all (i,j) in the range [1,m].[1,n]. (See Figure 2.2). Several examples will now be described.

Intensity shift [acn]

$$c(i,j) \Leftarrow \begin{vmatrix} 0 & a(i,j) + k < 0 \\ a(i,j) + k & 0 \leq a(i,j) + k \leq W \\ W & W < a(i,j) + k \end{vmatrix}$$

k is a constant, set by the system user. Notice that this definition was carefully designed to maintain c(i,j) within the same range as the input, viz. [0,W]. This is an example of a process referred to as *intensity normalisation*. Normalisation is important because it permits iterative processing by this and other operators in a machine having a limited precision for arithmetic (e.g. 8-bits). Normalisation will be used frequently throughout this chapter.

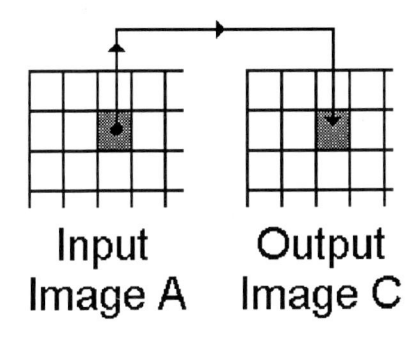

Figure 2.2 Monadic point-by-point operator. The (i,j)th pixel in the input image has intensity a(i,j). This value is used to calculate c(i,j), the intensity of the corresponding pixel in the output image.

Intensity multiply [mcn]

$$c(i,j) \Leftarrow \begin{vmatrix} 0 & a(i,j) \cdot k < 0 \\ a(i,j) \cdot k & 0 \leq a(i,j) \cdot k \leq W \\ W & W < a(i,j) \cdot k \end{vmatrix}$$

Logarithm [log]

$$c(i,j) \Leftarrow \begin{vmatrix} 0 & a(i,j) = 0 \\ W \cdot \left[\dfrac{Log(a(i,j))}{Log(W)} \right] & \text{otherwise} \end{vmatrix}$$

This definition arbitrarily replaces the infinite value of log(0) by zero, and thereby avoids a difficult rescaling problem.

Antilogarithm (exponential) [exp] $c(i,j) \Leftarrow W \cdot \exp(a(i,j)) / \exp(W)$

Negate [neg] $c(i,j) \Leftarrow W - a(i,j)$

Threshold [thr]

$$c(i,j) \Leftarrow \begin{vmatrix} W & k1 \leq a(i,j) \leq k2 \\ 0 & \text{otherwise} \end{vmatrix}$$

This is an important function, which converts a grey-scale image to a binary format. Unfortunately, it is often difficult, or even impossible to find satisfactory values for the parameters *k1* and *k2*.

Highlight [hil]

$$c(i,j) \Leftarrow \begin{cases} k3 & k1 \leq a(i,j) \leq k2 \\ a(i,j) & \text{otherwise} \end{cases}$$

Squaring [sqr] $\qquad\qquad c(i,j) \Leftarrow [\,a(i,j)\,]^2 / W$

2.2.2 Dyadic, Point-by-point Operators

Dyadic operators have a characteristic equation of the form:

$c(i,j) \Leftarrow h(a(i,j), b(i,j))$

There are two input images: $A = \{a(i,j)\}$ and $B = \{b(i,j)\}$ (Figure 2.3), while the output image is $C = \{c(i,j)\}$. It is important to realise that $c(i,j)$ depends upon only $a(i,j)$ and $b(i,j)$. Here are some examples of dyadic operators.

Add [add] $\qquad\qquad c(i,j) \Leftarrow [\,a(i,j) + b(i,j)\,] / 2.$

Subtract [sub] $\qquad\qquad c(i,j) \Leftarrow [\,(a(i,j) - b(i,j)) + W\,] / 2$

Multiply [mul] $\qquad\qquad c(i,j) \Leftarrow [\,a(i,j).b(i,j)\,] / W$

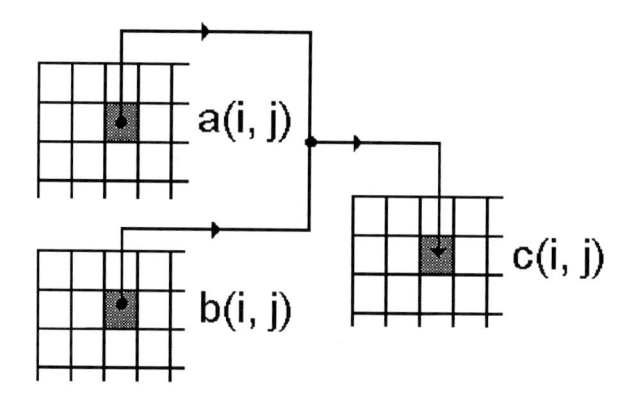

Figure 2.3 Dyadic point-by-point operator. The intensities of the $(i,j)^{th}$ pixels in the two input images (i.e. $a(i,j)$ and $b(i,j)$) are combined to calculate the intensity, $c(i,j)$, at the corresponding address in the output image.

Maximum [max] $c(i,j) \Leftarrow MAX[\,a(i,j), b(i,j)\,]$

When the maximum operator is applied to a pair of binary images, the *union* (OR function) of their white areas is computed. This function may also be used to *superimpose* white writing onto a grey-scale image.

Minimum [min] $c(i,j) \Leftarrow MIN[\,a(i,j), b(i,j)\,]$

When A and B are both binary, the *intersection* (AND function) of their white areas is calculated.

2.2.3 Local Operators

Figure 2.4 illustrates the principle of the operation of local operators. Notice that the intensities of several pixels are combined together, in order to calculate the intensity of just one pixel. Amongst the simplest of the local operators are those which use a set of 9 pixels arranged in a 3x3 square. These have a characteristic equation of the following form:

$c(i,j) \Leftarrow \quad g(\,a(i-1, j-1), a(i-1, j), a(i-1, j+1), a(i, j-1), a(i, j), a(i, j+1),$
$a(i+1, j-1), a(i+1, j), a(i+1, j+1)\,)$

where g(.) is a function of 9 variables. This is an example of a local operator which uses a *3x3 processing window*. (That is, it computes the value for one pixel on the basis of the intensities within a region containing 3x3 pixels. Other local operators employ larger windows and we shall discuss these briefly later.) In the simplified notation which we introduced earlier, the above definition reduces to: $E \Leftarrow g(A, B, C, D, E, F, G, H, I)$.

2.2.4 Linear Local Operators

An important sub-set of the local operators is that group which performs a linear weighted sum, and which are therefore known as *linear local operators*. For this group, the characteristic equation is:

$E \Leftarrow k1.(A.W1 + B.W2 + C.W3 + D.W4 + E.W5 + F.W6 + G.W7 + H.W8 + I.W9) + k2$

where W1, W2,...,W9 are weights, which may be positive, negative or zero. Values for the normalisation constants, k1 and k2 are given later. The matrix illustrated below is termed the *weight matrix* and is important, because it determines the properties of the linear local operator.

W1	W2	W3
W4	W5	W6
W7	W8	W9

(vii) In order to perform normalisation, the following values are used for k1 and k2.

$$k1 \leftarrow 1 / \sum_{p,q} | w_{p,q} |$$

$$k2 \leftarrow [1 - \sum_{p,q} w_{p,q} / \sum_{p,q} | w_{p,q} |].W/2$$

(viii) A filter using the following weight matrix performs a *local averaging function* over an 11x11 window [*raf(11,11)*].

1	1	1	1	1	1	1	1	1	1	1
1	1	1	1	1	1	1	1	1	1	1
1	1	1	1	1	1	1	1	1	1	1
1	1	1	1	1	1	1	1	1	1	1
1	1	1	1	1	1	1	1	1	1	1
1	1	1	1	1	1	1	1	1	1	1
1	1	1	1	1	1	1	1	1	1	1
1	1	1	1	1	1	1	1	1	1	1
1	1	1	1	1	1	1	1	1	1	1
1	1	1	1	1	1	1	1	1	1	1
1	1	1	1	1	1	1	1	1	1	1

This produces quite a severe 2-directional blurring effect. Subtracting the effects of a blurring operation from the original image generates a picture in which spots, streaks and intensity steps are all emphasised. On the other hand, large areas of constant or slowly changing intensity become uniformly grey. This process is called *high-pass filtering*, and produces an effect similar to unsharp masking, which is familiar to photographers.

2.2.5 Non-linear Local Operators

Largest intensity neighbourhood function [lnb]

$$E \Leftarrow MAX(A, B, C, D, E, F, G, H, I)$$

This operator has the effect of spreading bright regions and contracting dark ones.

Edge detector [command sequence: lnb, sub]

$$E \Leftarrow MAX(A, B, C, D, E, F, G, H, I) - E$$

This operator is able to highlight edges (i.e. points where the intensity is changing rapidly).

Median filter [mdf(5)]
$$E \Leftarrow \textbf{FIFTH\_LARGEST} (A,B,C,D,E,F,G,H,I)$$

This filter is particularly useful for reducing the level of noise in an image. (Noise is generated from a range of sources, such as video cameras and x-ray detectors, and can be a nuisance if it is not eliminated by hardware or software filtering.)

Crack detector[1] [lnb, lnb, neg, lnb, lnb, neg]
This operator is equivalent to applying the above Prolog+ sequence of operations and then subtracting the result from the original image. This detector is able to detect thin dark streaks and small dark spots in a grey-scale image; it ignores other features, such as bright spots and streaks, edges (intensity steps) and broad dark streaks.

Roberts edge detector [red]
The Roberts gradient is calculated using a 2x2 mask. This will determine the edge gradient in two diagonal directions (i.e. the cross-differences).

$$E \Leftarrow \sqrt{(A - E)^2 + (B - D)^2}$$

The following approximation to the Roberts gradient magnitude is called the Modified Roberts operator. This is simpler and faster to implement and it more precisely defines the Prolog+ operator *red*. It is defined as

$$E \Leftarrow \{ |A - E| + |B - D| \} / 2$$

Sobel edge detector [sed]
This popular operator highlights the edges in an image; points where the intensity gradient is high are indicated by bright pixels in the output image. The Sobel edge detector uses a 3x3 mask to determine the edge gradient.

$$E \Leftarrow \sqrt{[(A + 2.B + C) - (G + 2.H + I)]^2 + [(A + 2.D + G) - (C + 2.F + I)]^2}$$

The following approximation is simpler to implement in software and hardware and more precisely defines the Prolog+ operator *sed*:

$$E \Leftarrow \{ |(A + 2.B + C) - (G + 2.H + I)| + |(A + 2.D + G) - (C + 2.F + I)| \} / 6$$

---

[1] This is an example of an operator that can be described far better using computer notation rather than mathematical notation.

See Figure 2.5, for a comparison of the Roberts and Sobel edge detector operators when applied to a sample monochrome image. Note that, while the Roberts operator produces thinner edges, these edges tend to break up in regions of high curvature. The primary disadvantage of the Roberts operator is its high sensitivity to noise, since fewer pixels are used in the calculation of the edge gradient. There is also a slight shift in the image, when the Roberts edge detector is used. The Sobel edge detector does not produce such a shift.

(a)

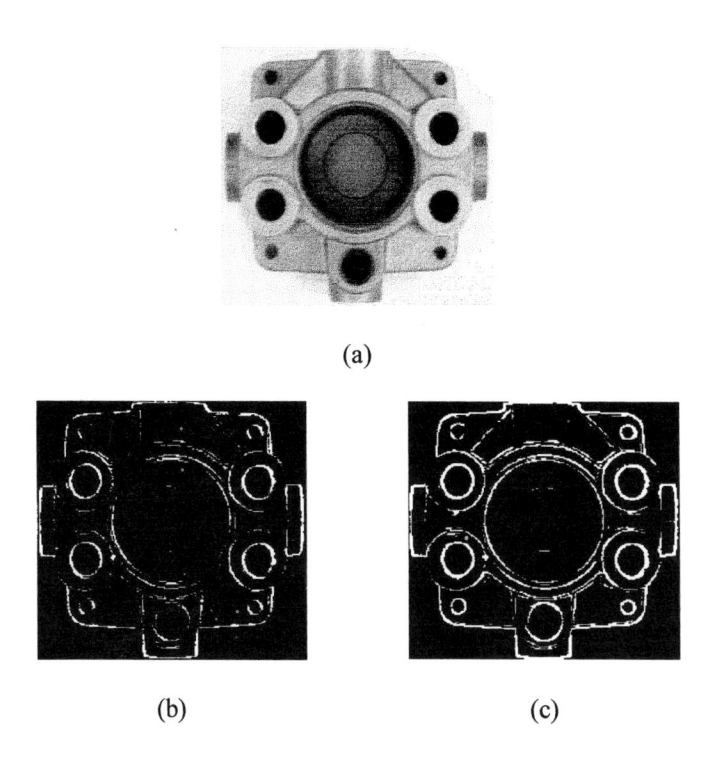

(b)   (c)

Figure 2.5 Edge detection. (a) Original image. (b) Roberts edge gradient (after thresholding). (c) Sobel edge gradient (after thresholding).

Prewitt edge detector
The Prewitt edge-detector is similar to the Sobel operator, but is more sensitive to noise as it does not possess the same inherent smoothing. This operator uses the two 3x3 shown below to determine the edge gradient,

-1	-1	-1
0	0	0
1	1	1

P_1

-1	0	1
-1	0	1
-1	0	1

P_2

where P_1 and P_2 are the values calculated from each mask respectively. The Prewitt gradient magnitude is defined as: $E \Leftarrow \sqrt{P_1^2 + P_2^2}$

Frei and Chen edge detector
This operator uses the two 3x3 masks shown below to determine the edge gradient,

-1	-√2	-1
0	0	0
1	√2	1

F_1

-1	0	1
-√2	0	√2
-1	0	1

F_2

where F_1 and F_2 are the values calculated from each mask respectively. The Frei and Chen gradient magnitude is defined as: $E \Leftarrow \sqrt{F_1^2 + F_2^2}$

Rank filters [mdf, rid]
The generalised 3x3 rank filter is:

c(i, j) ⇐ k1.(A′.W1 + B′.W2 + C′.W3 + D′.W4 + E′.W5 + F′.W6 + G′.W7 + H′.W8 + I′.W9) + k2

where
        A′ = **LARGEST** (A, B, C, D, E, F, G, H, I)
        B′ = **SECOND_LARGEST** (A, B, C, D, E, F, G, H, I)
        C′ = **THIRD_LARGEST** (A, B, C, D, E, F, G, H, I)
        ..
        I′ = **NINTH_LARGEST** (A, B, C, D, E, F, G, H, I)

and *k1* and *k2* are the normalisation constants defines previously. With the appropriate choice of weights (W1, W2,...,W9), the rank filter can be used for a range of operations including edge detection, noise reduction, edge sharping and image enhancement.

Direction codes [dbn]
This function can be used to detect the *direction* of the intensity gradient. A direction code function DIR_CODE is defined thus:

$$\text{DIR\_CODE}(A,B,C,D,F,G,H,I) \Leftarrow \begin{cases} 1 & \text{if } A \geq \text{MAX}(B,C,D,F,G,H,I) \\ 2 & \text{if } B \geq \text{MAX}(A,C,D,F,G,H,I) \\ 3 & \text{if } C \geq \text{MAX}(A,B,D,F,G,H,I) \\ 4 & \text{if } D \geq \text{MAX}(A,B,C,F,G,H,I) \\ 5 & \text{if } F \geq \text{MAX}(A,B,C,D,G,H,I) \\ 6 & \text{if } G \geq \text{MAX}(A,B,C,D,F,H,I) \\ 7 & \text{if } H \geq \text{MAX}(A,B,C,D,F,G,I) \\ 8 & \text{if } I \geq \text{MAX}(A,B,C,D,F,G,H) \end{cases}$$

Using this definition the operator *dbn* may be defined as:

E ⇐ DIR_CODE(A,B,C,D,F,G,H,I)

2.2.6 N-tuple Operators

The N-tuple operators are closely related to the local operators and have a large number of linear and non-linear variations. N-tuple operators may be regarded as generalised versions of local operators. In order to understand the N-tuple operators, let us first consider a *linear* local operator which uses a large processing window, (say $r.s$ pixels) with most of its weights equal to zero. Only N of the weights are non-zero, where $N \ll r.s$. This is an N-tuple filter. (See Figure 2.6.) The N-tuple filters are usually designed to detect specific patterns. In this role, they are able to locate a simple feature, such as a corner, annulus, the numeral "2", in any position etc. However, they are sensitive to changes of orientation and scale. The N-tuple can be regarded as a sloppy template, which is convolved with the input image.

Non-linear tuple operators may be defined in a fairly obvious way. For example, we may define operators which compute the average, maximum, minimum or median values of the intensities of the N pixels covered by the N-tuple. An important class of such functions is the *morphological operators*. (See Sections 2.4 and 2.5.) Figure 2.7 illustrates the recognition of the numeral '2' using an N-tuple. Notice how the goodness of fit varies with the shift, tilt, size, and font. Another character ('**Z**' in this case) may give a score that is close to that obtained from a '**2**', thus making these two characters difficult to distinguish reliably.

2.2.7 Edge Effects

All local operators and N-tuple filters are susceptible to producing peculiar effects around the edges of an image. The reason is simply that, in order to calculate the intensity of a point near the edge of an image, we require information about pixels outside the image, which of course are simply not present. In order to make some attempt at calculating values for the edge pixels, it is necessary to make some assumptions, for example that all points outside the image are black, or have the same values as the border pixels. This strategy, or whatever one we adopt, is perfectly arbitrary and there will be occasions when the edge effects are so pronounced that there is nothing that we can do but to remove them by masking [*edg*]. Edge effects are important because they require us to make special provisions for them when we try to patch several low-resolution images together.

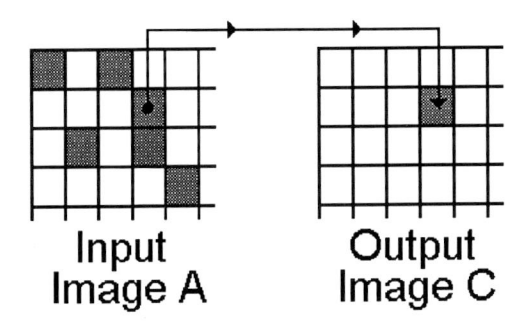

Figure 2.6 An N-tuple filter operates much like a local operator. The only difference is that the pixels whose intensities are combined together do not form a compact set. A linear N-tuple filter can be regarded as being equivalent to a local operator which uses a large window and in which many of the weights are zero.

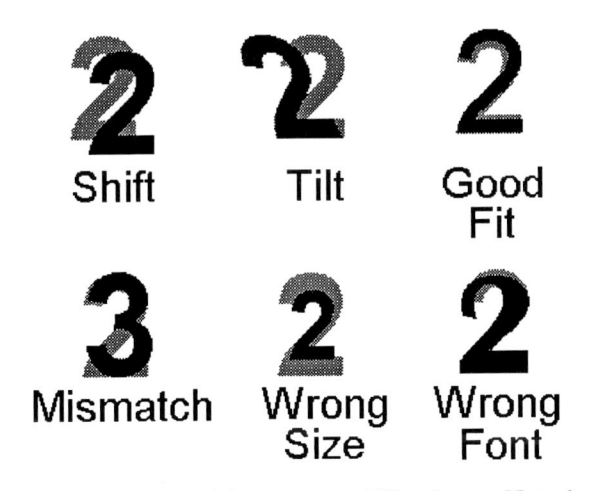

Figure 2.7 Recognising a numeral '2' using an N-tuple.

2.2.8 Intensity Histogram *[hpi, hgi, hge, hgc]*

The intensity histogram is defined in the following way:

(a) Let

$$s(p,i,j) \leftarrow \begin{cases} 1 & a(i,j) = p \\ 0 & \text{otherwise} \end{cases}$$

(b)   Let h(p) be defined thus: $h(p) \leftarrow \sum_{i,j} s(p,i,j)$

It is not, in fact, necessary to store each of the $s(p,i,j)$, since the calculation of the histogram can be performed as a serial process in which the estimate of $h(p)$ is updated iteratively, as we scan through the input image. The *cumulative histogram*, $H(p)$, can be calculated using the following recursive relation:

$H(p) = H(p-1) + h(p)$, where $H(0) = h(0)$.

Both the cumulative and the standard histograms have a great many uses, as will become apparent later. It is possible to calculate various intensity levels which indicate the occupancy of the intensity range [*pct*]. For example, it is a simple matter to determine that intensity level, $p(k)$, which when used as a threshold parameter ensures that a proportion k of the output image is black, $p(k)$ can be calculate using the fact that $H(p(k)) = m.n.k$. The *mean intensity* [*avg*] is equal to:

$$\sum_p (h(p).p) / (m.n)$$

while the *maximum intensity* [*gli*] is equal to $MAX(p \mid h(p) > 0)$ and the *minimum intensity* is equal to $MIN(p \mid h(p) > 0)$.

One of the principal uses of the histogram is in the selection of threshold parameters. It is useful to plot $h(p)$ as a function of p. It is often found from this graph that a suitable position for the threshold can be related directly to the position of the "foot of the hill" or to a "valley" in the histogram.

An important operator for image enhancement is given by the transformation:

$c(i,j) \Leftarrow [W.H(a(i,j))] / (m.n)$

This has the interesting property that the histogram of the output image $\{c(i,j)\}$ is flat, giving rise to the name *histogram equalisation* [*heq*] for this operation. Notice that histogram equalisation is a *data-dependent* monadic, point-by-point operator.

An operation known as "*local area histogram equalisation*" relies upon the application of histogram equalisation within a small window. The number of pixels in a small window that are darker than the central pixel is counted. This number defines the intensity at the equivalent point in the output image. This is a powerful filtering technique, which is particularly useful in texture analysis applications. (See Section 2.7.)

2.3 Binary Images

For the purposes of this description of binary image processing, it will be convenient to assume that a(i,j) and b(i,j) can assume only two values: 0 (black) and 1(white). The operator "+" denotes the Boolean **OR** operation, "•" represents the **AND** operation and where '⊗' denotes the Boolean **Exclusive OR** operation. Let #(i,j) denote the number of white points addressed by N(i,j), including (i,j) itself.

Inverse [not] $\quad\quad\quad\quad\quad\quad\quad\quad\quad$ c(i,j) ⇐ **NOT**(a(i,j))

AND white regions [and, min] $\quad\quad$ c(i,j) ⇐ a(i,j) • b(i,j)

OR [ior, max] $\quad\quad\quad\quad\quad\quad\quad\quad$ c(i,j) ⇐ a(i,j) + b(i,j)

Exclusive OR [xor] (Find differences between white regions.)
$$c(i,j) \Leftarrow a(i,j) \otimes b(i,j)$$

Expand white areas [exw]

c(i,j) ⇐ a(i-1, j-1) + a(i-1, j) + a(i-1, j+1) + a(i, j-1) + a(i, j) + a(i, j+1) + a(i+1, j-1) + a(i+1, j) + a(i+1, j+1)

Notice that this is closely related to the local operator *lnb* defined earlier. This equation may be expressed in the simplified notation: E ⇐ A + B + C + D + E + F + G + H + I

Shrink white areas [skw]

c(i,j) ⇐ a(i-1, j-1) • a(i-1, j) • a(i-1, j+1) • a(i, j-1) • a(i, j) • a(i, j+1) • a(i+1, j-1) • a(i+1, j) • a(i+1, j+1)

or more simply c(i,j) ⇐ A • B • C • D • E • F • G • H • I

Edge detector [bed] $\quad$ c(i,j) ⇐ E • **NOT**(A • B • C • D • F • G • H • I)

Remove isolated white points [wrm]

$$c(i,j) \Leftarrow \begin{cases} 1 & a(i,j) \cdot (\#(i,j) > 1) \\ 0 & \text{otherwise} \end{cases}$$

Count white neighbours [cnw] $c(i,j) \Leftarrow \#(a(i,j) = 1)$.

Where $\#(Z)$ is the number of times Z occurs. Notice that $\{c(i,j)\}$ is a grey-scale image.

Connectivity detector [cny]. Consider the following pattern:

1	0	1
1	X	1
1	0	1

If X=1, then all of the 1's are 8-connected to each other. Alternatively, if X=0, then they are not connected. In this sense, the point marked X is critical for connectivity. This is also the case in the following examples:

1	0	0
0	X	1
0	0	0

1	1	0
0	X	0
0	0	1

0	0	1
1	X	0
1	0	1

However, those points marked X below are not critical for connectivity, since setting X=0 rather than 1 has no effect on the connectivity of the 1's.

1	1	1
1	X	1
0	0	1

0	1	1
1	X	0
1	1	1

0	1	1
1	X	0
0	1	1

A connectivity detector shades the output image with 1's to indicate the position of those points which are critical for connectivity and which were white in the input image. Black points, and those which are not critical for connectivity, are mapped to black in the output image.

Euler number [eul]. The Euler number is defined as the number of connected components (blobs) minus the number of holes in a binary image. The Euler number represents a simple method of counting blobs in a binary image, provided they have no holes in them. Alternatively, it can be used to count holes in a given object, providing they have no "islands" in them. The reason why this approach is used to count blobs, despite the fact that it may seem a little awkward to use, is that the Euler number is very easy and fast to calculate. It is also a useful means of classifying shapes in an image. The Euler number can be computed by using three local operators. Let us define three numbers N1, N2 and N3, where Nα indicates the number of times that one of the patterns in the pattern set α (α = 1, 2 or 3) occur in the input image.

0	0
0	1

0	0
1	0

1	0
0	0

0	1
0	0

Pattern set 1 (N1)

0	1
1	0

1	0
0	1

Pattern set 2 (N2)

1	1
1	0

1	1
0	1

0	1
1	1

1	0
1	1

Pattern set 3 (N3)

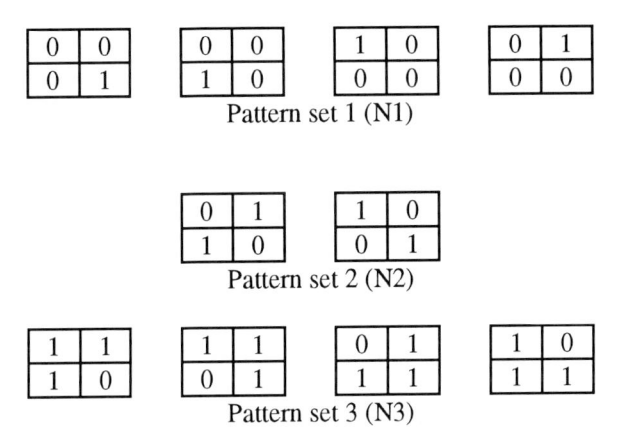

The *8-connected* Euler number, where holes and blobs are defined in terms of 8-connected figures, is defined as: *(N1-2.N2-N3)/4*. It is possible to calculate the *4-connected* Euler number using a slightly different formula, but this parameter can give results which seem to be anomalous when we compare them to the observed number of holes and blobs.

Filling holes [blb]. Consider a white blob-like figure containing a hole (lake), against a black background. The application of the hole-filling operator will cause all of the holes to be *filled-in*; by setting all pixels in the holes to white. This operator will not alter the outer edge of the figure.

Region labelling [ndo]. Consider an image containing a number of separate blob-like figures. A region-labelling operator will shade the output image so that each blob is given a separate intensity value. We could shade the blobs according to the order in which they are found, during a conventional raster scan of the input image. Alternatively, the blobs could be shaded according to their areas; the biggest blobs becoming the brightest. This is a very useful operator, since it allows objects to be separated and analysed individually. (Figure 2.8) Small blobs can also be eliminated from an image using this operator. Region labelling can also be used to count the number of distinct binary blobs in an image. Unlike the Euler number, counting based on region labelling is not effected by the presence of holes.

Other methods of detecting/removing small spots. A binary image can be represented in terms of a grey-scale image in which only two grey levels, 0 and W, are allowed. The result of the application of a conventional low-pass (blurring) filter to such an image is a grey-scale image in which there is a larger number of possible intensity values. Pixels which were well inside large white areas in the input image are mapped to very bright pixels in the output image. Pixels which were well inside black areas are mapped to very dark pixels in the output image. However, pixels which were inside small white spots in the input image are mapped to mid-grey intensity levels (Figure 2.9). Pixels on the edge of large white areas are also mapped to mid-grey intensity levels. However, if there is a cluster of small spots, which are closely spaced together, some of them may also disappear.

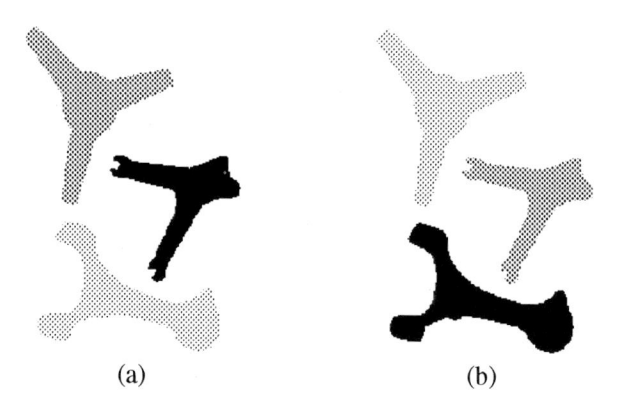

Figure 2.8 Shading blobs in a binary image (a) according to their areas and (b) according to the order in which they are found during a raster scan (left to right; top to bottom).

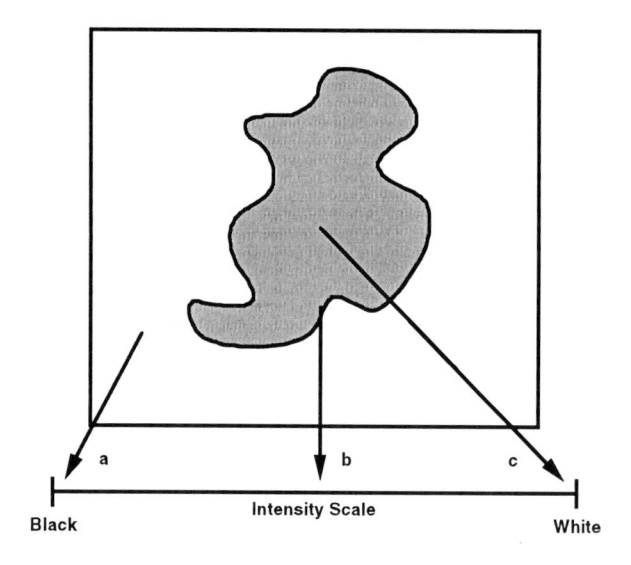

Figure 2.9 Using a grey-scale blurring filter to remove noise from a binary image. (a) Background points are mapped to black. (b) Edge points are mapped to the central part of the intensity range. Thresholding at mid-grey has the effect of smoothing the edge of large blobs. (c) Central areas of large white blobs are mapped to white.

Based on these observations, the following procedure has been developed. It has been found to be effective in distinguishing between small spots and, at the same time, achieving a certain amount of edge smoothing of the large bright blobs which remain:

```
raf(11,11),    % Low-pass filter using a 11x11 local operator
thr(128),      % Threshold at mid-grey
```

This technique is generally easier and faster to implement than the blob shading technique described previously. Although it may not achieve the desired result exactly, it can be performed at high speed.

An N-tuple filter having the weight matrix illustrated below, can be combined with simple thresholding to distinguish between large and small spots. Assume that there are several small white spots within the input image and that they are spaced well apart. All pixels within a spot which can be contained within a circle of radius three pixels will be mapped to white by this particular filter. Pixels within a larger spot will become darker than this. The image is then thresholded at white to separate the large and small spots.

			-1	-1	-1			
		-1				-1		
	-1						-1	
-1								-1
-1				20				-1
-1								-1
	-1						-1	
		-1				-1		
			-1	-1	-1			

Grass-fire transform and skeleton [gfa, mdl, mid]. Consider a binary image containing a single white blob, Figure 2.10. Imagine that a fire is lit at all points around the blob's outer edge and the edges of any holes it may contain. The fire will burn inwards, until at some instant, advancing fire lines meet. When this occurs, the fire becomes extinguished locally. An output image is generated and is shaded in proportion to the time it takes for the fire to reach each point. Background pixels are mapped to black.

The importance of this transform, referred to as the *grass-fire* transform, lies in the fact that it indicates distances to the nearest edge point in the image [BOR-86]. It is therefore possible to distinguish thin and fat limbs of a white blob. Those points at which the fire lines meet are known as *quench* points. The set of quench points form a "match-stick" figure, usually referred to as a *skeleton or medial axis transform*. These figures can also be generated in a number of different ways [GON-87] (Figure 2.11).

One such approach is described as *onion-peeling*. Consider a single white blob and a "bug" which walks around the blob's outer edge, removing one pixel at a time. No edge pixel is removed, if by doing so we would break the blob into two disconnected parts. In addition, no white pixel is removed, if there is only one white pixel amongst its 8-neighbours. This simple procedure leads to an undesirable effect in those instances when the input blob has holes in it; the skeleton which it produces has small loops in it which fit around the holes like a tightened noose. More sophisticated algorithms have been devised which avoid this problem.

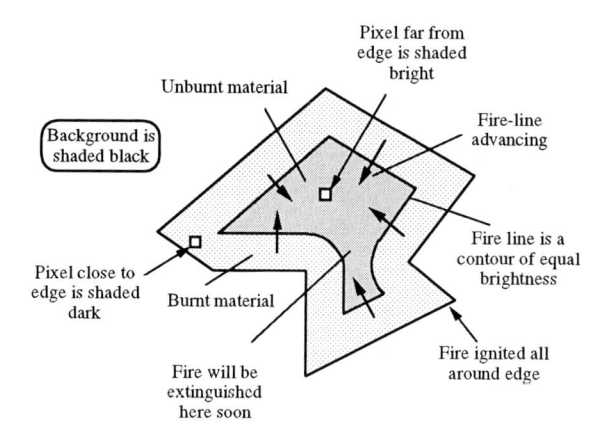

Figure 2.10 Grass-fire transform.

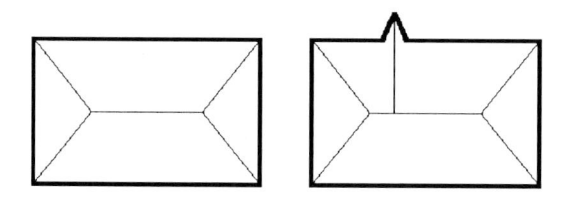

Figure 2.11 Application of the Medial Axis Transform.

Edge smoothing and corner detection. Consider three points B1, B2 and B3 which are placed close together on the edge of a single blob in a binary image. (See Figure 2.12.) The perimeter distance between B1 and B2 is equal to that between B2 and B3. Define the point P to be that at the centre of the line joining B1 and B3. As the three points now move around the edge of the blob, keeping the spacing between them constant, the locus of P traces a smoother path than that followed by B2 as it moves around the edge. This forms the basis of a simple edge smoothing procedure.

A related algorithm, for corner detection, shades the edge according to the distance between P and B2. This results in an image in which the corners are highlighted, while the smoother parts of the image are much darker.

Many other methods of edge smoothing are possible. For example, we may map white pixels which have fewer than, say, three white 8-neighbours to black. This has the effect of eliminating "hair" around the edge of a blob-like figure. One of the techniques described previously for eliminating small spots offers another possibility. A third option is to use the processing sequence: [*exw, skw, skw, exw*], where *exw* represents expand white areas and *skw* denotes shrink white areas.

Convex hull [chu]. Consider a single blob in a binary image. The convex hull is that area enclosed within the smallest convex polygon which will enclose the shape (Figure 2.13). This can also be described as the region enclosed within an elastic string, stretched around the blob. The area enclosed by the convex hull, but not within the original blob is called the *convex deficiency*, which may consist of a number of disconnected parts, and includes any holes and indentations. If we regard the blob as being like an *island*, we can understand the logic of referring to the former as *lakes* and the latter as *bays*.

2.3.1 Measurements on Binary Images

To simplify the following explanation, we will confine ourselves to the analysis of a binary image containing a single blob. The area of the blob can be measured by the total number of object (white) pixels in the image. However, we must first define two different types of edge points, in order to measure an object's perimeter.

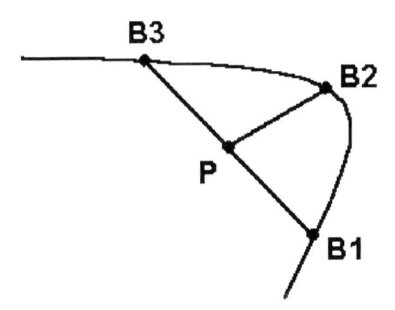

Figure 2.12 Edge smoothing and corner detection.

Figure 2.13 Convex hull of a 'club' shape. The lightly shaded region indicates the shape's convex deficiency.

The *4-adjacency* convention (Figure 2.14) only allows the four main compass points to be used as direction indicators, while *8-adjacency* uses all eight possible directions. If 4-adjacency convention is applied to the image segment given in Figure 2.14(c), then none of the four segments (two horizontal and two vertical) will appear as touching, i.e. they are not connected. Using the 8-adjacency convention, the segments are now connected, but we have the ambiguity that the inside of the shape is connected to the outside. Neither convention is satisfactory, but since 8-adjacency allows diagonally-connected pixels to be represented, it leads to a more faithful perimeter measurement.

Assuming that the 8-adjacency convention is used, we can generated a coded description of the blob's edge. This is referred to as the *chain code* or *Freeman code* [*fcc*]. As we trace around the edge of the blob, we generate a number, 0-7, to indicate which of the eight possible directions we have taken (i.e. from the centre, shaded pixel in Figure 2.14(b)). Let N_o indicate how many *odd-numbered* code values are produced as we code the blob's edge, and N_e represent the number of *even-numbered* values found. The *perimeter* of the blob is given *approximately* by the formula: $N_e + \sqrt{2}.N_o$

This formula will normally suffice for use in those situations where the perimeter of a smooth object is to be measured. The *centroid* of a blob [*cgr*] determines its position within the image and can be calculated using the formulae:

$$I \leftarrow \sum_j \sum_i (a(i,j).i) / N_{i,j} \quad \text{and} \quad J \leftarrow \sum_j \sum_i (a(i,j).j) / N_{i,j}$$

where $N_{i,j} \leftarrow \sum_j \sum_i a(i,j)$

Although we are considering images in which the a(i,j) are equal to 0 (black) or 1 (white), it is convenient to use a(i,j) as an ordinary arithmetic variable as well.

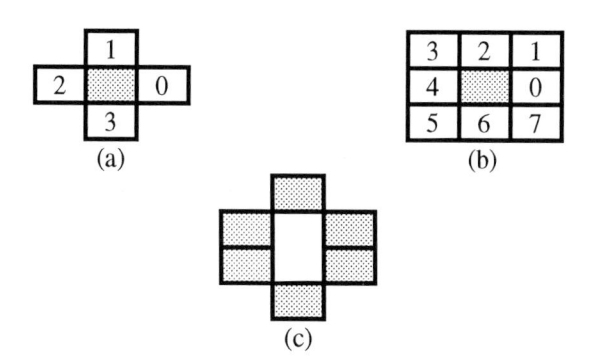

Figure 2.14 Chain code. (a) 4-adjacency coding convention. (b) 8-adjacency coding convention. (c) Image segment.

2.3.2 Shape Descriptors

The following are just a few of the numerous shape descriptors that have been proposed:

(a) The distance of the furthest point on the edge of the blob from the centroid.
(b) The distance of the closest point on the edge of the blob from the centroid.
(c) The number of protuberances, as defined by that circle whose radius is equal to the average of the parameters measured in (a) and (b).
(d) The distances of points on the edge of the blob from the centroid, as a function of angular position. This describes the silhouette in terms of polar co-ordinates. (This is not a single-valued function.)
(e) Circularity = Area / Perimeter^2. This will tend to zero for irregular shapes with ragged boundaries, and has a maximum value (=$1/4\pi$) for a circle.
(f) The number of holes.(Use *eul* and *ndo* to count them.)
(g) The number of bays.
(h) Euler number.
(i) The ratio of the areas of the original blob and that of its convex hull.
(j) The ratio of the areas of the original blob and that of its circumcircle.
(k) The ratio of the area of the blob to the square of the total limb-length of its skeleton.
(l) Distances between joints and limb ends of the skeleton.
(m) The ratio of the projections onto the major and minor axes.

2.4 Binary Mathematical Morphology

The basic concept involved in mathematical morphology is simple: an image is probed with a template shape, called a structuring element, to find where the structuring element fits, or does not fit within a given image. [DOU-92] (Figure 2.15) By marking the locations where the template shape fits, structural information, can be gleaned about the image. The structuring elements used in practice are usually geometrically simpler than the image they act on, although this is not always the case. Common structuring elements include points, point pairs, vectors, lines, squares, octagons, discs, rhombi and rings. Since shape is a prime carrier of information in machine vision applications, mathematical morphology has an important role to play in industrial systems [HAR-87b].

The language of binary morphology is derived from that of set theory [HAR-92b]. General mathematical morphology is normally discussed in terms of Euclidean N-space, but in digital image analysis we are only interested in a discrete or digitised equivalent in two-space. The following analysis is therefore restricted to binary images, in a digital two-dimensional integer space, Z^2. The image set (or scene) under analysis will be denoted by A, with elements $a = (a1, a2)$. The shape parameter, or structuring element, that will be applied to scene A

will be denoted by *B*, with elements *b = (b1, b2)*. The primary morphological operations that we will examine are dilation, erosion, opening and closing.

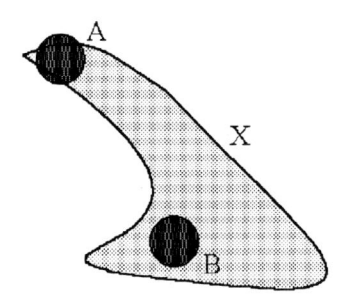

Figure 2.15 A structuring element fitting, *B*, and not fitting, *A*, into a given image scene *X* [DOU-92].

Dilation (also referred to as filling and growing) is the expansion of an image set *A* by a structuring element *B*. It is formally viewed as the combination of the two sets using vector addition of the set elements. The dilation of an image set *A* by a structuring element *B*, will be denoted $A \oplus B$, and can be represented as the union of translates of the structuring element *B* [HAR-92b]:

$$A \oplus B = \bigcup_{a \in A} B_a$$

where $\bigcup$ represents the union of a set of points and the translation of *B* by point *a* is given by, $B_a = \{ c \in Z^2 \mid c = b + a \text{ for some } b \in B \}$. This is best explained by visualising a structuring element *B* moving over an image *A* in a raster fashion. Whenever the origin of the structuring element touches one of the image pixels in *A*, then the entire structuring element is placed at that location. For example, in Figure 2.16 the grid image is dilated by a cross-shaped structuring element, contained within a 3x3 pixel grid.

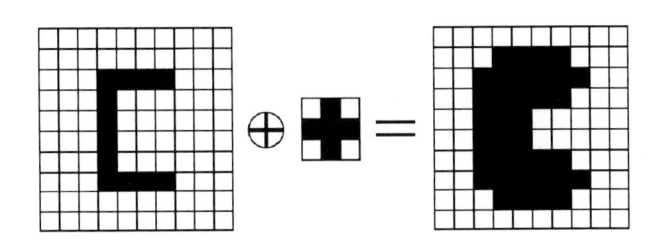

Figure 2.16 Dilation of a grid image by a cross structuring element.

Erosion is the dual morphological operation of dilation and is equivalent to the shrinking (or reduction) of the image set A by a structuring element B. This is a morphological transformation which combines two sets using vector subtraction of set elements [HAR-92b]. The erosion of an image set A by a structuring element B, denoted $A \ominus B$, can be represented as the intersection of the negative translates:

$$A \ominus B = \bigcap_{b \in B} A_{-b}$$

where $\bigcap$ represents the intersection of a set of points. Erosion of the image A by B is the set of all points for which B translated to a point x is contained in A. This consists of sliding the structuring element B across the image A, and where B is fully contained in A (by placing the origin of the structuring element at the point x) then x belongs to the eroded image $A \ominus B$. For example, in Figure 2.17 the grid image is eroded by a cross-shaped structuring element, contained within a 3x3 pixel grid.

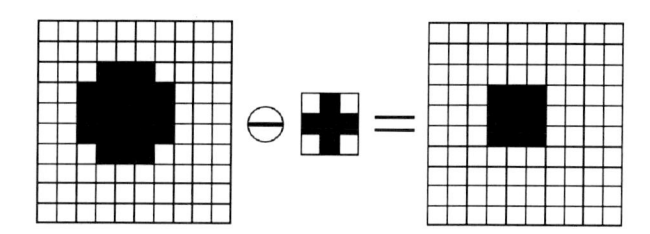

Figure 2.17 Erosion of a grid image by a cross structuring element.

A duality relationship exists between certain morphological operators, such as erosion and dilation. This means that the equivalent of such an operation can be performed by its dual on the complement (negative) image and by taking the complement of the result [VOG-89]. Although duals, erosion and dilation operations are not inverses of each other. Rather they are related by the following duality relationships:

$$(A \ominus B)^c = A^c \oplus \overline{B} \text{ and } (A \oplus B)^c = A^c \ominus \overline{B}$$

Where A^c refers to the complement of the image set A and,

$$\overline{B} = \{ x \mid \text{for some } b \in B, x = -b \}$$

refers to the reflection of B about the origin. (Serra [SER-82; SER-86] refers to this as the *transpose* of the structuring element.)

2.4.1 Opening and Closing Operations

Erosion and dilation tend to be used in pairs to extract, or impose, structure on an image. The most commonly found erosion-dilation pairings occur in the *opening* and *closing* transformations.

Opening is a combination of erosion and dilation operations that have the effect of removing isolated spots in the image set A that are smaller than the structuring element B and those sections of the image set A narrower than B. This is also viewed as a geometric *rounding* operation. (Figure 2.18) The opening of the image set A by the structuring element B, is denoted $A \bigcirc B$, and is defined as $(A \ominus B) \oplus B$.

Closing is the dual morphological operation of opening. This transformation has the effect of filling in holes and blocking narrow valleys in the image set A, when a structuring element B (of similar size to the holes and valleys) is applied. (Figure 2.18) The closing of the image set A by the structuring element B, is denoted $A \bullet B$, and is defined as $(A \oplus B) \ominus B$.

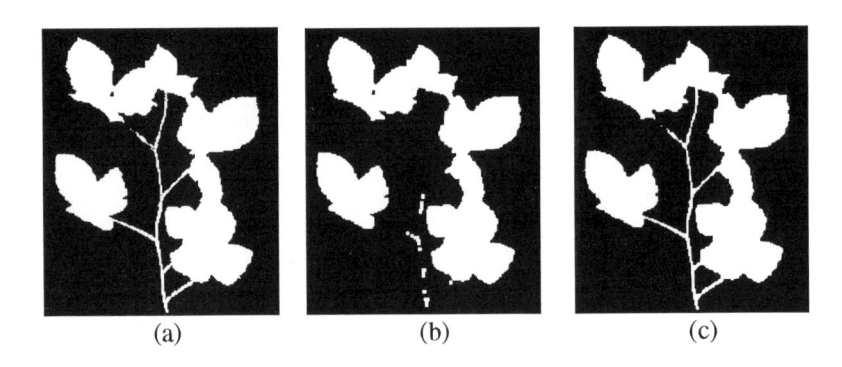

Figure 2.18 Application of a 3x3 square structuring element to a binary image of a small plant. (a) Original image. (b) Result of morphological opening. (c) Result of morphological closing.

One important property that is shared by both the opening and closing operations is *idempotency*. This means that successful reapplication of the operations will not change the previously transformed image [HAR-87b]. Therefore, $A \bigcirc B = (A \bigcirc B) \bigcirc B$ and $A \bullet B = (A \bullet B) \bullet B$.

Unfortunately, the application of morphological techniques to industrial tasks, which involves complex operations on "real-world" images, can be difficult to implement. Practical imaging applications tend to have structuring elements that are unpredictable in shape and size. In practice, the ability to manipulate arbitrary structuring elements usually relies on their decomposition into component parts.

2.4.2 Structuring Element Decomposition

Some vision systems [DUF-73; STE-78; WAL-88b] can perform basic morphological operations very quickly in a parallel and/or pipelined manner. Implementations that involve such special purpose hardware tend to be expensive, although there are some notable exceptions [WAL-94]. Unfortunately, some of these systems impose restrictions on the shape and size of the structuring elements that can be handled. Therefore, one of the key problems involved in the application of morphological techniques to industrial image analysis is the generation and/or decomposition of large structuring elements. Two main strategies are used to tackle this problem.

The first technique is called *dilation* or *serial decomposition*. This decomposes certain large structuring elements into a sequence of successive erosion and dilation operations, each step operating on the preceding result. Unfortunately, the decomposition of large structuring elements into smaller ones is not always possible. Also, those decompositions that are possible are not always easy to identify and implement.

If a large structuring element B can be decomposed into a chain of dilation operations, $B = B_1 \oplus B_2 \oplus \ldots \oplus B_N$ (Figure 2.19), then the dilation of the image set A by B is given by:

$$A \oplus B = A \oplus (B_1 \oplus B_2 \oplus \ldots \oplus B_N) = (((A \oplus B_1) \oplus B_2) \ldots) \oplus B_N.$$

Similarly, using the so-called chain rule [ZHU-86], which states that $A \ominus (B \oplus C) = (A \ominus B) \ominus C$, the erosion of A by B is given by:

$$A \ominus B = A \ominus (B_1 \oplus B_2 \oplus \ldots \oplus B_N) = (((A \ominus B_1) \ominus B_2) \ldots) \ominus B_N.$$

A second approach to the decomposition problem is based on "breaking up" the structuring element, B, into a union of smaller components, $B_1, \ldots, B_N$. We can think of this approach as 'tiling' of the structuring element by sub-structuring elements. (Figure 2.20) Since the 'tiles' do not need to be contiguous or aligned, any shape can be specified without the need for serial decomposition of the structuring element, although the computational cost of this approach is proportional to the area of the structuring element [WAL-88b]. This is referred to as *union* or *parallel decomposition*. Therefore, with B decomposed into a union of smaller structuring elements, $B = B_1 \cup B_2 \cup \ldots \cup B_N$, then the dilation of an image A by the structuring element B can be rewritten as:

$$\begin{aligned} A \oplus B &= A \oplus (B_1 \cup B_2 \cup \ldots \cup B_N) \\ &= (A \oplus B_1) \cup (A \oplus B_2) \cup \ldots \cup (A \oplus B_N) \end{aligned}$$

Likewise, the erosion of A by the structuring element B can be rewritten as:

$$A \ominus B = A \ominus (B_1 \cup B_2 \cup \ldots \cup B_N)$$
$$= (A \ominus B_1) \cap (A \ominus B_2) \cap \ldots \cap (A \ominus B_N)$$

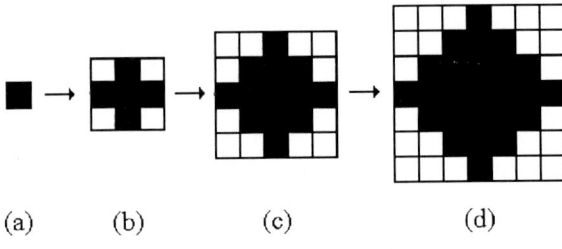

Figure 2.19 Construction of a 7x7 structuring element by successive dilation of a 3x3 structuring element. (a) Initial pixel. (b) 3x3 structuring element and the result of the first dilation. (c) Result of the second dilation. (d) Result of the third dilation [WAL-88b].

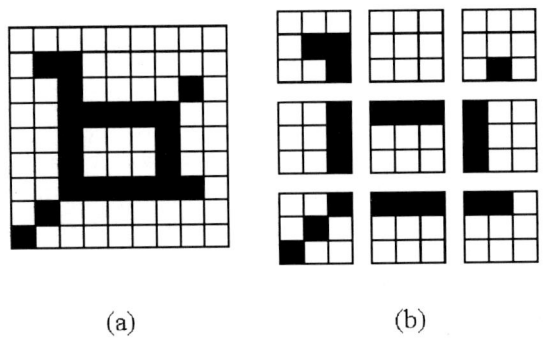

Figure 2.20 Tiling of a 9x9 arbitrary structuring element. (a) The initial 9x9 structuring element. (b) Tiling with nine 3x3 sub-structuring elements [WAL-88b].

This makes use of the fact that $A \ominus (B \cup C) = (A \ominus B) \cap (A \ominus C)$ [HAR-87b]. Due to the nature of this decomposition procedure, it is well suited to implementation on parallel computer architectures.

Waltz [WAL-88b] compared these structural element decomposition techniques, and showed that the serial approach has a 9:4 speed advantage over its parallel equivalent. (This was based on an arbitrarily specified 9x9 pixel structuring element, when implemented on a commercially available vision system.) However, the parallel approach has a 9:4 advantage in the number of degrees of freedom. (Every possible 9x9 structuring element can be achieved with the *parallel decomposition*, but only a small subset can be realised with the serial approach.) Although slower than the serial approach, it has the advantage that there is no need for serial decomposition of the structuring element.

Classical parallel and serial methods mainly involve the numerous scanning of image pixels and are therefore inefficient when implemented on conventional computers. This is so, because the number of scans depends on the total number of pixels (or edge pixels) in the shape to be processed by the morphological operator. Although the parallel approach is suited to some customised (parallel) architectures, the ability to implement such parallel approaches on *serial* machines is discussed by Vincent [VIN-91].

2.5 Grey Scale Morphology

Binary morphological operations can be extended naturally to process grey scale imagery, by the use of neighbourhood minimum and maximum functions [HAR-87b]. Heijmans [HEI-91], presents a detailed study of grey scale morphological operators, in which he outlines how binary morphological operators and thresholding techniques can be used to build a large class of useful grey scale morphological operators. Sternberg [STE-86], discusses the application of such morphological techniques to industrial inspection tasks.

In Figure 2.21, a one-dimensional morphological filter, operates on an analogue signal (equivalent to a grey scale image). The input signal is represented by the thin curve and the output by the thick black curve. In this simple example, the structuring element has an approximately parabolic form. In order to calculate a value for the output signal, the structuring element is pushed upwards, from below the input curve. The height of the top of the structuring element is noted. This process is then repeated, by sliding the structuring element sideways. Notice how this particular operator attenuates the intensity peak but follows the input signal quite accurately everywhere else. Subtracting the output signal from the input would produce a result in which the intensity peak is emphasised and all other variations would be reduced.

The effect of the basic morphological operators on two-dimensional grey scale images can also be explained in these terms. Imagine the grey scale image as a landscape, in which each pixel can be viewed in 3-D. The extra height dimension represents the grey scale value of a pixel. We generate new images by passing the structuring element above/below this landscape. (See Figure 2.21.)

Grey scale dilation. This is computed as the maximum of translations of the grey surface. Grey level dilation of image A by the structuring element B produces an image C defined by:

$$C(r,c) = Max_{(i,j)}\{ A(r-i, c-j) + B(i,j) \} = (A \oplus B)(r,c)$$

where *A, B* and *C* are grey level images. Commonly used grey level structuring elements include rods, disks, cones and hemispheres. This operation is commonly used to smooth small negative contrast grey level regions in an image.

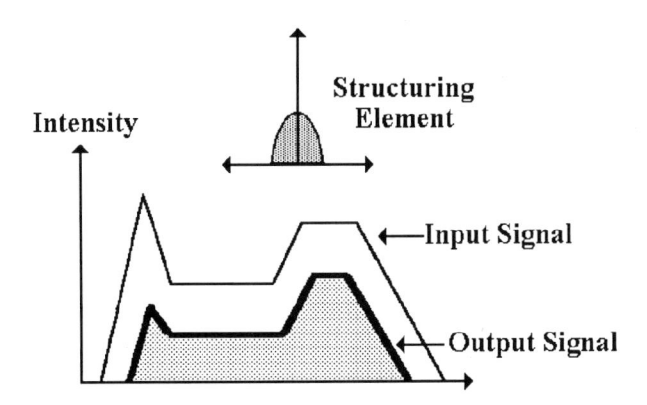

Figure 2.21 A 1-dimensional morphological filter, operating on an analogue signal.

Grey scale erosion. The grey value of the erosion at any point is the *maximum* value for which the structuring element centred at that point, still fits entirely within the foreground under the surface. This is computed by taking the *minimum* of the grey surface translated by all the points of the structuring element. (Figure 2.21). Grey level erosion of image *A* by the structuring element *B* produces an image *C* defined by:

$$C(r,c) = Min_{(i,j)}\{ A(r+i, c+j) - B(i,j) \} = (A \ominus B)(r,c)$$

This operation is commonly used to smooth small positive contrast grey level regions in an image.

Grey scale opening. This operation is defined as the grey level erosion of the image followed by the grey level dilation of the eroded image. That is, it will cut down the peaks in the grey level topography to the highest level for which the elements fit under the surface.

Grey scale closing. This operation is defined as the grey level dilation of the image followed by the grey level erosion of the dilated image. Closing fills in the valleys to the maximum level for which the element fails to fit above the surface. For a more detailed discussion on binary and grey scale mathematical morphology, see Haralick and Shapiro [HAR-92b] and Dougherty [DOU-92].

2.6 Global Image Transforms

An important class of image processing operators is characterised by an equation of the form B ⇐ f(A), where A = {a(i,j)} and B = {b(p,q)}. Each element in the output picture, B, is calculated using all or, at least a large proportion of the pixels in A. The output image, B, may well look quite different from the input image, A. Examples of this class of operators are: lateral shift, rotation, warping, Cartesian to Polar co-ordinate conversion, Fourier and Hough transforms.

Integrate intensities along image rows [rin]. This operator is rarely of great value when used on its own, but can be used with other operators to good effect, for example detecting horizontal streaks and edges. The operator is defined recursively:

b(i,j) ⇐ b(i,j-1) + a(i,j)/n where b(0,0) = 0

Row maximum [rox]. This function is often used to detect local intensity minima. c(i,j) ⇐ MAX(a(i,j), c(i,j-1))

Geometric transforms. Algorithms exist by which images can be shifted [*psh*], rotated [*tur*], undergo axis conversion [*ctr, rtc*], magnified [*pex* and *psq*] and warped. The reader should note that certain operations, such as rotating a digital image, can cause some difficulties because pixels in the input image are not mapped exactly to pixels in the output image. This can cause smooth edges to appear stepped. To avoid this effect, interpolation may be used, but this has the unfortunate effect of blurring edges. (See [BAT-91] for more details.)

The utility of axis transformations is evident when we are confronted with the examination of circular objects, or those displaying a series of concentric arcs, or streaks radiating from a fixed point. Inspecting such objects is often made very much easier, if we first convert from *Cartesian* to *Polar* co-ordinates. Warping is also useful in a variety of situations. For example, it is possible to compensate for *barrel*, or *pin-cushion* distortion in a camera. Geometric distortions introduced by a wide-angle lens, or trapezoidal distortion due to viewing the scene from an oblique angle can also be corrected. Another possibility is to convert simple curves of known shape into straight lines, in order to make subsequent analysis easier.

2.6.1 Hough Transform

The Hough transform provides a powerful and robust technique for detecting lines, circles, ellipses, parabolae, and other curves of pre-defined shape, in a binary image. Let us begin our discussion of this fascinating topic, by describing the simplest version, the basic Hough Transform, which is intended to detect straight lines. Actually, our objective is to locate *nearly linear* arrangements of disconnected white spots and *"broken" lines*. Consider that a straight line in the input image is defined by the equation $r = x.\text{Cos } \phi + y.\text{Sin } \phi$, where r and ϕ are two unknown parameters, whose values are to be found. Clearly, if this line intersects the point (x_i, y_i), then $r = x_i.\text{Cos } \phi + y_i.\text{Sin } \phi$ can be solved for many different values of (r, ϕ). So, each white point (x_i, y_i) in the input image may be associated with a *set* of (r, ϕ) values. Actually, this set of points forms a *sinusoidal curve* in (r, ϕ) space. (The latter is called the *Hough Transform (HT)* image.) Since each point in the input image generates such a sinusoidal curve, the whole of that image creates a multitude of overlapping sinusoids, in the HT image. In many instances, a large number of sinusoidal curves are found to converge on the same spot in the HT image. The (r, ϕ) address of such a point indicates the slope, ϕ, and position, r, of a straight line that can be drawn through a large number of white spots in the input image.

The implementation of the Hough transform for line detection begins by using a two-dimensional accumulator array, $A(r, \phi)$, to represent quantised (r, ϕ) space. (Clearly, an important choice to be made is the step size for quantising r and ϕ. However, we shall not dwell on such details here.) Assuming that all the elements of $A(r, \phi)$ are initialised to zero, the Hough Transform is found by computing a set $S(x_i, y_i)$ of (r, ϕ) pairs satisfying the equation $r = x_i.\text{Cos } \phi + y_i.\text{Sin } \phi$. Then, for all (r, ϕ) in $S(x_i, y_i)$, we increment $A(r, \phi)$ by one. This process is then repeated for all values of i such that the point (x_i, y_i) in the input image is white. We repeat that bright spots in the HT image indicate "linear" sets of spots in the input image. Thus, line detection is transformed to the rather simpler task of finding local maxima in the accumulator array, $A(r, \phi)$. The co-ordinates (r, ϕ) of such a local maximum give the parameters of the equation of the corresponding line in the input image. The HT image can be displayed, processed and analysed just like any other image, using the operators that are now familiar to us.

The robustness of the HT techniques arises from the fact that, if part of the line is missing, the corresponding peak in the HT image is simply darker. This occurs because fewer sinusoidal curves converge on that spot and the corresponding accumulator cell is incremented less often. However, unless the line is almost completely obliterated, this new darker spot can also be detected. In practice, we find that "near straight lines" are transformed into a cluster of points. There is also a spreading of the intensity peaks in the HT image, due to noise and quantisation effects. In this event, we may conveniently threshold the HT image and then find the centroid of the resulting spot, to calculate the parameters of the straight line in the input image. Pitas [PIT-93] gives a more detailed description of this algorithm. Figure 2.22 illustrates how this approach can be used to find a line in a noisy binary image.

The Hough transform can also be generalised to detect groups of points lying on a curve. In practice, this may not be a trivial task, since the complexity increases very rapidly with the number of parameters needed to define the curve. For circle detection, we define a circle parametrically as: $r^2 = (x - a)^2 + (y - b)^2$

where, (a, b) determines the co-ordinates of the centre of the circle and r is its radius. This requires a three-dimensional parameter space, which cannot, of course, be represented and processed as a single image. For an arbitrary curve, with no simple equation to describe its boundary, a look-up table is used to define the relationship between the boundary co-ordinates an orientation and the Hough transform parameters. (See [SON-93] for more details.)

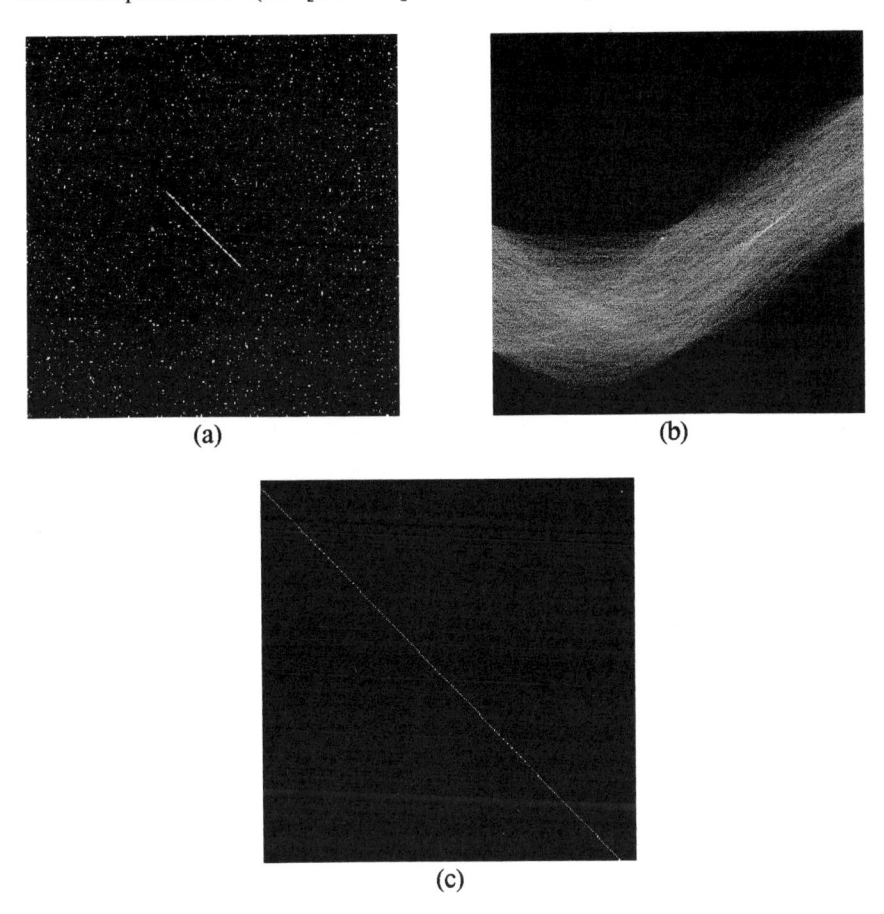

Figure 2.22 Hough transform. (a) Original image. (b) Hough transform. (c) Inverse Hough transform applied to a single white pixel located at the point of maximum intensity in (b). Notice how accurately this process locates the line in the input image, despite the presence of a high level of noise.

2.6.2 Two-dimensional Discrete Fourier Transform

We have just seen how the transformation of an image into a different domain can sometimes make the analysis task easier. Another important operation to which this remark applies is the Fourier Transform. Since we are discussing the processing of images, we shall discuss the *two-dimensional Discrete Fourier Transform*. This operation allows spatial periodicities in the intensity within an image to be investigated, in order to find, amongst other features, the dominant frequencies. The two-dimensional Discrete Fourier Transform of an N.N image f(x,y) is defined as follows: [GON-87]

$$F(u,v) = \frac{1}{N} \sum_{x=0}^{N-1} \sum_{y=0}^{N-1} f(x,y) \exp[-j2\pi(ux+vy)/N]$$

where $0 \leq u,v \leq N-1$. The inverse transform of F(u,v) is defined as:

$$f(x,y) = \frac{1}{N} \sum_{u=0}^{N-1} \sum_{v=0}^{N-1} F(u,v) \exp[j2\pi(ux+vy)/N]$$

where $0 \leq x,y \leq N-1$.

Several algorithms have been developed to calculate the two-dimensional Discrete Fourier Transform. The simplest makes use of the observation that this is a *separable transform* which can be computed as a sequence of two one-dimensional transforms. Therefore, we can generate the two-dimensional transform by calculating the one-dimensional Discrete Fourier Transform along the image rows and then repeating this on the resulting image but, this time, operating on the columns. [GON-87] This reduces the computational overhead when compared to direct two-dimensional implementations. The sequence of operations is as follows:

f(x,y) → Row Transform → $F_1(x,v)$ → Column Transform → $F_2(u,v)$

Although this is still computationally slow compared to other many shape measurements, the Fourier transform is quite powerful. It allows the input to be represented in the frequency domain, which can be displayed as a *pair* of images. (It is not possible to represent both amplitude and phase using a single monochrome image.) Once the processing within the frequency domain is complete, the inverse transform can be used to generate a new image in the original, so-called, *spatial* domain.

The Fourier power, or amplitude, spectrum plays an important role in image processing and analysis. This can be displayed, processed and analysed as an intensity image. Since the Fourier transform of a real function produces a complex function: $F(u,v) = R(u,v) + i.I(u,v)$, the frequency spectrum of the image is the magnitude function

$$|F(u,v)| = \sqrt{R^2(u,v) + I^2(u,v)}$$

and the power spectrum (spectral density) is defined as $P(u,v) = |F(u,v)|^2$

Figure 2.23 illustrates how certain textured features can be highlighted using the two-dimensional Discrete Fourier Transform. The image is transformed into the frequency domain and an ideal band-pass filter (with a circular symmetry) is applied. This has the effect of limiting the frequency information in the image. When the inverse transform is calculated, the resultant textured image has a different frequency content which can then be analysed. For more details on the Fourier transform and its implementations, see [PIT-93] and [GON-87].

(a)

(b)

(c)

(d)

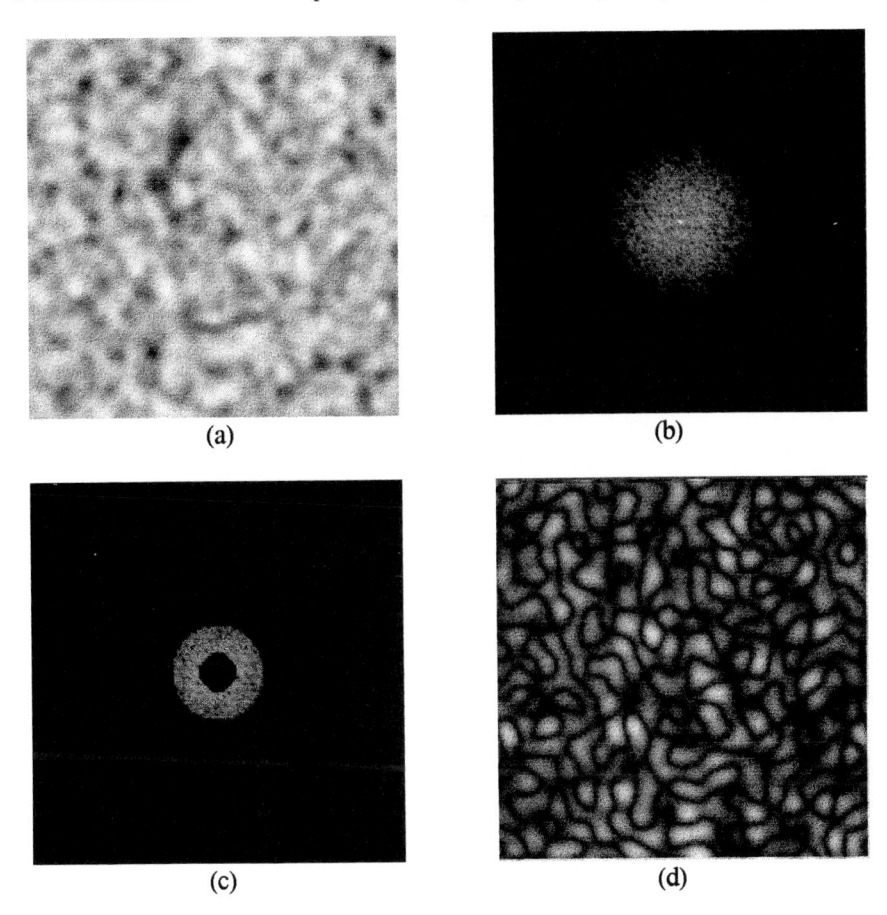

Figure 2.23 Filtering a textured image in the frequency domain. (a) Original textured image. (b) Resultant transformed image in the frequency domain after using the two-dimensional Discrete Fourier Transform. (The image is the frequency spectrum shown as an intensity function.) (c) Resultant frequency domain image after an ideal band-pass filter is applied to image. (d) The resultant spatial domain image after the inverse two-dimensional discrete Fourier transform is applied to the band-pass filtered image in (c).

2.7 Texture Analysis

Texture is observed in the patterns of a wide variety of synthetic and natural surfaces (e.g. wood, metal, paint and textiles). If an area of a textured image has a large *intensity* variation then the dominant feature of that area would be *texture*. If this area has little variation in *intensity* then the dominant feature within the area is *tone*. This is known as the *tone-texture concept*. Although a precise formal definition of texture does not exist, it may be described subjectively using terms such as *coarse, fine, smooth, granulated, rippled, regular, irregular* and *linear*, and of course these features are used extensively in manual region segmentation. There are two main classification techniques for texture: *statistical* and *structural*.

2.7.1 Statistical Approaches

The statistical approach is well suited to the analysis and classification of random or natural textures. A number of different techniques have been developed to describe and analyse such textures [HAR-79], a few of which are outlined below.

Auto-Correlation Function (ACF)
Auto-correlation derives information about the basic 2-D tonal pattern that is repeated to yield a given periodic texture. Although useful at times, the ACF has severe limitations. It cannot always distinguish between textures, since many subjectively different textures have the same ACF, which is defined as follows:

$$A(\delta x, \delta y) = (\sum_{i,j} [I(i,j).I(i+\delta x, j+\delta y)]) / \sum_{i,j} [I(i,j)]^2$$

where $\{I(i,j)\}$ is the image matrix. The variables (i, j) are restricted to lie within a specified window outside which the intensity is zero. Incremental shifts of the image are given by $(\delta x, \delta y)$. It is worth noting that the ACF and the power spectral density are Fourier transforms of each other.

Fourier spectral analysis
The Fourier spectrum is well suited to describing the directionality and period of repeated texture patterns, since they give rise to high energy narrow peaks in the power spectrum. (See Section 2.6 and Figure 2.23.) Typical Fourier descriptors of the power spectrum include: the location of the highest peak, mean, and variance and the difference in frequency between the mean and the highest value of the spectrum. This approach to texture analysis is often used in aerial/satellite and medical image analysis. The main disadvantage of this approach is that the procedures are not invariant even, under monotonic transforms of its intensity.

Edge Density

This is a simple technique in which an edge detector or high pass filter is applied to the textured image. The result is then thresholded and the edge density is measured by the average number of edge pixels per unit area. Two-dimensional, or directional filters/edge detectors, may be used as appropriate.

Histogram Features

This useful approach to texture analysis is based on the intensity histogram of all or part of an image. Common histogram features include: *moments, entropy dispersion, mean* (an estimate of the average intensity level), *variance* (this second moment is a measure of the dispersion of the region intensity), *mean square value* or *average energy, skewness* (the third moment which gives an indication of the histograms symmetry) and *kurtosis* (cluster prominence or "peakness"). For example a narrow histogram indicates a low contrast region, while two peaks with a well-defined valley between them indicates a region that can readily be separated by simple thresholding.

Texture analysis, based solely on the grey scale histogram, suffers from the limitation that it provides no information about the relative position of pixels to each other. Consider two binary images, where each image has 50% black and 50% white pixels. One of the images might be a checkerboard pattern, while the second one may consist of a salt and pepper noise pattern. These images generate exactly the same grey level histogram. Therefore, we cannot distinguish them using first order (histogram) statistics alone. This leads us naturally to the examination of the co-occurrence approach to texture measurement.

2.7.2 Co-occurrence Matrix Approach

The co-occurrence matrix technique is based on the study of *second-order grey level spatial dependency statistics*. This involves the study of the grey level spatial interdependence of pixels and their spatial distribution in a local area. Second order statistics describe the way grey levels tend to occur together, in pairs and therefore provide a description of the type of texture present. A *two-dimensional* histogram of the spatial dependency of the various grey level picture elements within a textured image is created. While this technique is quite powerful, it does not describe the shape of the primitive patterns making up the given texture.

The co-occurrence matrix is based on the estimation of the second order joint conditional probability density function, $f(p,q,d,a)$, for angular displacements, a, equal to *0, 45, 90* and *135* degrees. Let $f(p,q,d,a)$ be the probability of going from one pixel with grey level p to another with grey level q, given that the distance between them is d and the direction of travel between them is given by the angle a. (For N_g grey levels - the size of the co-occurrence matrix will be $N_g.N_g$.) For example, assuming the intensity distribution shown in the sub-image given below, we can generate the co-occurrence matrix for $d = 1$ and a is taken as 0 degrees.

2	3	3	3
1	1	0	0
1	1	0	0
0	0	2	2
2	2	3	3

Sub-image with 4 grey-levels.

Grey Scale	0	1	2	3
0	6	2	1	0
1	2	4	0	0
2	1	0	4	2
3	0	0	2	6

Co-occurrence matrix $\{f(p,q,1,0)\}$ for the sub-image.

A co-occurrence distribution that changes rapidly with distance, d, indicates a fine texture. Since the co-occurrence matrix also depends on the image intensity range, it is common practice to normalise the textured image's grey scale prior to generating the co-occurrence matrix. This ensures that first-order statistics have standard values and avoids confusing the effects of first- and second-order statistics of the image.

A number of texture measures (also referred to as texture attributes) have been developed to describe the co-occurrence matrix numerically and allow meaningful comparisons between various textures. [HAR-79] (See Figure 2.24.) Although these attributes are computationally intensive, they are simple to implement. Some sample texture attributes for the co-occurrence matrix are given below.

Energy, or angular second moment, is a measure of the homogeneity of a texture. It is defined thus,

$$Energy = \Sigma_p \Sigma_q [\, f(p,q,d,a)\,]^2$$

In a uniform image, the co-occurrence matrix will have few entries of large magnitude. In this case the *Energy* attribute will be large.

Entropy is a measure of the complexity of a texture and is defined thus:

$$Entropy = -\sum_p \sum_q [f(p,q,d,a).Log(f(p,q,d,a))]$$

It is commonly found that what a person judges to be a complex image tends to have a higher *Entropy* value than a simple one.

Inertia is the measurement of the moment of inertia of the co-occurrence matrix about its main diagonal. This is also referred as the *contrast* of the textured image. This attribute gives an indication of the amount of local variation of intensity present in an image.

$$Inertia = \sum_p \sum_q [(p-q)^2.f(p,q,d,a)]$$

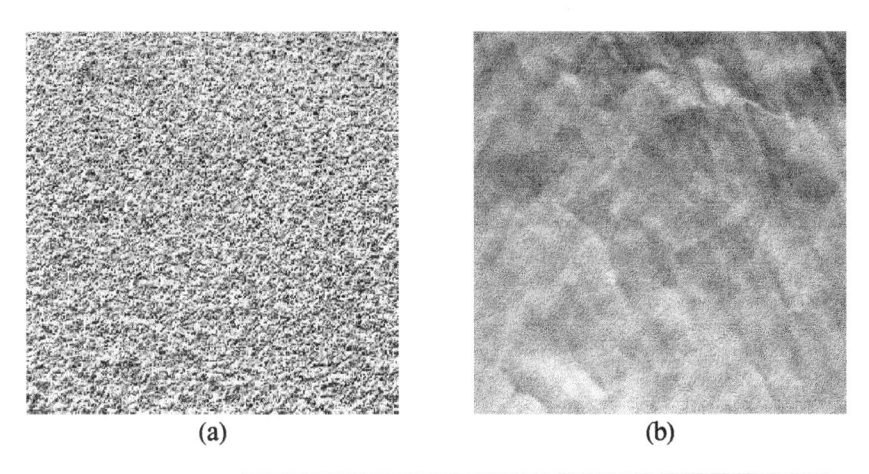

(a)

(b)

	Sand		Paper	
	f(p,q,1,0)	f(p,q,1,90)	f(p,q,1,0)	f(p,q,1,90)
Energy (x10^6)	1.63	1.7	3.49	3.42
Inertia (x10^8)	5.4	6.5	.181	.304

(c)

Figure 2.24 Co-occurrence based texture analysis. (a) Sand texture. (b) Paper texture. (c) Texture attributes.

2.7.3 Structural Approaches

Certain textures are deterministic in that they consist of identical *texels* (basic texture element), which are placed in a repeating pattern according to some well-defined but unknown placement rules. To begin the analysis, a texel is isolated by identifying a group of pixels having certain invariant properties, which repeat in the given image. A texel may be defined by its: grey level, shape, or homogeneity of some local property, such as size or orientation. Texel spatial relationships may be expressed in terms of adjacency, closest distance and periodicities.

This approach has a similarity to language; with both image elements and grammar, we can generate a syntactic model. A texture is labelled *strong* if it is defined by deterministic placement rules, while a *weak* texture is one in which the texels are placed at random. Measures for placement rules include: edge density, run lengths of maximally connected pixels and the number of pixels per unit area showing grey levels that are locally maxima or minima relative to their neighbours

2.7.4 Morphological Texture Analysis

Textural properties can be obtained from the erosion process (Sections 2.4 and 2.5) by appropriately parameterizing the structuring element and determining the number of elements of the erosion as a function of the parameters value [DOU-92]. The number of white pixels of the morphological opening operation as a function of the size parameter of the structuring element, H, can determine the size distribution of the *grains* in an image. Granularity of the image F is defined as:

$$G(d) = 1 - (\#[F \bigcirc H_d] / \#F)$$

Where H_d is a disk structuring element of diameter d or a line structuring element of length d, and $\#F$ is the number of elements in F. This measures the proportion of pixels participating in grains smaller than d.

2.8 Implementation Considerations

Of course, all of the image processing and analysis operators that have been mentioned above can be implemented using a conventional programming language, such as C or Pascal. However, it is important to realise that many of the algorithms are time-consuming when realised in this way. The monadic, dyadic and local operators can all be implemented in time $K.m.n$ seconds, where K is a constant that is different for each function and (m,n) define the image resolution. However, some of the global operators require $O(m^2.n^2)$ time. With these points in mind, we see that a low-cost, slow but very versatile image processing system can be assembled, simply by embedding a *frame-store* into a conventional desk-top computer. (A *frame-store* is a device for digitising video images and displaying computer-processed/generated images on a monitor.)

The monadic operators can be implemented using a look-up table, which can be realised simply in a ROM or RAM. The dyadic operators can be implemented using a straightforward *Arithmetic and Logic Unit* (ALU), which is a standard item of digital electronic hardware. The linear local operators can be implemented, nowadays, using specialised integrated circuits. One manufacturer sells a circuit board which can implement an 8x8 linear local operator in real-time on a standard video signal. Several companies market a broad range of image processing modules that can be plugged together, to form a very fast image processing system that can be tailored to the needs of a given application. Specialised architectures have been devised for image processing. Among the most successful are parallel processors, which may process one row of an image at a time (vector processor), or the whole image (array processor). Competing with these are *systolic array*, *neural networks* and *transputer* networks. See Dougherty and Laplante [DOU-95] for a discussion on the considerations that need to be examined in the development of real-time imaging systems.

2.8.1 Morphological System Implementation

While no single image processing operation is so important that all others can be ignored, it is interesting to consider the implementation of the morphological operators, since it reflects the range of hardware and software techniques that can be applied to achieve high speed.

There are two classical approaches to the implementation of morphological techniques on computer architectures, parallel and sequential (serial) methods. (See Section 2.4.) Klien and Serra [KLI-72], discuss an early example of one of the many commercial computer architectures for digital image processing which implement the basic morphological operations: erosion and dilation. Morphological operations with 3x3 pixel structuring elements, are easily implemented by array architectures, such as CLIP [DUF-73]. Other system implementations include Sternberg [STE-78]. Waltz [WAL-88b, WAL-94] describes examples of a near real-time implementation of binary morphological processing using large (up to 50x50 pixels), arbitrary structuring elements, based on commercially available image processing boards. The success of this approach, referred to as *SKIPSM* (*Seperated-Kernal Image Processing using Finite State Machines*), was achieved by reformulating the algorithm in such a way that it permitted high-speed hardware implementation. Similar algorithmic methods allow fast implementation of these operators in software.

A number of companies now manufacture industrial vision systems that incorporate video rate morphological operations, albeit with a limited range of structuring elements. These include Machine Vision Int., Maitre, Synthetic Vision Systems, Vicom, Applied Intelligence Systems and Leitz [HAR-87b].

2.9 Commercial Devices

In this section, we discuss generic types of computing sub-systems for machine vision, rather than giving details of existing commercial products, since any review of current technology would become out of date quite quickly. The

the mechanical and image acquisition and display interfaces is also a significant benefit when installing vision systems. However, it can be difficult to add further functionality at a later date without upgrading the system.

2.9.4 Turn-key Systems

Turn-key vision systems are self-contained machine vision systems, designed for a specific industrial use. While some such systems are custom designed, many turn-key systems contain commercially available plug-in cards. Turn-key systems tend to be designed for a specific market niche, such as can-end inspection, high-speed print recognition and colour print registration. So, not only is the hardware tuned for to deal with high-speed image analysis applications, it is also optimised for a specific imaging task. While the other systems discussed usually require significant development to produce a final solution for an imaging application, turn-key systems are fully developed, although they need to be integrated into the industrial environment. This should not be taken lightly, as this can often be a difficult task. It may not be possible to find a turn-key system for a specific application.

While we have avoided the discussion of any specific commercial devices, there are a number of valuable information sources available, some of these are provided by commercial organisations but some of the most valuable are free ! (See the Internet resource list in Appendix C.) One commercial resource that is well worth considering is *Opto*Sense®* [WHI-94]. This is machine vision database that gives details of a large number of machine vision vendors and their products and services.

2.9.5 Software

As was mentioned earlier, there is a large number of image processing, and analysis, packages available, for a wide range of computing platforms. Several of these packages are freely available over the Internet. (See Appendix C.) Some of these packages are tightly tied to a given vision system, while others are compiled for a number of host computers and operating systems. The majority of the software packages have interactive imaging tools that allow ideas to be tested prior to coding the for efficient operation. For more information on the hardware and software aspects of real-time imaging, including a survey of commonly used languages, see [DOU-95].

2.10 Further Remarks

The image processing operators described in this chapter have all found widespread use in industrial vision systems. Other areas of application for image processing may well use additional algorithms to good effect. Two key features of industrial image processing systems are the cost and speed of the target system (i.e. the one installed in a factory). It is common practice to use a more versatile and slower system for problem analysis and prototyping. While the target system must continue to operate in an extremely hostile environment. (It may be hot, greasy, wet and or dusty.) It must also be tolerant of abuse and neglect. As far as possible, the target system should be self-calibrating and able to verify that it is "seeing" appropriate images. It should provide enough information to ensure that the factory personnel are able to trust it; no machine system should be built that is a viewed by the workers as a mysterious black box. Consideration of these factors is as much a part of the design process as writing the software. (See Appendices A and B.)

3

Intelligent Image Processing

3.1 Interactive Image Processing

Simply reading the previous chapter, or any other text on image processing, does not, on its own, equip a person to design an effective industrial vision system. A person cannot examine a picture by eye and decide what algorithms are necessary for filtering, analysing and measuring it. Proof of this is not hard to find: try it, but be prepared for disappointment! In the past, many people have adopted this approach, only to discover later that, what they were convinced would be an effective image processing algorithm, was not reliable and effective in practice, or did not work at all. Inspection of an object or scene by eye, followed by introspective self-analysis is now totally discredited as a method of choosing image processing algorithms for machine vision systems. Over the last two decades, this has gradually been accepted as one of the central tenets of the machine vision systems development process.

It is now widely accepted that an interactive "tool-box" is needed, by which a person can experiment with image processing algorithms. Facilities of this kind were originally developed in the 1970s. One of the earliest interactive image processing systems was called *SUSIE* (*Southampton University System for Image Evaluation*. [BAT-79]) From that, other systems were spawned, notably *Autoview* ([BAT-92b], sold in USA by 3M Company, under the name *System 77*) and *VCS*. [VCS] These systems are all related to one another, in having similar command repertoires, although they use different mnemonic command names. The set of image processing functions used in the following pages is listed in Appendix E, and forms the basis of the Prolog+ language, which we shall describe in this chapter. The appropriate mnemonics have also been listed beside the image processing functions described in Chapter 2.

A detailed account of the techniques and benefits of interactive image processing is given elsewhere. [BAT-92b]

3.1.1 Modus Operandi

A person working with an interactive image processor sits in front of a computer terminal, with either one large, high-resolution screen, or several medium-resolution ones. Displayed on the terminal are various text messages and at least two images, called the *current* and *alternate images*. It is bad practice, though all too common, to superimpose the textual data onto the pictures, since neither can be seen properly.

As the user types each command, he sees the results of the corresponding operation, displayed on the terminal screen almost instantaneously. For example, the user might type the command *neg*, followed by the *"Return"* key. (The result of this particular operation is that the negative of the current image is computed.) The original current image replaces the alternate image, which is discarded. (Figure 3.1) The user might type another command, e.g. *thr(97,126)*, which specifies numeric parameters, in order to perform thresholding.

Measurements on images can also be obtained. For example, the user might type *avr* in order to measure the average intensity. The average intensity value is printed on the text display for the user to read. Dyadic operations (such as *add, sub, mul, max* and *min*) use both the current and alternate images. Again, the current image is transferred to the alternate image display and the result is displayed as the new current image. The image digitise operation (*grb*), reading a picture from archive (disc) store (*rea*), and image generation (e.g. *wgx*) all move the current image into the alternate image display and put the new image into the current image display. At any time during the interactive process, the user can press the *"Return"* key, to interchange the current and alternate images. Thus, the user can always undo the last operation performed, since the previous (current) image is always retained. (Of course, the dyadic functions, are not exactly reversible, because the alternate image is lost.) The user can also examine the results of each processing step, since the "before" and "after" images are displayed side by side and can be compared directly. As each new command is typed, it is recorded in the text display, thereby providing the user with a record of the interactive session.

3.1.2 Prototyping Inspection Systems

Provided *most* of the interactive image processing functions take less than one second to execute, the user is able to maintain a high level of concentration during quite a long session, since he feels as though he is in complete control of the machine. Longer execution times than this simply reduce the effectiveness of the interaction, although most users can tolerate a few (exotic and infrequently used) image processing operators having execution times of 30 seconds. It is good practice, when adding new functions, to keep execution times less than 10 seconds, if at all possible.

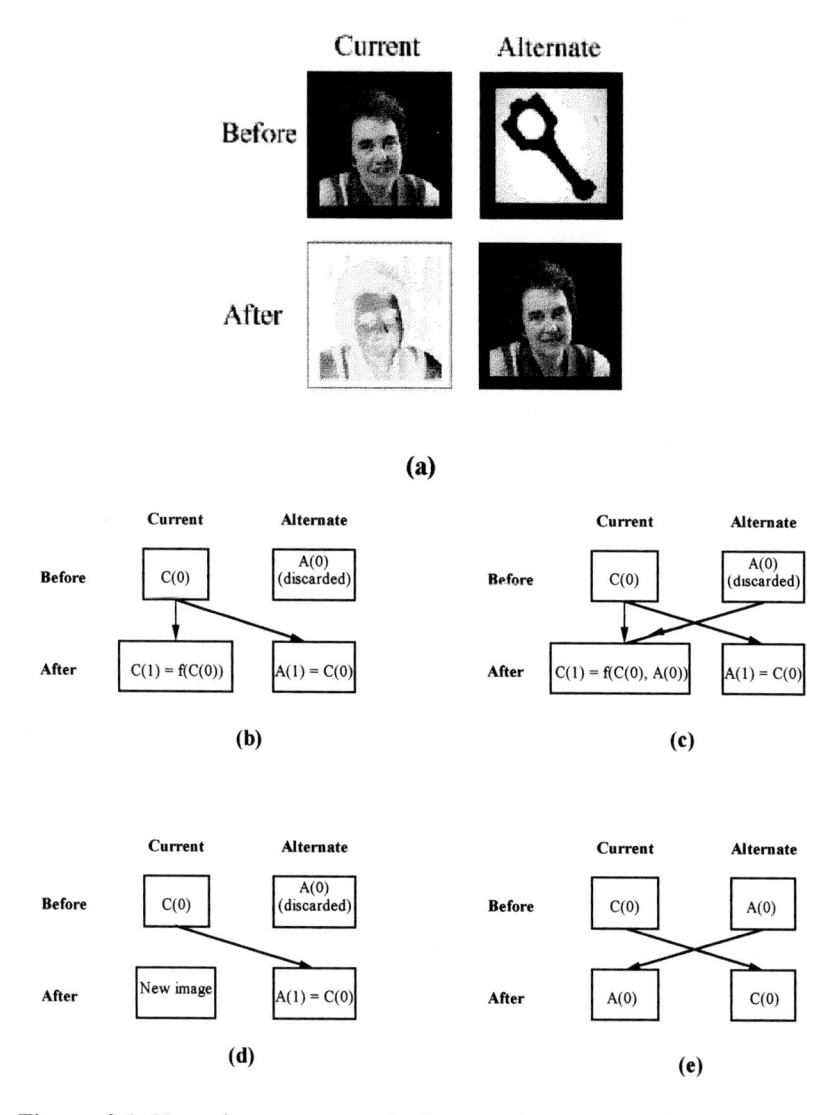

Figure 3.1 How the current and alternate images are changed during interactive processing. (a) Command *neg*. The picture on the top-left, will be denoted by P, and is displayed in the *current image* area, while that on the top-right will be called Q and is displayed in the *alternate image* area. After *neg* has been performed, the negative of P is displayed as the current image. Image Q has been discarded and image P is now displayed in the alternate image area. (b) Diagrammatic representation of monadic and local operators. Examples: *neg, thr, lpf, lnb, exw, bed.* (c) Dyadic operators. Examples: *add, sub, mul, max, min.* (d) Image digitisation (*grb*), reading images from disc (*rea*) and image generation (e.g. *zer, wgx, hic*). (e) Switch images (*swi*).

An experienced user of an interactive system can very often identify what processing steps are appropriate for a given application task, in less than an hour. We assume here, of course, that a satisfactory solution exists. If it does not, the user will in a short time discover that fact. Numerous times during the last two decades, representatives of a manufacturing company have visited the authors laboratories, bringing with them samples of a certain product that is causing them particular concern at that time. Typical product faults reported during such a visit are scratches, cracks, stains, malformations, swarf, foreign bodies etc. Very often such defects can only be detected visually, although human inspectors are unreliable. The purpose of the visit is, of course, to explore the possibility of using a machine vision system to inspect the objects, either during the manufacturing process, or just after they have been made. After an initial period of experimentation, when the lighting and viewing system is set up, the interactive image processor is used to find procedures that are able to perform the inspection. In the experience of the authors and many of their colleagues, working in different companies, the process of designing / choosing an algorithm using an interactive image processor takes only a short time, typically a few hours, or even minutes! At the end of the exercise, the experimenter can, of course, be *certain* that the algorithm he has discovered actually works!

3.1.3 Building Simple Programs

One important feature of many of the interactive image processing system mentioned above is that they allow *macros* to be written. By combining several commands in a macro, new image processing procedures can often be written, without the need to program at the pixel level. For example, drawing isopohotes (intensity contours), or plotting the intensity profile along a given line, are tasks that are ideally suited for implementation using macros. Just as a small repertoire of arithmetic operators / functions (+, -, *, /, sin, cos, log, exp, etc.) is sufficient to represent a large class of mathematical functions, so a modest set of image processing operator can be used to define a much larger collection of macros. *SUSIE, Autoview* and *VCS* all have macro facilities, which are described in detail in [BAT-91]. Although macros have been found to be very useful in the past, they do not provide the full range of facilities needed for certain applications.

Interactive image processing is regarded, by the authors, as providing an essential prelude to the more conventional approach to writing and developing vision programs. It must be made clear, however, that interaction cannot fully replace the conventional approach to program-writing. Nor is the latter approach sufficient on its own, as a means of developing software for vision systems. For these reasons, it was decided, some years ago, to search for a way of combining the interactive and conventional approach to writing programs, in an attempt to provide the benefits of both. As we shall see in the following pages, considerable success has been achieved by embedding a collection of image processing commands within the Artificial Intelligence language, Prolog. Some systems allow programs to be written using a more conventional language, such as Basic or C, with image processing operations being made available in the form of a library of sub-routines. Naturally, the same mnemonics are used for interactive and program-mode operation, to make program writing as easy a possible. The following is a brief VCS program, which shows image processing operations being performed within a Basic program.

```
100     for i = 1 to 100     % Begin outer loop
200     grb                  % Digitise an image
300     for j = 1 to 5       % Begin inner loop
400     lpf                  % Low pass filter
500     next j               % Terminate inner loop
600     gli min max          % Find the maximum and minimum
                             % intensities
700     X = max - min        % Compute intensity range
800     print "Intensity range =", X
                             % Print result
900     next i               % Terminate outer loop
1000    stop                 % Program finished
```

It is normal practice for the VCS [VCS] user to develop an image processing sequence, interactively, then include the newly discovered function sequence within a Basic-like program. Several other image processing languages have been developed using the same basic idea, but with software founded upon Pascal, Fortran, or C. Image processing systems have also been developed around other languages, including APL, Forth, Smalltalk and Lisp.

3.1.4 Interaction and Prolog

In this chapter, we shall follow the same general route: embedding image processing operations within Prolog. As we hope to show throughout the remainder of this book, Prolog provides a powerful and natural mode of programming for controlling image processing. Compared to the more conventional "imperative" computer languages, such as Basic, Pascal, Fortran, and C, the benefits of Prolog are not widely known amongst the general population of computer users.

It was decided very early in the search for a suitable language as a "host" for image processing that the benefits of interaction must not be lost. Despite their great love of the Macintosh / Windows-style of operating environment, the authors are convinced that interactive image processing is still best performed using the *command-line* mode of operation. They have found no real alternative to that. Pull-down menus and customised dialogue boxes are undoubtedly useful for novice users but are of more limited value for more experienced users, involved in prototyping image processing. To appreciate this fully, one has to witness a command-line interactive image processor in the hands of an expert user. Better still, one has to become an expert user. Selecting items from a pull-down menu is slower and generally less convenient when values for arguments have to be specified. Of course, a much greater investment of time and effort must be made to learn to use a command-line system but the authors maintain that this is well worthwhile. Their experience does not support the approach taken in the design of many of the user interfaces built into many modern commercial image processing packages. Some of these rely almost exclusively on pull-down menus and customised dialogue boxes. *Image analyst, IP-Lab* [IPL], *OptiLab* [OPT], *Photoshop* [PHO], *Image* [IMA] and *Visilog* [VIS] are all in this category. Prolog+, the extension to standard Prolog that we shall describe below, permits the use of command-line, pull-down menus, customised dialogue boxes and screens, and other modes of operation, not yet mentioned. Its *primary* mode of interaction is via command line control, although each user needs, at times, to use different interaction tools. Moreover, an expert user can program these himself, thereby creating new interface tools, to suit his own or the application requirements.

3.2 Introducing Prolog+

The image processing techniques described in the previous chapter are not particularly intelligent, in the sense that many potential applications of machine vision require greater reasoning power than they can provide. One of the main themes of this book is that an intelligent program-controlled image processing system can be made, by the simple expedient of embedding image manipulation and measurement routines within standard Prolog. For the moment, we may think of the image processor as being a hardware peripheral device that is attached to a computer running a Prolog program, via a bi-directional, low-bandwidth, serial line. (Figure 3.2) The top-level control language for this system will be called Prolog+[1] and, as we shall see, is able to solve many of the problems that do not yield to standard image processing techniques operating alone. Prolog+ can also be implemented using software, with no specialised hardware whatsoever. Many high-performance desk-top computers are provided with a video input. For example, many of the Macintosh family of computers can be fitted with an *"AV"* card, to provide the ability to digitise images. In addition, a small, cheap, medium-quality camera [QUC] is available which plugs into the serial (RS422) port of virtually any Macintosh computer. A Prolog+ system can be built around these or other standard hardware platforms, needing only software to run. It is also possible to build a Prolog+ system which controls dedicated electronic hardware for fast image processing.

Since Prolog is a rather unusual computer language, having been devised originally for AI work, we shall describe it briefly in the following pages. However, to obtain a more comprehensive understanding of this fascinating

---

[1] Prolog+ should not be confused with Prolog++, which is an extension of LPA MacProlog [MAC], providing Object Oriented Programming (OOP) facilities. Hence, it is a trivial matter to extend Prolog+ further, so that it permits OOP image processing programs to be written. To date, the authors have not yet developed this idea and are not aware of any other work in this area.

language, the uninitiated reader is urged to refer to the standard textbooks. [CLO-87, BRA-90] Throughout this book, we shall attempt to justify the choice of Prolog as the basis for intelligent image processing, by describing and listing Prolog+ programs which together solve a range of diverse machine vision problems. For the moment, let it suffice to say that many major benefits have been obtained through the use of Prolog that could not have been obtained nearly so easily using the other languages listed above, except possibly Lisp. The authors can report that, in their experience, which now stretches over ten years, no serious shortcomings of this approach have come to light.

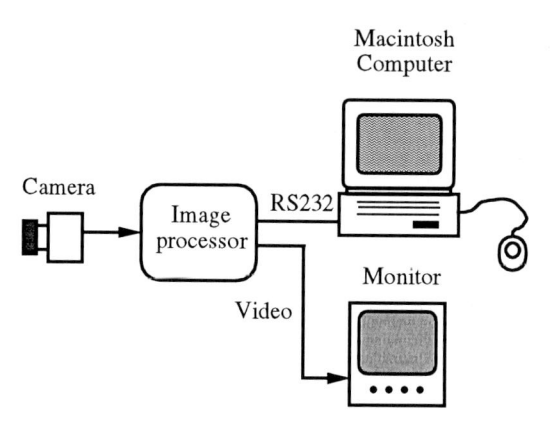

Figure 3.2 A simple approach to implementing Prolog+. The Macintosh computer is programmed in Prolog.

Prolog+ was originally intended as a vehicle for developing intelligent image processing procedures. Prolog+ programs can be written, edited and run, just as one would expect using Prolog. It is a superset of standard (Edinburgh) Prolog and, is also intended to provide the same interactive image processing facilities as *SUSIE* [BAT-79], *Autoview* [BAT-92b], *SuperVision* [INT] and *VCS* [VCS]. Thus, Prolog+ provides facilities for both programmed and interactive image processing and is well placed to assist in prototype development and problem analysis for industrial vision applications. Hardware implementations of Prolog+ are described later in this chapter. (See Appendix D for a discussion on software implementation of Prolog+.) In addition, a comprehensive operating environment, centred on a version of Prolog+ implemented using LPA MacProlog [MAC], has been devised and will be discussed in detail in the following chapter. This includes auto-starting, user-extendible pull-down menus, cursor, on-line HELP, automatic script generation, replaying recorded speech, speech synthesis, speech recognition, advisory programs ("expert systems") for machine vision system designers, and various demonstration packages. In Chapter 4, we shall also discuss the use of Prolog+ for Natural Language (NL) understanding, interfacing to other languages / software packages and building networks of multi-camera / multi-processor machine vision systems. In addition to its ability to perform image processing, Prolog+ also provides facilities for controlling a range of external hardware devices, such as robots, an (X,Y,θ)-table, cameras (pan, tilt, focus and zoom), relays, solenoids, computer-controlled lighting, etc. and this will be discussed in Chapter 5. Apart from analysing data from a video source, Prolog+, used in conjunction with an appropriate hardware interface unit, can sense and act upon information derived from a range of other devices: proximity sensors, thermocouples, pressure gauges, instruments measuring optical activity, pH and salinity, ultra-sonic range sensors, digital micrometers, etc. We shall say more about this in Chapter 5, while in Chapter 6, we shall concentrate upon the specialised topic of recognising colours specified by name. Again, we shall use the Prolog+ language. However, for the moment, we must concentrate upon first principles, i.e. the ability of Prolog+ to perform complex image processing operations.

The repertoire of image processing commands embedded within Prolog+ is evident from the discussion in the previous chapter and also from Appendix E. The list of image processing operators is constantly growing, as new primitives are being added.

3.3 Review of Prolog

It took one of the authors (B.G.B.) two years to understand *why* Prolog is important and just two weeks to become proficient at using it! The reason why Prolog is worth further study is simply that it allows a novel and very natural mode of programming to be used. It permits a person to state the nature of a solution, rather than *how* to find one. To understand this, consider the task of finding a marriage partner for a given hypothetical man. It is relatively easy to specify the basic "requirements", to a dating agency, in terms such as those listed below[2]:

Sex:	Female
Age:	[45,55]
Height:	[150,180]
Weight:	[45,75]
Language:	English
Personality:	A long list of desirable characteristics might be included here.

Clearly, this list is incomplete but it is sufficiently detailed to allow us to illustrate the general principles of what is known as *Declarative Programming*. Writing a Prolog program to find a wife / husband is straightforward and has

---

[2] This list is intended merely to illustrate a point about Prolog; it does not make any statement about the absolute desirability of any type of personality, or racial characteristics.

actually been used by dating agencies. Here is the program to find a wife, using the very small number of criteria specified above:

```
find_wife(X) :-
   person(X),              % Find a person called X
   sex(X,female),          % Is person X female?
   age(X,A),               % Find the age, A, of person X
   A ≥ 45,                 % Is person ≥ 45 years old?
   A ≤ 55,                 % Is person ≤ 55 years old?
   height(X, H),           % Find the height, H, of person X
   H ≥ 150,                % Is person at least 150cm tall
   H ≤ 180,                % Is person at most 180cm  tall
   weight(X,W),            % Find the weight of person X
   W ≥ 45,                 % Does person weigh at least 45 kg
   W ≤ 75,                 % Does person weigh at most 75 kg
   speaks(X,English).      % Does X speak English?
   must_be(X, [kind, truthful, generous, loving, loyal]).
                           % Obvious meaning[3].
```

Given such a program and a set of stored data about a collection of people, Prolog will search the database, to find a suitable match. There is no need to tell Prolog how to perform the search. The reader is urged to consider rewriting the program using Basic, C, Fortran, or Pascal. The program will take much longer to write, will consist of many more lines of code and the result will be altogether much less satisfactory, because a person programming a computer in one of these languages has to impose an unnatural mode of thought on the problem. Prolog programming is very much more natural in its style, since it allows a person to think more directly about the type of solution required, rather than how to find it. Prolog differs from most other computer languages, such as Pascal, C, Forth, APL, Occam, Lisp, Fortran and assembly code, in several very important ways.

(a) Firstly, a Prolog "program"[4] does *not* consist of a sequence of instructions, as routines written in these other languages do. Instead, it is a description of (part of) the world. For this reason, Prolog is referred to as a *declarative* language, whereas most other computer languages, military orders, knitting patterns, automobile repair manuals, musical scores and culinary recipes are all examples of *imperative* languages. This is a vital difference, which distinguishes Prolog (and a very small group of related languages) from the well-known conventional languages of Computer Science.

(b) The "flow of control" in a Prolog program does not follow the normal convention of running from top to bottom. We shall see later that the flow is

---

[3] A suitable definition of *must_be* might be as follows:
```
       must_be(_,[]).
       must_be(A, [B|C]) :-
              personality(A,B), !, must_be(A,C).
```

[4] The correct term is "application", since a program is, strictly speaking, a sequence of instructions. However, we shall continue to use the term "program", since this is more familiar to most readers.

just as likely to be in the reverse direction, through a control mechanism called *back-tracking*.

(c) Through the use of back-tracking, it is possible to make and subsequently revise temporary assignments of values to variables. This process is called *instantiation / de-instantiation*[5] and is akin to re-evaluating assumptions made earlier in life. Instantiation is performed, in order to try and prove some postulate, theorem or statement, which may or may not be true. As far as Prolog is concerned, theorem proving is the equivalent process to running or executing an imperative language program.

(d) It is possible to make very general statements in Prolog in a way that is not possible in most other languages, such as those listed above. We shall see more of this feature later, but for the moment, let us illustrate the point with a simple example. In Prolog it is possible to define a relationship, called *right* in terms of another relationship, called *left*.

| In English: | "A is to the *right* of B if B is to the *left* of A." |
| In Prolog: | `right(A,B) :- left(B,A).` |

(Read ":-" as "*can be proved to be true if*" or more succinctly as "*if*".) Notice that neither A nor B have yet been defined. In other words, we do not need to know what A and B are in order to define the relationship *right*.[6] For example, A and B might be features of an image such as blob centres or corners. Alternatively, A and B may be political "objects", either people (*Hitler* and *Lenin*) or policies (*National Socialism* and *Marxism*).The point to note is that Prolog+ allows the relationships *left* and *right* to be applied to any such objects, with equal ease.[7]

(e) Prolog makes very extensive use of *recursion*. While, Pascal, C and certain other imperative languages also allow recursion, in Prolog it forms an essential control mechanism.

Prolog was devised specifically for and has subsequently been applied extensively in Artificial Intelligence. It is, for example, one of the prime tools for research in Natural Language Understanding [GAZ-89] and has been used for such tasks as planning a complex sequence of actions, given applications constraints. It excels as a basis for writing rule-based programs for decision-making and it is a straightforward matter to write expert systems in Prolog. However, Prolog is not suitable for writing programs requiring a large amount of

---

[5] The word *"instantiation"* is derived from the same linguistic root as *"instance"*. Prolog tries to find an *instance* of some variable(s) which cause the given predicate to be true.

[6] *left* is defined in Section 3.5.5.

[7] Is *hitler* to the right of *lenin*? In the political sense, "yes", while the answer is "no", when we consider at the layout of words in the preceding sentence.

numerical manipulation. Nor is Prolog appropriate for real-time control, or other computational processes requiring frequent processing of interrupts.

3.3.1 Sample Program

This section is intended to refresh the memories of readers who have previously encountered Prolog; it is not intended as an introduction for people who have never seen the language before. The following program deals with the ancestry and ages of members of two fictitious families.

```
/* The following facts specify in which years certain people were
born.
Interpretation: born(roger,1943) means that
      "roger was born in 1943".
*/
born(roger,1943).
born(susan,1942).
born(pamela,1969).
born(graham,1972).
born(thomas,1953).
born(angela,1954).
born(elizabeth,1985).
born(john,1986).
born(marion,1912).
born(patricia,1911).
born(gertrude,1870).
born(david,1868).

/* These facts describe the parent-child relationships which exist
in the families.
Interpretation: parent(X,Y) means that
      "X is a parent of Y".
*/
parent(roger,pamela).
parent(roger,graham).
parent(patricia,roger).
parent(anne,patricia).
parent(david,patricia).
parent(marion,susan).
parent(susan,graham).
parent(susan,pamela).
parent(thomas,john).
parent(angela,john).
parent(thomas,elizabeth).
parent(angela,elizabeth).

/* Defining a relationship called "child". Read this as follows:
      "A is a child of B if
            B is a parent of A."
*/
child(A,B) :- parent(B,A).
```

```prolog
/* Defining a relationship called "older". Read this as follows:
      "A is older than B if
              the age of A is X AND
              the age of B is Y AND
              X > Y". */
older(A,B) :-
      age(A,X),
      age(B,Y),
      X > Y.

/* Defining a relationship "age". Read this as follows:
      "A has age B if
              A was born in year X AND
              it is now year Y AND
              X ≤ Y AND
              B is equal to Y - X". */
age(A,B) :-
      born(A,X),
      date(Y,_,_),
      X ≤ Y,
      B is Y - X.

/* The definition of "ancestor" has two clauses. Prolog always
tries to satisfy the top one first. If this fails, it then tries to
satisfy the second clause.
Interpretation: ancestor(A,B) means that
      "A is an ancestor of B."
The first clause should be interpreted as follows:
      "A is an ancestor of B if A is a parent of B".
*/
ancestor(A,B) :- parent(A,B).

/* The second clause should be interpreted as follows:
      "A is an ancestor of B if
              A is a parent of Z AND
              Z is an ancestor of B."
      Notice the use of recursion here. */
ancestor(A,B) :-
      parent(A,Z),
      ancestor(Z,B).

/* Definition of "print_descendents".
      This uses backtracking to find all possible solutions. The
      first clause always fails but in doing so it prints the
      descendants and their dates of birth.
*/
print_descendents(A) :-
      nl,                        % New line
      write('The known descendants of '),
                                 % Print a message
      write(A),                  % Print value of A
      write(' are:'),            % Print a message
      ancestor(A,Z),             % Find Z such that A is ancestor of Z
      born(Z,Y),                 % Z was born in year Y
      nl,                        % New line
      tab(10),                   % 10 white spaces
      write(Z),                  % Print value of Z
      write(', born '),          % Print a message
      write(Y),                  % Print value of Y
      fail.                      % Force back-tracking

% The second clause always succeeds and prints a new line
print_descendents(_) :- nl.
```

3.3.2 Sample Queries

```
Query:   born(susan, 1942)
YES

Query:   born(susan, X)
X =  1942
YES

Query:   born(X, 1942)
X =  susan
YES

Query:   born(X, Y)
X =  roger
Y =  1943

X =  susan
Y =  1942

X =  pamela
Y =  1969

X =  graham
Y =  1972

X =  thomas
Y =  1953

X =  angela
Y =  1954

X =  elizabeth
Y =  1985

X =  john
Y =  1986

X =  marion
Y =  1912

X =  patricia
Y =  1911

X =  gertude
Y =  1870

X =  david
Y =  1868

NO MORE SOLUTIONS
(Notice the alternative
 solutions generated by
 this general query.)
```

```
Query:   age(marion, Z)
Z =  77
YES

Query:   older(marion, susan)
YES

Query:   older(susan, marion)
NO
(This really means NOT PROVEN)

Query:   child(susan, Z)
Z =  marion
YES

Query:   ancestor(susan, Z)
Z =  graham
Z =  pamela

NO MORE SOLUTIONS
(Notice the alternative solutions.)

Query:   ancestor(Z, graham)
Z =  roger
Z =  susan
Z =  patricia
Z =  anne
Z =  david
Z =  marion

NO MORE SOLUTIONS

Query:   print_descendents(marion)

The known descendants of marion are:
          susan, born 1942
          graham, born 1972
          pamela, born 1969
YES

Query:   print_descendents(anne)

The known descendants of anne are:
          patricia, born 1911
          roger, born 1943
          pamela, born 1969
          graham, born 1972
YES

Query:   print_descendents(wilfred)
The known descendants of wilfred are:
YES
(There are no known descendents of
 wilfred.)
```

3.4 The Nature of Prolog+

The reader is reminded that Prolog+ is an extension of standard Prolog, in which a rich repertoire of image processing functions is available as a set of built-in predicates. The way that the image processing predicates operate follows the standard pattern established for printing in Prolog (*c.f. nl, write, tab*). That is, they always succeed but are never resatisfied on back-tracking. For example, as a "side effect" of trying to satisfy the goal *neg*, Prolog+ calculates the negative of the current image. (Figure 3.1) The blurring operator (low-pass filter, *lpf*) also acts on the current image. The goal *thr(125,193)* performs thesholding, setting all pixels in the current image in the range [125,193] to white and all others to black. The goal *cwp(Z)* always succeeds and instantiates Z to the number of white pixels in the current image. However, the goal *cwp(15294)* will only succeed if there are exactly 15294 white pixels in the current image. While Prolog+ is trying to prove the compound goal *[avr(Z),thr(Z)]*, Z is instantiated to the average intensity within the current image (calculated by *avr*). This value is then used to define the threshold parameter used by *thr*. With these points in mind, we are ready to examine our very first Prolog+ program.

```
grab_and_threshold :-
      grb,              % Digitise an image
      lpf,              % Blur (Low-pass filter)
      avr(Z),           % Calculate average intensity
      thr(Z).           % Threshold at average intensity level
```

Since each sub-goal in this simple program succeeds, the effect is the same as we had written a sequence of commands using a conventional (i.e. imperative) computer language. The goal *grab_and_threshold* always succeeds. However, the following Prolog+ program is a little more complicated.

```
big_changes(A) :-
      repeat,           % Always succeeds on backtracking
      grb,              % Digitise an image from the camera
      lpf,              % Perform lpf (low-pass filter)
      lpf,              % Perform lpf
      lpf,              % Perform lpf
      sca(3),           % Retain only 3 bits of each intensity value
      rea,              % Read image stored during previous cycle
      swi,              % Switch current & alternate images
      wri,              % Save the image for the next cycle
      sub,              % Subtract the two images
      abs,              % Compute "absolute value" of intensity
      thr(1),           % Threshold at intensity level 1.
      cwp(A),           % A is number of white pixels in image
      A > 100.          % Are differences between images significant?
```

The operator *repeat* succeeds, as does each of the image processing operators, [*grb, lpf, ... , cwp(A)*]. If the test *A > 100* then fails, the program back-tracks to *repeat*, since none of the image processing predicates is resatisfied on backtracking. (Remember that *repeat* is always resatisfied on backtracking.) Another image is then captured from the camera and the whole image processing sequence is repeated. The loop terminates when *A* exceeds 100. When this happens, the goal *big_changes(A)* succeeds and *A* is instantiated to the number of white pixels in the difference image. The goal *big_changes(A)* performs the processing *sequence [grb,... , cwp(A)]* an indefinitely large number of times and only succeeds when two consecutive images are found that are significantly different from one another. This program could be used as the basis for an intruder alarm, which signals when a person enters a restricted area. By adjusting the number in the final sub-goal (i.e. *A>100*), it is possible to tolerate between small changes (e.g. a cat wandering in front of the camera) while still being able to detect larger objects, such as a person being in view.

Although image processing commands, such as *grb*, *thr*, *cwp* etc. are always satisfied, errors will be signalled if arguments are incorrectly specified. Since *thr(X)* requires one numeric argument, the goal will fail if X is uninstantiated. On the other hand, the following compound goal is satisfied [*X is 197, thr(X)*], as is

```
gli(A,B),          % Get lower & upper limits of intensity
C is (A+B)/2,      % Average them
thr(C).            % Use average value as threshold⁸
```

The compound goal [*X is 186, Y is 25, thr(X,Y)*] fails, since *thr* fails when its second argument is less than the first one,[9] while [*X is 1587, thr(X)*] fails, because the parameter (X) is outside the range of acceptable values, i.e. [0,255]. Notice that *thr* has already been used with different numbers of arguments (different *arities*). Throughout this book, we shall use *thr* with 0, 1 or 2 arguments. Other image processing predicates will be treated in the same way. For example, we have already used *wri* (write image to disc) and *rea* (read image from disc) without arguments. We shall also encounter them with a single argument, which is instantiated to a string of alpha-numeric characters. Such arguments may be generated according to the usual Prolog conventions. For example, the Prolog symbol generator, *gensym*, may be used to create a series of file-names, *image_file1, image_file2,...*, as the following illustration shows:

```
process_image_sequence :-
    grb,                        % Digitise an image
    process_image,              % Process the image
    gensym(image_file,X),       % Generate new symbol name
    wri(X),
    process_image_sequence.     % Repeat processing
```

---

[8] The observant reader will notice that C can be instantiated to an integer ((A+B) is even) or a decimal value ((A+B) is odd). The effect of the latter is the same as if C is first rounded *down*, before trying to satisfy the goal *thr(C)*.

[9] The image processor signals an error, which causes the failure of the Prolog goal, *thr(186,25)*.

Notice here that we have "condensed" almost all of the image processing into the subsidiary predicate, *process_image*. Using this approach, a simpler, revised version of *big_changes* may be defined using the subsidiary predicate, *process*.

```
big_changes(A) :-
    process       % Listed below
    cwp(A),       % Instantiate A to number of white pixels in image
    A > 100.      % Test to see whether differences between images
                  % are large. If not, back-track to process.
```

where *process* is defined thus:

```
process :-   grb, lpf, lpf, lpf, sca(3), rea, swi, wri,
             sub, abs, thr(1).
```

The use of subsidiary predicates, such as *process_image* and *process*, allows the programmer to think at a higher conceptual level and to defer deciding what image processing is to be performed until later.

3.5 Prolog+ Programs

Now that we have illustrated the basic principles of Prolog+, we are in a position to be able to consider more complex programs. In this section, we shall present a range of more advanced programs, which illustrate various features of Prolog+. It is important to realise that we must always use Prolog+ to describe the *image* generated by the camera, not the object / scene being inspected. The importance of this point cannot be over-emphasised. Additional points of general interest will be discussed as they arise.

3.5.1 Recognising Bakewell Tarts

Consider Figure 3.3, which shows diagrammatic side and plan views of a small cake, popular in Britain and which is called a Bakewell tart. Now, let us use Prolog+ to describe the image obtained by viewing such a cake from above.

```
bakewell_tart :-
    segment_image,    % Convert image to form shown in Fig. 3.3
    outer_edge,       % Check the outer edge
    cherry,           % Check the cherry
    icing.            % Check the icing
```

Programs written in Prolog+ are almost invariably written from the top level downwards. In this instance, *bakewell_tart* was the first predicate to be defined. Notice that there are four obvious stages in verifying that the tart is a good one:

(a) Simplify the image (to a 4-level form), using *segment_image*.
(b) Check the integrity of the outer edge, using *outer_edge*.

(c) Check the presence, size and placing of the cherry, using *cherry*.
(d) Check the icing, using *icing*.

Even a novice Prolog programmer can understand that *bakewell_tart* is only satisfied if all four of the subsidiary tests succeed. The secondary predicates, though not defined yet, are not necessary for us to understand the process of recognising a Bakewell tart. Three of these are defined below. (*segment_image* is not given here, because it is problem specific and would distract us from the main point.)

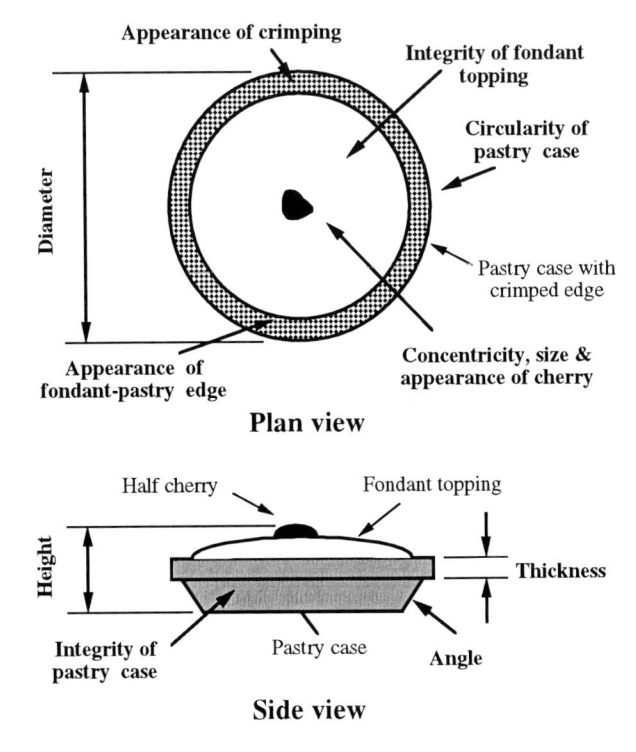

Figure 3.3 Critical dimensions of a Bakewell tart.

```
outer_edge :-
      thr(1),
      circular.          % Select outer edge
                         % Standard test for circularity. Defined below

cherry :-
      thr(1),            % Select outer edge
      cgr(X1,Y1),        % Centroid of outer edge
      swi,               % Switch images
      thr(200),          % Select cherry
      swi,               % Switch images - restore image for use later
      cgr(X2,Y2),        % Centroid of the cherry
      distance([X1,Y1,],[X2,Y2],D),
                         % D is distance [X1,Y1] to [X2,Y2],
      D < 20.            % Are cherry and outer edge nearly concentric?
```

```
icing :-
    thr(128),           % Select icing
    rea(mask),          % Read annular mask image from disc
    xor,                % Calculate differences between these images
    cwp(A),             % Calculate area of white region
    A > 50.             % Allow a few small defects in icing

circular :-
    cwp(A),             % Calculate area
    perimeter(P),       % Calculate perimeter.
    S is A/(P*P),       % Shape factor = Area/(Perimeter^2)
    S < 0.08.           % Min value for S is 1/4*π for a circle
```

Notice the highly modular approach to Prolog+ programming and the fact that it is possible to define what is an "acceptable" Bakewell Tart, in a simple and natural way. Apart from the predicate *segment_image*, whose definition depends upon the lighting and camera, the program *bakewell_tart* is complete. It is able to perform a simple yet effective means of inspecting Bakewell tarts.

3.5.2 Recognising Printed Letters

The top two layers of a Prolog+ program for recognising printed letters are given below:

```
% Top level predicate for recognising printed letters, which may
% be either upper or lower case and in any one of three fonts
letter(X) :- upper_case(X).      % Letter X may be upper case, ...
letter(X) :- lower_case(X).      % ... or X may be lower case

upper_case(X) :-
    font(Y),                              % Find what font we are using
    member(Y,[times,courier,helvetica]),
                                 % These 3 fonts are of interest to us
    recognise_upper_case(X,Y).   % X is upper case in font Y

lower_case(X) :-
    font(Y),                              % Find what font we are using
    member(Y,[times,courier,helvetica]),
                                 % These 3 fonts are of interest to us
    recognise_lower_case(X,Y).   % X is lower case in font Y
```

The complex task of recognising an upper- or lower-case letter in any of the three known fonts has been reduced to a total of 156 (=3*2*26) simpler sub-problems. (A simple declarative definition of the *sans serif* upper case letter A is presented later.) Now, let us consider what changes have to be made if a new font (e.g. *Palatino*) is to be introduced. Two changes have to be made:

(i) the second line in the body of *upper_case* and *lower_case* is changed to
    member(Y,[times,courier,helvetica,palatino])
(ii) two new clauses are added for each letter X, one for
    recognise_upper_case(X,palatino) and another for
    recognise_lower_case(X,palatino).

If we wanted to add recognition rules for the numeric characters, then 10 new clauses would be added, as in (ii). In other words, extending the scope of a Prolog+ program is conceptually simple, if rather tedious to accomplish. Here, as promised, is a naive but quite effective declarative definition of the *sans serif* upper-case letter A:

```
recognise_upper_case(a,sans_serif) :-
    apex(A),            % There is an apex called A[10].
    tee(B),             % There is a tee-joint called B
    tee(C),             % There is a tee-joint called C
    line_end(D),        % There is a line_end called D
    line_end(E),        % There is a line_end called E
    above(A,B),         % A is above B.
    above(A,C),
    about_same_vertical(B,C),
    about_same_vertical(D,E),
    above(B,D),
    above(C,E),
    connected(A,B),
    connected(A,C),
    connected(B,D),
    connected(C,E),
    connected(B,C),
    left(B,C),
    left(D,E).
```

The reader should be able to understand the above program without detailed knowledge about how the predicates *apex, tee, line_end above, about_same_vertical, above, connected* and *right* are defined. Figure 3.4 shows some of the objects which would be recognised by this program. Obviously, *recognise_upper_case* can be refined by adding further conditions, to eliminate some of the more bizarre objects that are recognised by the present definition and are shown in Figure 3.4.

3.5.3 Identifying Table Cutlery

The following Prolog+ program identifies items of table cutlery that are viewed in silhouette. (It is assumed, for the sake of brevity, that the input image can be segmented using simple thresholding.) The top-level predicate is *camera_sees(Z)* and is a general purpose utility that can find any object, given an appropriate definition for the subsidiary predicate *object_is*. In its present somewhat limited form, the program recognises only forks and knives; additional clauses for *object_is* are needed to identify other utensils, such as spoons, plates, mats, etc. When the query *camera_sees(Z)* is specified, Z is instantiated to the type of object seen by the camera. In practice, there may be many objects visible

---

[10] The name of a feature, such as an apex or tee-joint, may be the same as its address within the image. When using Prolog+, this is often very convenient.

to the camera and the program will progressively analyse each one in turn. Any objects visible to the camera that are not recognised are signalled by instantiating Z to the value *unknown_type*, see Figure 3.5.

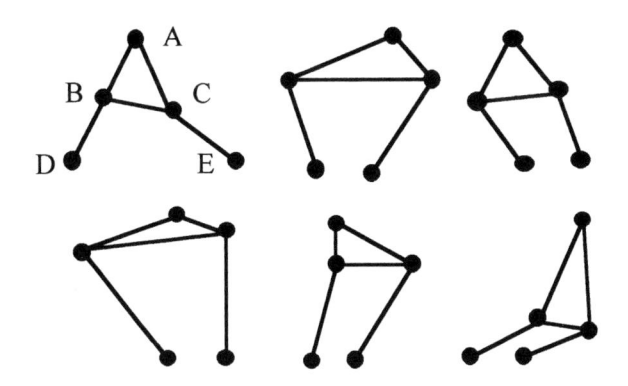

Figure 3.4 Objects which are inappropriately recognised as the *sans serif* letter **A** by the goal *recognise_upper_case(a,sans_serif)*.

(a)

(b)

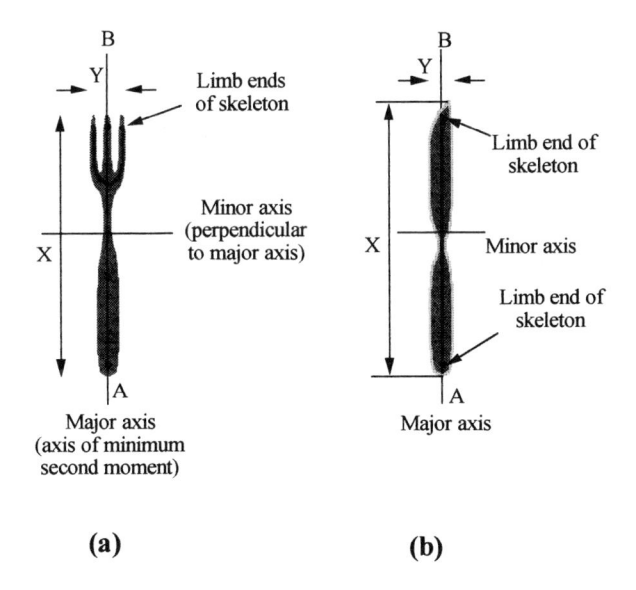

Figure 3.5 Explaining the operation of two object recognition programs (a) *object_is(fork)*. AB is the axis of minimum second moment. Criteria to be satisfied, before this object can be accepted as a fork: $150 \leq X \leq 450$; $25 \leq Y \leq 100$; $4 \leq X/Y \leq 10$; skeleton here has 3-5 limb ends. (b) *object_is(knife)*. AB is the axis of minimum second moment. Criteria to be satisfied, before this object can be accepted as a fork: $150 \leq X \leq 450$; $25 \leq Y \leq 100$; $6 \leq X/Y \leq 12$; skeleton must have exactly 2 limb ends.

```prolog
% Top level predicate for recognising individual items of table
% cutlery
camera_sees(Z) :-
      grb,               % Digitise an image from the camera
      segment_image,     % Example: [enc, thr(128)]
      ndo,               % Shade resulting binary image so that each
                         % blob has a different intensity
      wri(temp),         % Save image in disc file, named "temp"
      repeat,            % Begin loop - analyse all blobs in turn
      next_blob,         % Select a blob from the image saved in "temp"
      object_is(Z),      % Identify the blob as an object of type Z
      finished.          % Succeeds only when no more blobs to analyse

% Select one blob from the image stored in disc file "temp".
% Remove this blob from the stored image, so that it will not
% be considered next time.
next_blob :-
      rea(temp),         % Read image from disc file "temp"
      gli(_,A),          % Identify next blob - i.e. brightest
      hil(A,A,0),        % Remove it from stored image
      wri(temp),         % Save remaining blobs
      swi,               % Revert to previous version of stored image
      thr(A,A).          % Select one blob

% Recognises individual non-overlapping objects in a binary image.
object_is(fork) :-       % Figure 3.5(a) shows how this clause works
      mma(X,Y),          % Find lengths along major & minor axes (X,Y)
      X ≥ 150,           % Length must be ≥ 150 pixels
      X ≤ 450,           % Length must be ≤ 450 pixels
      Y ≥ 25,            % Width must be ≥ 25 pixels
      X ≤ 100,           % Width must be ≤ 100 pixels
      Z is X/Y,          % Calculate aspect ratio - whatever
                         % orientation
      Z ≤ 10,            % Aspect ratio must ≤ 10
      Z ≥ 4,             % Aspect ratio must be ≥ 4
      count_limb_ends(N),
                         % Instantiate N to number of limb ends
      N ≥ 3,             % Skeleton of fork must have ≥ 3 limb ends
      N ≤ 5.             % Skeleton of fork must have ≤ 5 limb ends

% Add as many clauses here as are needed to recognise each possible
% type of object.
object_is(knife) :-      % Figure 3.5(b) shows how this clause works
      mma(X,Y),          % Find lengths along major & minor axes (X,Y)
      X ≥ 150,           % Length must be ≥ 150 pixels
      X ≤ 450,           % Length must be ≤ 450 pixels
      Y ≥ 25,            % Width must be ≥ 25 pixels
      X ≤ 100,           % Width must be ≤ 100 pixels
      Z is Y/X,          % Calculate aspect ratio - whatever
                         % orientation
      Z ≤ 12,            % Aspect ratio must ≤ 12
      Z ≥ 6,             % Aspect ratio must be ≥ 6
      count_limb_ends(2).
                         % Skeleton of a knife has exactly 2 limb ends
                         % Catch-all clause. Object is not recognised

object_is(unknown_type).
```

```
% Search is finished.  Image is black everywhere.
finished :-
       rea(temp),      % Read image from disc
       thr(1),         % Threshold stored image
       cwp(0).         % Succeeds if number of white points = 0

% Count the limb ends on a skeleton ("match-stick") figure
count_limb_ends(N) :-
       mdl,            % Generate skeleton of the blob
       cnw,            % Count white neighbours for 3*3 window
       min,            % Ignore back-ground points
       thr(2,2),       % Select limb ends
       eul(N).         % Instantiate N to number of limb ends
```

3.5.4 Analysing all Visible Objects

The list of all objects that are visible to the camera can be found using the predicate *list_all_objects* defined thus:

```
list_all_objects(A) :-
       grb,                        % Same pre-processing ...
       segment_image,              % ... as is used in ...
       ndo,                        % ... the predicate ...
       wri(temp),                  % ... "camera_sees"
       find_object_list([],A).     % Generate list of objects seen

find_object_list(A,A) :-  % Terminating recursion
       finished.          % Succeeds when no more blobs to
                          % analyse

find_object_list(A,B) :-  % Analyse all blobs in the image
       next_blob,         % Select a blob from the image saved in
                          % "temp"
       object_is(C),      % Identify the blob as an object of
                          % type C
       !,                 % Do not want to back-track, so include
                          % cut (!) here to make recursion more
                          % efficient
       find_object_list([C|A],B).
                          % Recursive call to analyse all blobs
```

This makes use of the fact that *object_is* is able to recognise an object in the presence of other objects, provided that they do not touch or overlap. We will now make good use of *list_all_objects* in performing a much more difficult task, namely that of recognising a well-laid table place setting.

3.5.5 Recognising a Table Place Setting

The following Prolog+ program can recognise a table place setting with the cutlery laid out as shown in Figure 3.6(a). For this program to work properly, we must first define additional clauses for the predicate *object_is*, so that appropriate types of object, such as *small_knife, tea_spoon, dinner_fork, plate, mat*, etc., can be recognised. Notice that *table_place_setting* is defined in standard Prolog, without any further use of the image processing built-in predicates.

Figure 3.6 Table place settings: (a) Ideal arrangement. *Key:* 1, *dinner_fork*; 2, *mat*; 3, *dinner_knife*; 4, *small_knife*; 5, *soup_spoon*; 6, *small_fork*; 7, *desert_spoon*. (b) Scene that is misrecognised by *table_place_setting* with the original definitions of *left* and *below* but which is correctly rejected with the revised versions. (c) A scene that is incorrectly recognised by the revised version of the program.

```
table_place_setting :-
      list_all_objects(A),
      equal_sets(A, [mat, plate, dinner_knife, small_knife,
      dinner_fork, small_fork, soup_spoon, desert_spoon]),
      left(dinner_fork, mat),          % Defined below
      left(mat, dinner_knife),
      left(dinner_knife, small_knife),
      left(small_knife, soup_spoon),
      below(mat, small_fork),
      below(small_fork, desert_spoon).
```

For completeness, we now define the predicates *left*, *below* and *equal_sets*. These and many other useful "general purpose" predicates like them form part of a Prolog+ Library, which augments the basic language.

```
left(A,B) :-
      location(A,Xa,_),       % Horizontal position of A is Xa
      location(B,Xb,_),       % Horizontal position of B is Xb
      !,                      % Inhibit backtracking
      Xa < Xb.                % Compare horizontal positions

below(A,B) :-
      location(A,_,Ya),       % Vertical position of A is Ya
      location(B,_,Yb),       % Vertical position of B is Yb
      !,                      % Inhibit backtracking
      Ya < Yb.                % Compare vertical positions

equal_sets([],[]).            % Terminate recursion
```

```
equal_sets([A|B],C) :-         % Checking two non-empty lists are
                               % equal
        member(A,C),           % A is a member of C
        cull(A,C,D),           % Delete A from list C. Result is D
        !,                     % Improve efficiency of recursion
        equal_sets(B,D).       % Recursion. Are sets B and D also
                               % equal?

cull(_,[],[]).                 % Cannot delete anything from empty list

cull(A,[A|B],C) :-             % Delete A from list if A is at its head
        !,                     % Improve efficiency of recursion
        cull(A,B,C).           % Repeat until the lists A and B are both
                               % empty

cull(A,[B|C],[B|D]) :-
                               % A is not head of "input" list so work on
                               % tails
        !,                     % Improve efficiency of recursion
        cull(A,C,D).           % Repeat until the lists A & B are both empty
```

Using these simple definitions, a range of unusual configurations of cutlery and china objects is accepted. (See Figure 3.6(b).) To improve matters, we should refine our definitions of *left* and *below*. When redefining *left(A,B)*, we simply add extra conditions, for example that A and B must be at about the same vertical position:

```
left(A,B) :-
        location(A,Xa,Ya),        % Horizontal position of A is Xa
        location(B,Xb,Yb),        % Horizontal position of B is Xb
        !,                        % Inhibit backtracking
        Xa < Xb,                  % Compare horizontal positions
        about_same(Ya,Yb, 25).    % Tolerance level is specified by 3rd
                                  % argument

about_same(A,B,C) :- A ≤ B + C.
about_same(A,B,C) :- A ≥ B - C.
```

The predicate *below* is redefined in a similar way:

```
below(A,B) :-
        location(A,Xa,Ya),        % Vertical position of A is Ya
        location(B,Xb,Yb),        % Vertical position of B is Yb
        !,                        % Inhibit backtracking
        Ya < Yb.                  % Compare vertical positions
        about_same(Xa,Xb, 25).    % Tolerance level is specified by 3rd
                                  % argument
```

In practical application, it would probably be better for the tolerance parameter required by *about_same* (third argument) to be related to the size of the objects to be recognised. This would make a program such as *table_place_setting* more robust, by making it size-independent. This modification to our program improves matters, but it still recognises certain cutlery arrangements as being valid place settings, even though we would probably want to exclude them in while *inside* can be defined thus.

```
inside(A,B) :-
      isolate(A),      % Isolate object A
      wri,             % Save image until later
      isolate(B),      % Isolate object B
      rea,             % Recover saved image during "wri" operation
      sub,             % Subtract images
      thr(0,0),        % Find all black pixels
      cwp(0).          % There are exactly zero black pixels.
```

Notice that *inside(A,B)* creates two binary images, containing objects A and B. These images are then compared. If all white pixels in one image (i.e. the one containing object B) are also white in the other (containing object A), then we can conclude that A is inside B and the goal *inside(A,B)* succeeds.

There are, of course, many other abstract relationships that we must define. Here are the definitions of a few of them:

```
% Are objects A and B concentric ?
concentric(A,B) :-
      location(A,Xa,Ya),     % Could use centroid to define
                             % "location"
      location(B,Xb,Yb),     % Could use centroid to define "location"
      near([Xa,Ya],[Xb,Yb],10).
                             % Is distance [Xa,Ya] to [Xb,Yb] ≤ 10

% Are A and B in about the same vertical position ?
about_same_vertical(A,B) :-
      location(A,Xa,Ya),     % Could use centroid to define
                             % "location"
      location(B,Xb,Yb),     % Could use centroid to define "location"
      about_same(Ya,Yb,10).  % Is difference in vertical position ≤
                             % 10.

% Test whether object A in the top part of object B (i.e. the
% bottom-most point in A is above B's centroid). A must be entirely
% contained inside B.
top_of(A,B) :-
      isolate(A),      % Isolate object A
      dim(_,Ya,_,_),   % Bottom of object A
      isolate(B),      % Isolate object B
      cgr(B,_,Yb),     % Could use centroid to define "location"
      Ya ≤ Yb,         % Is bottom point in A above centre of B
      inside(A,B).

% Are the points [X1,Y1] and [X2,Y2] connected by a continuous set
% of white pixels ?
connected(X1,Y1,X2,Y2) :-
      ndo,                   % Shade blobs.
      pgt(X1,Y1,Z),          % Z is intensity at [X1,Y1]
      pgt(X2,Y2,Z),          % Z is intensity at [X2,Y2]
      Z = 255.               % Both pixels are white

% Are regions A and B adjacent ?
adjacent(A,B) :-
      isolate(A),            % Isolate region A
      exw,                   % Expand region by 1 pixel
```

```
            wri,              % Save image for use later
            isolate(B),       % Isolate region A
            rea,              % Read image A expanded
            min,              % Logical AND of the 2 images
            cwp(N),           % Count white points in both images
            N > 0.            % Are there some?
```

The reader may like to consider how a predicate can be defined which can test whether an object in a binary image is vertical. (Figure 3.8. The relevent image processing predicate is *lmi*.) This predicate could be used to prevent *table_place_setting* from detecting scenes like that shown in Figure 3.6(c).

3.6.3 Geometric Figures

The predicate *rectangle*, which is used in the definition of *picture*, seems to refer to a geometric figure, that can be defined in precise mathematical terms. (Figure 3.9) In fact, this is not so; human beings use the term "rectangles" much more loosely than this. Most people, including mathematicians in their everyday lives, would describe a rectangle using the *fuzzy concepts* of *"straight line"* and *"right angle"*. Most people are prepared to use these terms to describe a wide variety of carelessly drawn objects. (Figure 3.10) We cannot use the familiar equation for a straight line ($y = m.x + c$) directly, when we want to test whether a "broken" polygonal arc, or series of spots, could reasonably be regarded as forming a straight line. We could use the Hough transform to do this. Another approach is exemplified by the predicate *straight_line*, defined below. This provides a simple heuristic for testing whether a 1-pixel wide arc, whose end points are [X1,Y1] and [X2,Y2], could reasonably be accepted as being a straight line. (See Figure 3.11.)

```
straight_line([X1,Y1], [X2,Y2]) :-
        wri,              % Save image for use later
        zer,              % Create black image
        vpl((X1,Y1, X2,Y2, 255),
                          % Draw digital straight line
        neg,              % Negate the image
        gft,              % Grass fire transform
        rea,              % Recover saved image
        min,              % Use original image as mask on grass fire
                          % picture
        gli(_,Z),         % Find maximum intensity (i.e. distance from
                          % line joining [X1,Y1] and [X2, Y2])
        Z ≤ 10.           % Is whole arc ≤10 pixels away from this line?
```

Testing for right angles is easy, once the corner points have been identified. (See Section 2.3.)

```
right_angle(A,B,C,D) :-
        angle(A,B,P),         % Find angle of line joining A and B
        angle(C,D,Q),         % Find angle of line joining C and D
        S is Q - P,           % Find difference in angles
        about_same(S,90,5).   % Check angle difference is in range
                              % [85,95]
```

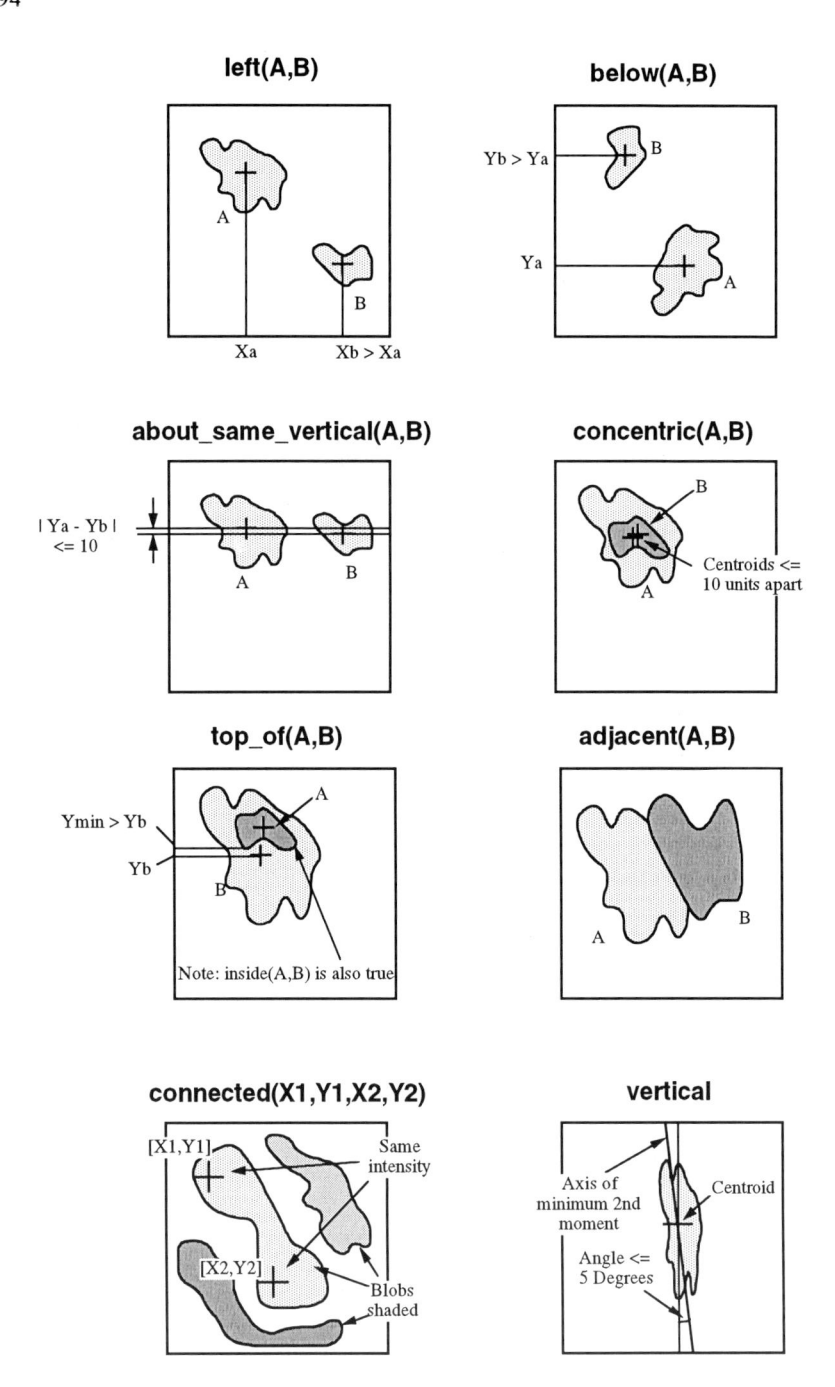

Figure 3.8 Spatial relationships in Prolog+.

Here, at last, is one possible test for rectangles, using the ideas just described.

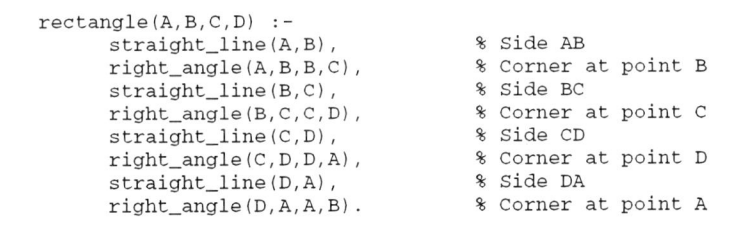

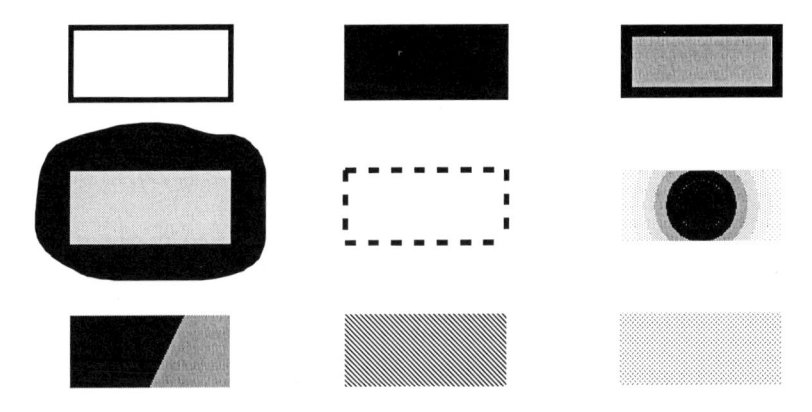

```
rectangle(A,B,C,D) :-
      straight_line(A,B),        % Side AB
      right_angle(A,B,B,C),      % Corner at point B
      straight_line(B,C),        % Side BC
      right_angle(B,C,C,D),      % Corner at point C
      straight_line(C,D),        % Side CD
      right_angle(C,D,D,A),      % Corner at point D
      straight_line(D,A),        % Side DA
      right_angle(D,A,A,B).      % Corner at point A
```

Figure 3.9 "Rectangles".

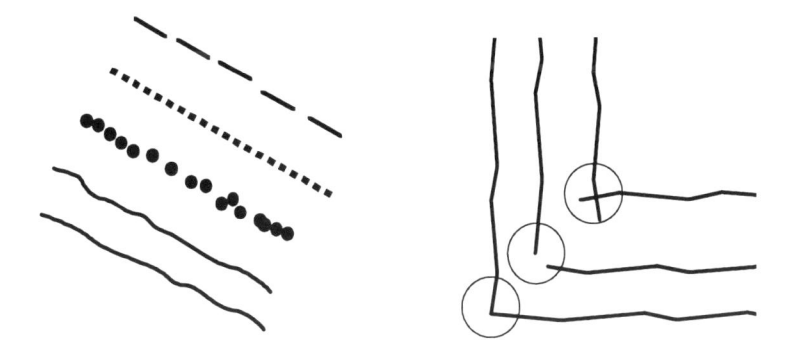

Figure 3.10 Some of the objects that people call "straight lines" and "right angles".

However, this is not the complete story! There are many other types of object that could legitimately be called rectangles. Some of these are illustrated in Figure 3.9. Clearly, a general definition of rectangles requires a lot more sophistication than we have space to explain here. Let it suffice to say that each of the objects shown in Figure 3.9 could be recognised by a separate Prolog+ clause.

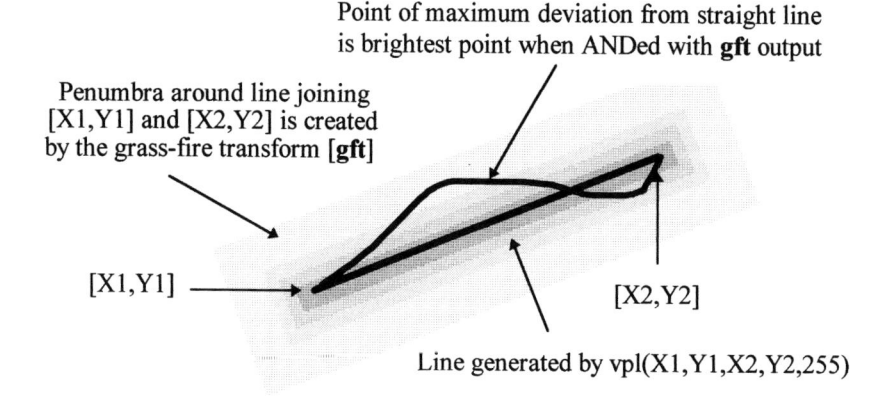

Figure 3.11 Operation of the predicate *straight_line*.

3.7 Implementation of Prolog+

Consider Figure 3.12 which shows the block diagram of a system that offers one possible implementation of Prolog+. This represents one of the most recent stages in a period of evolutionary development that has taken place since 1985. A number of other systems combining image processing with Prolog have been built. (See Figure 3.13.)

3.7.1 The # Operator

The essential feature of Figure 3.12 is that it shows a Prolog program controlling a proprietary image processor, the Intelligent Camera [INT]. Figure 3.14 shows the action that follows, when Prolog encounters a goal of the form # *en*. The # operator performs the following actions:

1. The character string *"en"* is sent, via the Macintosh computer's modem port, to the Intelligent Camera.
2. The Intelligent Camera interprets the character string *"en"* as a command and performs the appropriate image processing operation (negating all intensities in the image.)

3. When the Intelligent Camera completes the operation *en*, it signals that it has done so, via its serial port, to the control computer running Prolog.

4. Prolog receives the "command done" signal from the Intelligent Camera and interprets it so that the Prolog goal [# *en*] succeeds.

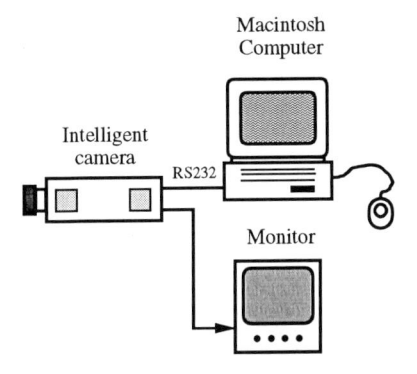

Figure 3.12 One possible implementation of Prolog+, using the Intelligent Camera [INT]. The Macintosh computer runs the Prolog+ software.

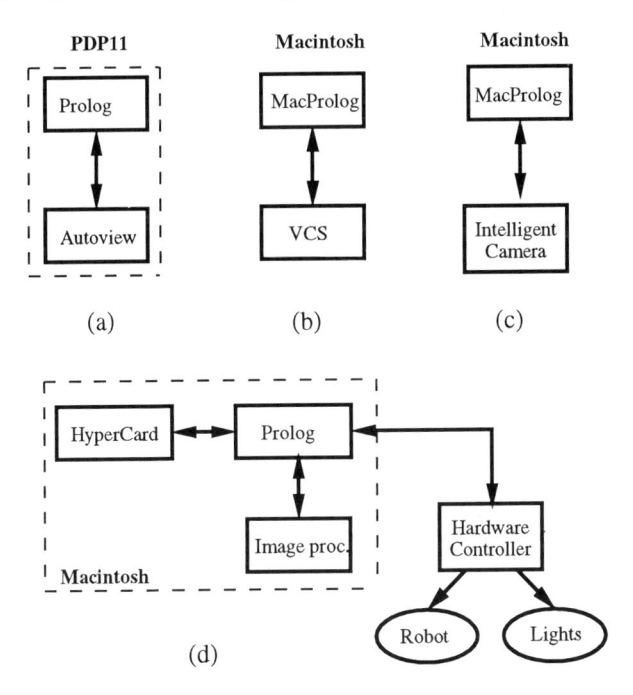

Figure 3.13 Other possible implementations of Prolog+. Each of these arrangements has been built, in the past. The earliest successful configuration to be used was (a). Most of the authors recent work has been based on (c) and (d).

The goal [# 'tf(128,W)'] is slightly more complex but is dealt with in a similar way. The only difference is that the character string "tf(128,W)" is sent to the Intelligent Camera. The # operator is defined in such a way that we may type "# tf(128,'W')" instead. (This alternative form is a little more convenient when we write Prolog+ programs.) The goal [# tf(X,'W')] where X is already instantiated to 123, is treated in just the same way; the character string transmitted to the Intelligent Camera is "tf(123,W)". However, both of the following goals [# tf(1234,'W')] and [X is 1234, # tf(X,'W')] will *fail*, because the Intelligent Camera cannot perform the operation. (The first parameter, X, is out of range.)

Two Prolog+ predicates are provided that can to receive data returned by the image processor. These access the local and global memories stored within the Intelligent Camera and have the form $m(R,V)$ and $g(R,V)$. The first of these instantiates the Prolog+ variable V to the value currently stored in the image processor's *local* memory register number R. The second instantiates V to the value stored in the *global* memory register R. Until recently, these were the only means that Prolog+ had of obtaining data about images.

What we have described thus far in this section will be referred to as *Very Simple Prolog+*, or *VSP*. Prolog+ contains far more facilities than this but the whole of Prolog+ as described so far in this book can be written in terms of VSP. The definition of the # operator is dependent upon the particular image processor being used. (See [BAT-91] for more details on this operator.) For the moment, the reader should note the following points:

(i) Repeating the # operator two or more times has the same effect as including it once in a program. In Prolog, this is achieved as follows: # # A :- # A.
Hence, # # # # en and # # en have the same effect as # en.
(ii) # is never resatisfied on back-tracking.
(iii) For the convenience of the user, the operator @ is defined to be synonymous with #, using the definition @ A :- # A.

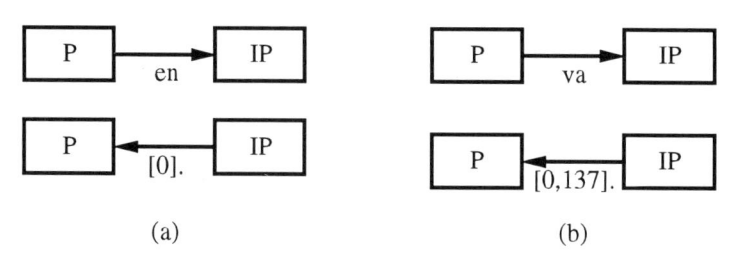

Figure 3.14 How the Prolog+ system shown in Figure 3.12 responds to two typical image processing goals. Data passes between the Prolog host computer and the image processor (Intelligent Camera). *Key:* P, Prolog host computer. IP, Image processor. (a) # en. The image processor returns the list [0]. to signal "task done". (b) # va. The image processor returns the list [0,137] to signal "task done" and that the average intensity is 137.

3.8 Comments

The idea of integrating image processing within Prolog was conceived in the mid-1980s. [BAT-91] Prior to that, a program providing a loose linkage between them had been developed. The tight connection between the image processing and Prolog that is implicit in Prolog+ has a distinctively beneficial effect, which is more significant than a mere cosmetic change. Indeed, the "ergonomics" of the combined software package are much improved, thereby enabling the Prolog+ programmer to concentrate better upon the high-level abstract concepts specific to the application. While it is possible to implement Prolog+ using only the '#' operator within Very Simple Prolog+ (VSP), closer integration between Prolog and the image processing software has distinct advantages, in terms of user acceptability. The software outlined in Appendix D was developed very recently and does not require the definition of the '#' operator as a prelude to implementing the full Prolog+ language. We should not confuse implementation details with the fundamental requirements of the language.

A very important aspect of Prolog+ is that it retains the ability to perform interactive image processing as a means of developing algorithms / heuristics for machine vision applications. Since interactive image processing has been found to be very effective in a wide variety of industrial applications, this was considered to be an essential feature of the language. We shall return to this theme in the following chapter, where we explain how interaction is actually achieved.

The mechanism for invoking image processing operators within Prolog+ programs has been described above. Each image processing function is executed as a side effect of evaluating a predicate that always succeeds but is never resatisfied on backtracking. In this respect, the image processing operators resemble the standard Prolog printing predicates: *write, nl, tab*, etc. Several Prolog+ programs have been presented. The ability to describe an object or scene that is to be recognised in future is an important aspect of Prolog+ programming. This allows a person to work in a natural way. The ability to use abstract symbolic relationships in these descriptions is particularly important. A person can, for example, define what spatial relationships must exist between objects within an image. The ability to define predicates such as *left, near, parallel* and from there develop definitions for general objects, such as "rectangles", are particularly valuable. The program for examining a table place setting (*table_place_setting*) is a good example of Prolog+ code that would be difficult to replace by software written in another language.

The ability of Prolog+ to define synonyms and to represent reciprocal relationships (e.g. *right* and *left*, *below* and *above*, etc.) is valuable, since it makes it a trivial, if tedious, task to write a program that is tolerant of human beings who are prone to forget / ignore rules about which word to use. A brief outline of the implementation of Prolog+ has been given in the pages above.

Several possible methods of implementing Prolog+ have been devised in the past:

(a) Using an external hardware device for image processing. This resides outside the Prolog host computer and is connected to it via a serial (RS232/RS422) line. This external device may, in fact, be another computer, a slow dedicated hardware unit, such as the Intelligent Camera, or consist of a fast hardware accelerator, controlled by a second computer.
(b) As (a), but using a fast parallel line.
(c) Using a full software implementation. More details are provided of the software implementation in Appendix D.

4

Enhanced Intelligent Systems

Thus far, Prolog+ has been presented simply as a language for intelligent image processing. However, this is *not* a book about image processing *per se*. Hence, we need to place Prolog+ in its proper context, as being just one element within a "tool box" containing a number of design aids for industrial vision systems engineers. The integrated operating environment, surrounding Prolog+ is just as important as the "core" language itself and forms one of four major topics discussed in this chapter.

The second major topic in this chapter is the use of speech to control a Prolog+ system. Speech input is seen as being attractive as a means of allowing the user of a vision system or robot, to control it, without using a keyboard. Speech input is faster and more natural for the user than is a keyboard. In addition, speech input keeps both hands free and is less prone to damage by dust and dirt in a factory environment. Of course, safeguards are needed to ensure that factory noise does not affect the speech recognition system and it may not be possible to use it at all in a very noisy plant. Strictly speaking, speech input and natural language understanding to accompany it, are part of the environment surrounding Prolog+. We have separated these issues in this chapter, simply for convenience.

In the third part of this chapter, we describe various design aids and discuss how they can be interfaced to a Prolog+ system. Among them is a so-called *Lighting Advisor*, which consists of two inter-linked HyperCard stacks and provides advice about a wide range of issues relating to the formation of a good image, prior to image digitisation and processing. Again, the Lighting Advisor effectively forms part of the Prolog+ environment.

The construction of loosely coupled networks of Prolog+ systems, controlled by a single Prolog program forms our final topic in this chapter. These networks can have as many as 32 host computers and up to 1024 cameras.

4.1 Prolog+ Environment: A Tool-box for Machine Vision

The Prolog+ environment provides a number of enhancements, extensions and utilities to the language described in the previous chapter and which have been designed as aids to the design process. The extensions to Prolog+ embodied within its operating environment include the following features:

- Library of useful programs, including several demonstrations. The latter are intended for education and training purposes.
- Auto-starting Prolog+ when the computer power is switched on.
- Interactive operation for prototyping industrial vision systems.
- Pull-down menus, which can be extended easily by the user, without any programming.
- Command keys, for rapid selection of certain important functions.
- Graphical display of the pose of a robot in a work cell.
- Speech Synthesis, for presenting symbolic data to a user in a natural way.
- On-line HELP and operating data display facilities.
- Cursor, used for both drawing and interactive investigation of image structure.
- Automatic Script Generation and editing.
- Linking to other programs, including a HyperCard controller for setting up a robot vision work cell.
- Speech Recognition, which can be used in conjunction with Prolog programs that are capable of understanding simple English sentences about a domain of limited scope. This topic is discussed in the following section.
- Design Aids, including a programs which gives advice about lighting and viewing techniques.

The ideas discussed in this chapter have all been studied and proved experimentally; Prolog+ programs have been written to substantiate all of these ideas. Integrating these utilities into a single, harmonious environment is a major task and, for this reason, is still under way.

4.1.1 Defining New Predicate Names

In Appendix E, we list the mnemonics for the image processing operators that are used throughout this book. However, the use of 3-letter mnemonics may not be to everyone's taste. Suppose that we wish to define a new name for the image processing operator *neg*. This can be accomplished very simply in the following way:

```
negate :- neg.
```

Either *neg* or *negate* may now be used. In the same way, a 2-letter mnemonic form may be defined:

```
ne :- neg.
```

When we wish to define a new name for an operator, such as *thr*, which has a variable arity[1], the situation is slightly more complicated:

```
threshold(X,Y) :- thr(X,Y).
threshold(X)   :- thr(X).
threshold      :- thr.
```

In the same way, it is possible to make up an entirely new image processing language, or to translate the terms into a foreign (natural) language. Here, for example, is a small portion of the translator for an image processor intended for use by Welsh speakers:

```
mwyaf      :- biggest.
ardal_gwyn(X) :- white_area(X).
anglir     :- blur.
```

4.1.2 Default Values for Arguments

It is a simple matter to redefine the default values for arguments of Prolog+ image processing operators. Suppose that we wish to define new default values for the operator *thr*:

```
thr(X) :- thr(X,200).     % Previous default value was 255
thr    :- thr(75,125).    % Not defined previously
thr    :- thr(128,255).   % Arbitrary but useful definition
```

Notice that it is necessary to include one clause for each case to be considered.

4.1.3 Useful Operators

A useful "program control" operator, '•', may be defined thus:

```
0•G.
N•G :-
     call(G),
     M is N -1,
     M•G.
```

This operator may be used like a FOR-loop in a conventional programming language, since it permits the programmer to order the repetition of an operation *G*. In formal Prolog terms, *N•G* is a goal, which either succeeds or fails. If *G* fails

---

[1] The arity measures the number of arguments.

on any of the N repetitions, then $N \cdot G$ will also fail. $N \cdot G$ succeeds if all N repetitions of G succeed. In order to understand the value of the '•' operator, notice that the two following definitions of *process* are identical in their effect:

```
% First definition, using the '•' operator
isophotes :-    % Draw smoothed intensity contours (isophotes)
     3•lpf,     % Repeat "lpf" three times
     sca(3),    % Discard 5 least significant bits (keep 3 bits)
     sed,
     thr(1).

% Second definition, repeating "lpf" three times
isophotes :-
     lpf, lpf, lpf,
     sca(3), sed,
     thr(1).
```

A useful non-linear filter for detecting thin dark streaks ("cracks") and small spots may be defined thus:

```
crack :-
     wri,            % Save image until later
     2•(3•lnb,neg),
     rea,            % Read image saved earlier
     sub.            % Subtract images
```

This is much more compact and rather more easily understood than the following equivalent version:

```
crack :-
     wri,
     lnb, lnb, lnb,
     neg,
     lnb, lnb, lnb,
     rea, sub.
```

The following provides a more general Prolog equivalent of the FOR-loop:

```
for(A,_,B,_) :-
     A > B,          % Terminate recursion
     !.              % Upper limit of variable exceeded
                     % Do not progress to next clause

for(A,B,C,D(_)) :-
     call(D(A)),     % Try to satisfy goal D(A)
     E is A + B,     % Increment counter, ready for next loop
     !, for(E,B,C,D(L)).   % Go on to next loop
```

A typical query using this definition is *for(3,2,27, process(_))*, which tries to satisfy *process(I)*, for I = 3, 5, 7, ..., 25, 27. A Prolog+ predicate, *case*, roughly equivalent to Pascal's CASE operator may be defined in the following way:

```
case(A,B) :-
     select_list_element(A,B,C),   % Instantiate C to A-th element
                                   % of list B
     call(C).                      % Satisfy goal C
```

```
% Fail for all negative values of first argument
select_list_element(N,_,fail) :-
        N ≤ 0, !, fail.

% Finish if N is 1.
select_list_element(1,[A],A).

% What to do if we have not found the N-th element yet
select_list_element(A,[B|C],D) :-
        length(C,N),    % Length of the list C is N
        N > 0,          % Is N > 1?
        X is A - 1, select_list_element(X,C,D).
                        % Repeat until we find the element we want

% Fail under all other circumstances
select_list_element(_,_,_) :- fail.
```

The conditional statement *if_then_else* may be defined thus:

```
if_then_else(A,B,C) :- A,!,B.      % If A succeeds, test B.
if_then_else (A,B,C) :- C.         % A has failed, so test C
```

The simpler *if_then* function could of course, be defined in the following way:

```
if_then(A,B) :- A,!,B.             % If A succeeds, test B.
```

However, an operator, ('->') may be defined as an alternative:

```
A -> B :- A,!,B.
```

The *AND* operator (&) may be defined in the following way:

```
op(900,xfy,'&'). % Precedence is 900 and '&' is left associative
A & B :- A, B.
```

The usefulness of this operator is simply that to, a naive user, '&' is easier to interpret than ','.

The *OR* operator (*or*) may be defined in the following way:

```
op(900,xfy,or).    % Precedence is 900 and is left associative
A or B :- A ; B.
```

The *if* operator can be helpful as an aid to user understanding. It can be defined thus

```
:- op(1200,xfx,if), term_expand( (A if B), A :- B) ).
```

and can be used *in lieu* of the Prolog ':-' operator. Here is the program *picture* (see Section 3.6.1) rewritten using the *&* and *if* operators.

```
picture if
    rectangle(A) &
    ellipse(B) &
    encloses(A,B) &
    text(C,'Qwerty') &
    inside(C,B) &
    print_texture(D) &
    encloses(A,D) &
    below(D,B).
```

4.1.4 Program Library

A large number of image processing functions, device control programs, and other utilities have been provided to augment the Prolog+ software [BAT-91]. This list is constantly being enlarged. A detailed explanation of some of these utilities is deferred until Chapters 5 and 6, where we shall discuss the control of lighting, electro-mechanical and various and other devices (Chapter 5) and colour image processing (Chapter 6). A series of demonstration programs has also been developed. (See Table 4.1.) Some of these are described in more detail in Chapter 7.

4.1.5 Auto-start

Prolog+ can be provided with an auto-start facility, by following two simple steps:

(i) Place the Prolog+ application software, or its alias, in the *"Startup Items"* folder. The MacProlog software can be located in another folder. When the computer power is switched on, the Prolog+ software will load automatically and will then be compiled.
(ii) A start-up goal (called *startup_goal*) will be initiated if we place the following statement in the program:
```
'<LOAD>'(_) :- startup_goal.
```

Prolog will try to satisfy the goal *startup_goal*, which may be used to good effect in a variety of ways:

(a) Automatic initialisation of a hardware implementation of Prolog+.
(b) Calibration of a robot vision system.
(c) Construction of pull-down menus. (see Section 4.1.7)
(d) Construction of a pull-down menu, used for speech recognition. (See Section 4.2.)
(e) Automatic entry into interactive mode.
(f) Automatic entry into a dialogue with a novice user.

So far, we have assumed that the full MacProlog software is loaded on start-up. It is possible, however, to build stand-alone applications which do not have the full MacProlog program development environment and these can be started automatically in the same way.

Demonstration	Description
Packing 2-D objects	Packs 2-D objects of arbitrary shape into a space also of arbitrary shape.
Telling the time	Analyses the image of a simple analogue clock and finds the time in the format *"five to nine"*, *"twenty past six"*, etc.
Dissecting small plants	Demonstrates the dissection of a small plant for micropropagation, given a silhouette of the plant.
Playing cards	Calculates the value and suit of both picture and non-picture playing cards, using a high magnification view of the top-left corner of each card.
Smart burglar alarm	Compares image obtained from camera with one digitised earlier. If the difference is large enough, the program signals an intruder.
Line drawings	Analyses simple line drawings and recognises parallel lines and squares.
Dominoes	Plays dominoes against a virtual robot (i.e. a person acting as a robot). (The domino rules are for the game 3s and 5s'.) Cheating is not allowed since the program is not clever enough to cope with devious behaviour.
Stacking boxes	Imagine a stack of boxes with sell-by dates on their sides. The program plans the actions of a virtual robot which restacks the boxes so that they are placed in date order.
Picking up objects	Analyses simple silhouette and decides where to place the gripper of a virtual robot. The program indicates whether or not it would be safe to lower the gripper, and whether the grip would be secure.
Learning shapes	Learns the shapes of simple objects, such as bottles, viewed in silhouette.
Learning colours	Learns the proportions of 8 colours plus black and white in a series of images. This program is suitable for recognising company logos, trademarks, etc. on printed cartons.

Table 4.1 Demonstration Programs. (See Chapter 7.)

4.1.6 Interactive Mode

We have already explained the role and importance of interactive image processing for prototyping industrial vision systems. The following program provides Prolog+ with an interactive facility:

```
interactive_mode :-
     prompt_read(['Please specify a Prolog+ goal'],X),
     not(X = end),                       % Exit interactive mode
     (( X = end_of_file, swi);  (call(X); # X)),
     interactive_mode.                   % Repeat
interactive_mode.                        % Force the goal to succeed
```

The third line in the body of the first clause of *interactive_mode* performs the following actions:

(a) If the user pressed the *RETURN* key in response to *prompt_read*, X is already instantiated to the value *end_of_file*. The effect is to switch the current and alternate images. (Uses the predicate *swi*.)
(b) If the user types a non-null character sequence (X), Prolog tries to satisfy the goal *call(X)*, which may or may not perform some image processing.
(c) Prolog tries to satisfy *call(X)* in the usual way, as if it were a normal Prolog goal. (That is, there is no attempt to do any image processing.) Hence, *call(X)* can either succeed or fail. If *call(X)* happens to succeed, the cycle repeats, beginning with the software issuing another invitation to the user to type a command.
(d) If *call(X)* fails, Prolog sends the character sequence defined by X to the image processor. Two actions can follow: *either* the image processor performs the operation X successfully, or it does not and the first clause fails.

4.1.7 User Extendible Pull-down Menus

Pull-down menus provide a convenient means by which the user of a computer system may order the execution of a range of useful functions. Pull-down menus are standard now on a wide range of popular computers and operating systems. They are unobtrusive when they are not in use, since they occupy only a narrow strip across the top of the computer screen. A single menu heading may allow access to a large number of "hidden" commands, which the user does not have to remember in order to use them properly. Since it is possible to organise hierarchical pull-down menus in MacProlog, it is conceivable that several hundred items could be made available via pull-down menus.

The following menu headings are normally visible in LPA MacProlog: *"File"*, *"Edit"*, *"Search"*, *"Windows"*, *"Fonts"* and *"Eval"*. Thus, there is plenty of room, across the standard Macintosh screen, to add several more menus to support Prolog+. As many as ten special-purpose menus (with single-character headings) can be fitted onto a computer with a standard-size screen. If space allows, the authors recommend the following menu headings for a Prolog+ system:

- *Utility* (for a variety of system-related functions).
- *Process* (for image-image mapping functions).
- *Analysis* (for image measurement and analysis functions).
- *Device* (for controlling external devices such as lights and a pick-and-place arm).
- *Table* (for controlling an (X,Y,θ)-table, or *Robot*).
- *Speech* - Optional. (Used in conjunction with the speech recognition system. See Section 4.2.)

To save space on small-screen computer monitors, it may be more convenient to use only the initial letters of the menu names, viz. *"U","P"*, *"A"*, *"D"* and *"T"* and these are the ones we shall use hereafter. Alternatively, special-purpose menus can be accommodated, simply by deleting one or more of the standard MacProlog menus. For example, the *"Fonts"* menu can safely be deleted, while the functions provided by the *"Windows"* menu can be performed in other ways. It is also possible to develop programs without the *"Search"* menu, though this is a little awkward in practice. It is unwise to delete or alter any of the other standard MacProlog menus. By using short names for menu headings (e.g. single letters), several more application-specific menus could be provided for Prolog+. To provide even greater choice, hierarchical menus can be used in MacProlog. For example, the *'Colour'* sub-menu discussed in Chapter 6 is a fixed sub-menu, appearing under the menu headed *"Utility"* or simply *"U"*. (Figure 4.1(b).)

In the standard Prolog+ software, the *"P", "A", "D"* and *"T"* menus are all empty initially. The *"U"* menu, though not empty, is quite short, with only a few entries. (Figure 4.1(a)) The *"Extend menu"* option under this menu allows the user to add items to any of the *"U", "P", "A", "D" and "T"* menus. The dialogue for doing this is illustrated in Figure 4.2. Figures 4.3 to 4.6 show menus that have been developed in this way.

Mechanism for Extending Menus

Adding new items to a menu can be achieved using the MacProlog built-in predicate *extend_menu*. In the Prolog+ software, the tedium of choosing suitable arguments for *extend_menu* is avoided by the use of a pull-down menu. The following program defines what action follows the user selecting the item "Extend menu" under the *"U"* menu. Using this facility, he is able to extend any one of the user-defined menus: *"U", "P", "A", "D"* or *"T"*. (The *"S"* menu is rather different in both form and function and is not intended primarily for use with the computer mouse. It is provided instead as a convenience, enabling the speech recogniser to operate properly. More will be said about the *"S"* menu in Section 4.2.)

```
% This clause is satisfied when the user selects "Extend menu"
% under the "U" menu.
'U'('Extend menu') :-
    scroll_menu(['Which menu do you wish to extend?'],
    ['U','P','A','D','T'],[],Z), [Y] = Z,
    prompt_read(['What is the name of the item you wish to add to
    the ', Y, ' menu?'],X),
    prompt_read(['What goal do you wish to be associated with the
    item ', X, ' in menu ',Y,' ?'],W), extend_menu(Y,[X]),
    assertz(menu_item(Y,X,W)).
```

(a)

(b)

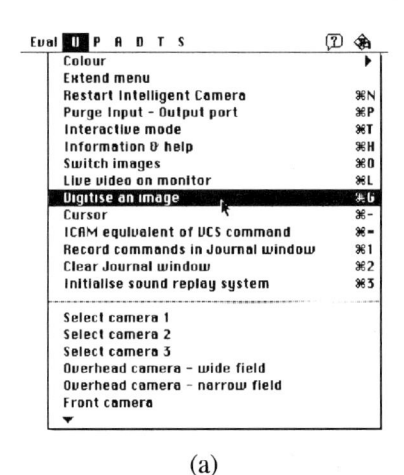

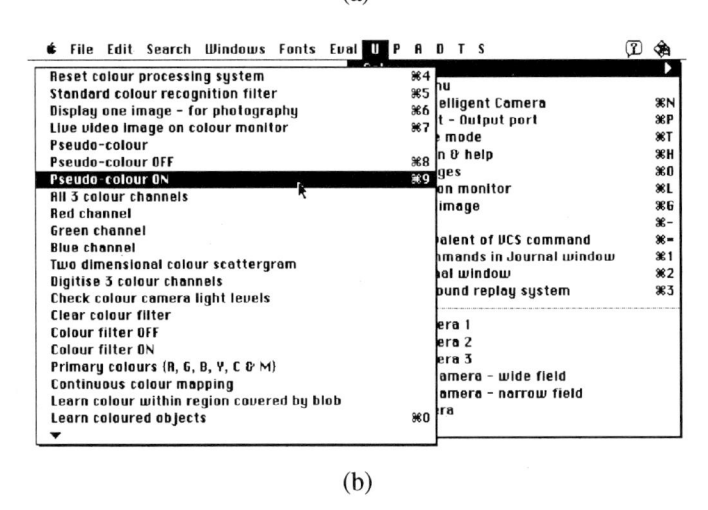

Figure 4.1 The *"U"* (Utility) menu. (a) The top part (above "Initialise sound replay system") is fixed. The lower part of the menu was added using the "Extend menu" facility. (b) The top part of the *"Colour"* sub-menu.

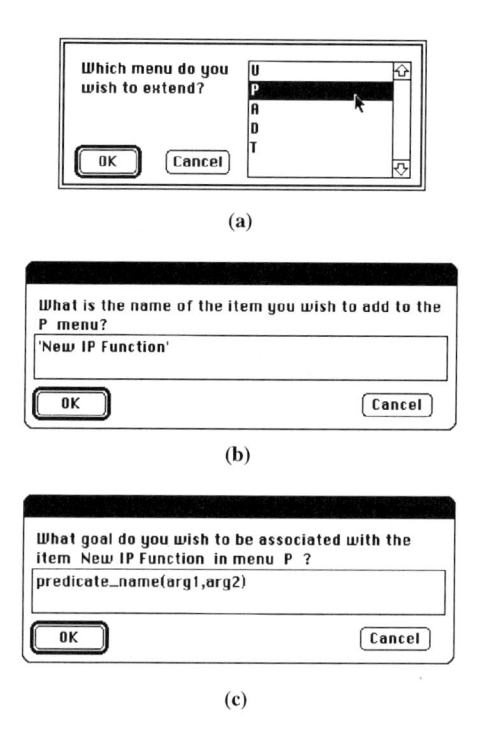

Figure 4.2 Dialogue for extending the menus. (a) User selects which menu is to be extended (b) User names the new menu item, (c) User defines what action is to follow when that menu item is selected.

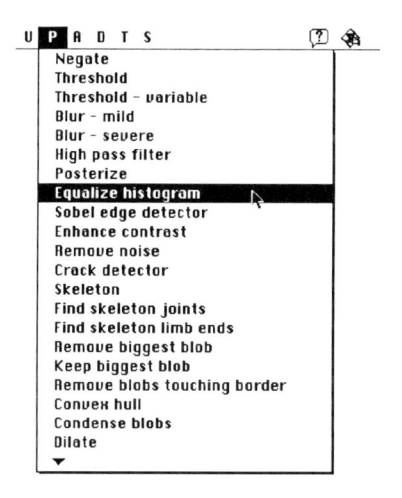

Figure 4.3 Extended *"P"* (Process) menu.

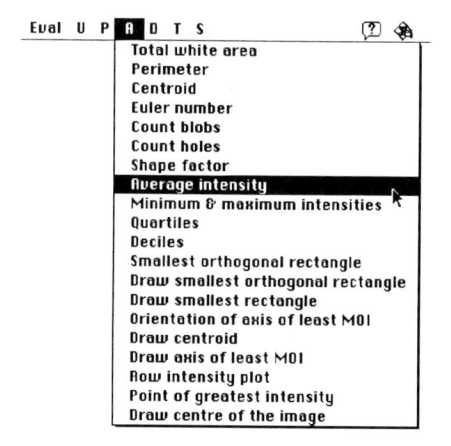

```
Eval  U  P  A  D  T  S                    ?
         Total white area
         Perimeter
         Centroid
         Euler number
         Count blobs
         Count holes
         Shape factor
         Average intensity
         Minimum & maximum intensities
         Quartiles
         Deciles
         Smallest orthogonal rectangle
         Draw smallest orthogonal rectangle
         Draw smallest rectangle
         Orientation of axis of least MOI
         Draw centroid
         Draw axis of least MOI
         Row intensity plot
         Point of greatest intensity
         Draw centre of the image
```

Figure 4.4 Extended *"A"* (Analysis) menu.

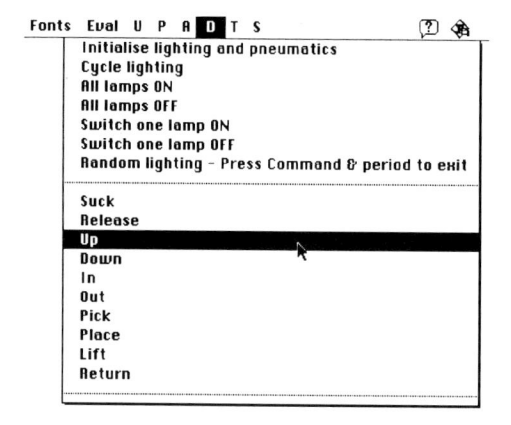

```
Fonts  Eval  U  P  A  D  T  S                    ?
         Initialise lighting and pneumatics
         Cycle lighting
         All lamps ON
         All lamps OFF
         Switch one lamp ON
         Switch one lamp OFF
         Random lighting - Press Command & period to exit
         Suck
         Release
         Up
         Down
         In
         Out
         Pick
         Place
         Lift
         Return
```

Figure 4.5 Extended *"D"* (Device) menu.

The effect of satisfying *'U' ('Extend menu')* is to *assert* one additional clause for *menu_item*. (The reader will recall that *assert* adds the new clause to the program immediately after all of the others already existing.) Notice that *menu_item* must be defined in a data window, otherwise it will not be retained when the user quits MacProlog. The following are typical entries in the "New Menus" window created by *'U' ('Extend menu')*.

```
menu_item('P', 'Equalize histogram', heq).
menu_item('P', 'Sobel edge detector', sed).
menu_item('A', 'Centroid', cgr(_,_)).  % Values are printed by "cgr"
```

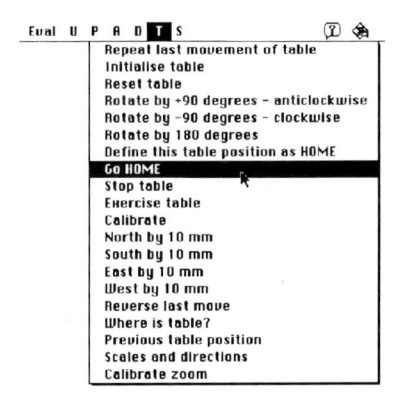

Figure 4.6 Extended *"T"* (Table) menu.

The newly added menu item is available for immediate use. However, it is necessary to build the user-defined menus whenever the Prolog+ system is (re)started. This can be achieved by satisfying the goal *build_menus*.

```
build_menus :-
      menu_item(A,B,C), extend_menu(A,[B]), fail.
build_menus.
```

To build the user-defined menus automatically, it is possible to use either

```
:- build_menus.           % Build the menus after compilation
```

or

```
'<LOAD>'(_) :- build_menus.
                          % Build menus after loading Prolog+ software
```

There is one final addition needed to the Prolog+ software: we must incorporate a set of program segments like that following, in order to define what action follows when a user-extended menu item is selected.

```
% What to do when item X is called from the "U" menu
'U'(X) :-
      menu_item('U',X,Y),  % Consult the database for new menu item
      call(Y).             % Satisfy the appropriate goal, Y
```

Similar definitions are needed for each of the other menus: *"P"*, *"A"*, *"D"* and *"T"*.

4.1.8 Command Keys

Pull-down menus are particularly useful for novices but are not always popular, particularly with very experienced computer users. To accommodate such people, the Prolog+ software is provided with a set of *Command Keys*. (See Table 4.2.)

Key	Function
A	Select all text in active window *
B	Balance parentheses *
C	Copy highlighted text to clipboard *
D	Find definition of predicate with name specified by highlighted text *
E	Find another copy of character sequence highlighted *
F	Find selected text *
G	Digitise (grab) an image (Standard MacProlog facility disabled)
H	Get HELP for Prolog+
I	Get information *
J	Plot colour scattergram (See Chapter 6)
K	Compile *
L	Display live video on monitor
M	Extend menu
N	Start / Restart Prolog+ system
O	Switch current and alternate images
P	Purge the I/O port of all data
Q	Not assigned
R	Replace and find text *
S	Select window *
T	Transparent (or interactive) mode
U	Repeat last query *
V	Paste *
W	What to find *
X	Cut highlighted text *
Y	Window details *
Z	Undo last editing change *
/	Help (MacProlog)
1	Record command in Journal window
2	Clear Journal window
3	Initialise sound replay system
4	Reset colour processing system (See Chapter 6)
5	Standard colour recognition filter (See Chapter 6)
6	Display one image (for photographic recording of results)
7	Display live image on colour monitor
8	Switch pseudo-colour display OFF (See Chapter 6)
9	Switch pseudo-colour display ON (See Chapter 6)
zero (0)	Learn coloured objects (See Chapter 6)
period (.)	Stop goal satisfaction
equals (=)	List named predicate
minus (-)	Cursor on

Table 4.2 Prolog+ command keys. To perform one of the operations listed above, the user presses the corresponding key, at the same time as holding the ⌘ key down. Notice that asterisk (*), denotes a standard MacProlog key command.

4.1.9 Graphical Display of a Robot Work Cell

Apart from image processing, Prolog+ is often used for controlling a robot or (X,Y,θ)-table. In Chapter 5, we shall discuss this topic again. For the moment, we shall simply assert that there is an outstanding requirement for a graphical display of a robot work cell. A drawing showing where the robot / table is and which lights are ON / OFF is all that is needed.

The use of HyperCard graphics for simulating a so-called *Flexible Inspection Cell* (FIC) is illustrated in Figure 4.7. The artwork was prepared using MacDraw and imported into HyperCard. It is possible to generate animated image sequences using HyperCard, which makes the construction of a controller for the FIC easy to achieve. HyperCard graphics does not, however, possess the ability to generate realistic-looking representations of 3D objects.

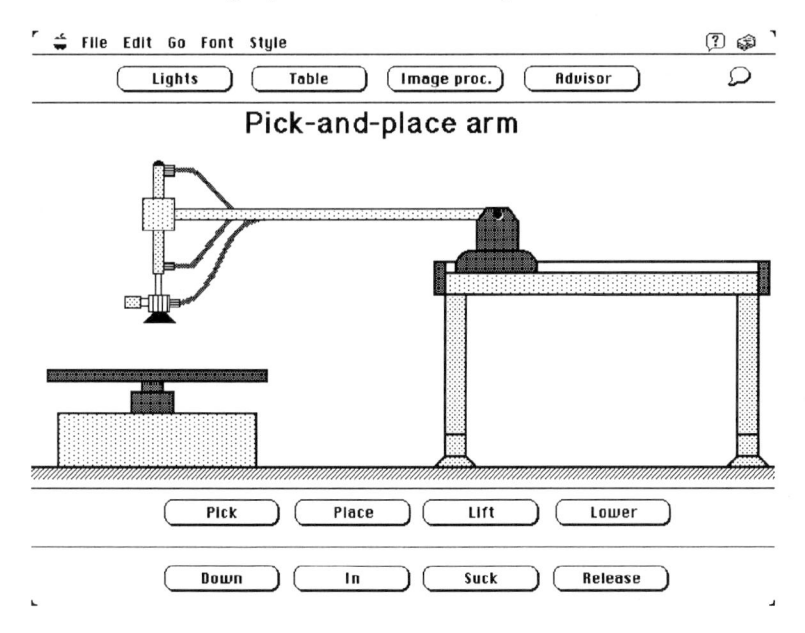

Figure 4.7 HyperCard graphics. The diagram shows a pneumatic pick-and-place arm, which forms part of the flexible inspection cell. (Also see Figure 4.8.) When the user clicks on one of the buttons, the arm moves to a new position and the diagram changes accordingly.

The particular arrangement shown in Figure 4.8 consists of an array of computer-controlled lights, a laser light-stripe generator, a pattern projector, a pneumatic pick-and-place arm, an (X,Y,θ)-table and four video cameras. Prolog+ programs are used to control the real FIC. Later, we shall see further examples of HyperCard graphics, used in the FIC controller. Throughout the remainder of this book, we shall discover that the FIC / Prolog+ combination provides a very versatile platform for studying and prototyping a wide range of inspection and robot guidance tasks. The Flexible Inspection Cell was developed in response to a realisation that traditional machine vision methods are not economic for inspecting products that are made in small batches (fewer than about 10^4 items / batch). The design and construction of the FIC hardware has taken place over a period of years, in parallel with and motivating the development of Prolog+. A detailed description of the function, control and use of the FIC will be deferred until Chapter 5.

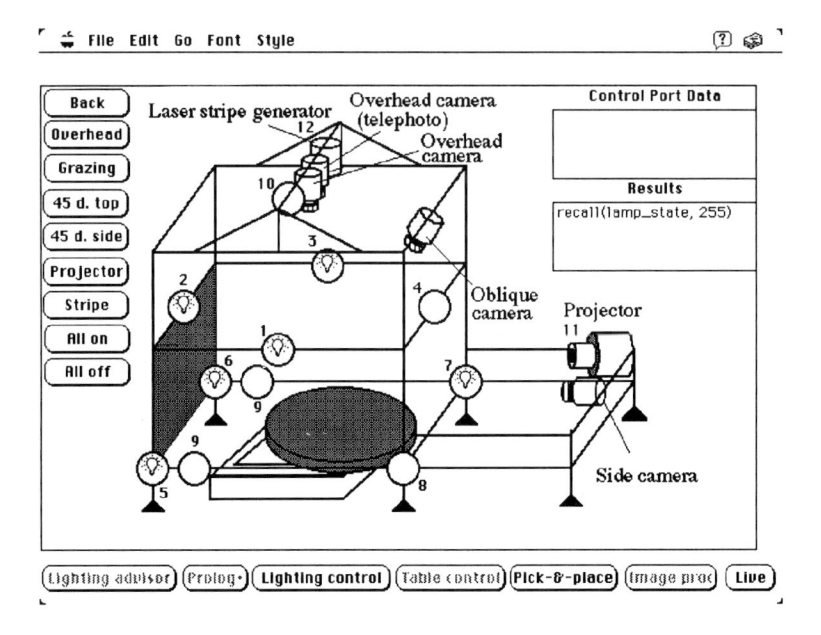

Figure 4.8 HyperCard controller for the Flexible Inspection Cell. Lamps are denoted by circles. Notice that some are ON (lamp symbol appears). The buttons on the left allow the user to set up a complex lighting pattern using a single mouse click. The dark button labelled "Pick & place" causes HyperCard to display one of a number of cards like that shown in Figure 4.7.

4.1.10 Speech Synthesis and Recorded Speech

Speech synthesis is valuable both as an aid to understanding the operation ("flow") of a Prolog program [BAT-91] and in providing a natural form for a human being to receive certain types of data. For some years, a speech synthesis package has been available for use with MacProlog. Another version, capable of working under the System 7 operating system, has been developed recently by McGowan, at Dublin City University. [MCG-94] The more recent development is based upon the standard Macintosh facility known as *Speech Manager*. This imposes certain restrictions: the most notable being that it can cope with strings of length not exceeding 256 characters. Several utilities for configuring the speech synthesiser were devised by McGowan and can conveniently be operated using a pull-down sub-menu, located beneath the "*U*" or "Utility" menu. Prior to using the speech synthesiser, it is necessary to choose certain voice characteristics, such as pitch and speaking rate.

Any synthesiser which generates speech for the English language must be able to cope with non-phonetic spelling. For this reason, a dictionary is often used to convert a phrase like *"Hello Sean. It is Tuesday, so you buy some bread"* to *"Hello Shawn. It is Chewsday, so you should by some bred"*.

An alternative, approach is to write all phrases that are to be spoken in phonetic form. This makes the program more difficult to write, since phrases that are to be spoken have to be to expressed in an unfamiliar "alphabet". McGowan's software provides a convenient facility which assists the user to convert standard English words into phonetic form and then generate utterances for both forms. In this way, the user can fairly quickly obtain a reasonably natural spoken phrase to convey the meaning he intends. A single Prolog+ predicate, *speak(X)* is all that is needed to operate the speech synthesiser, once it has been initialised. The variable X is instantiated to a text string.

An industrial vision system could make good use of speech output to report its status and results. However, this must be used with great care and consideration for the user, because any system (or person) that talks incessantly can be very irritating indeed. The situation is made even worse by the fact that a synthesised voice is imperfect, sounding "tinny", and may well have a disagreeable accent. An inspection system that tells the world, via a loudspeaker, that the objects it is examining have all been found to be satisfactory would quickly annoy anyone standing nearby. On the other hand, a system that can quickly and succinctly summarise its own performance, on demand, would be much more useful. Such a system should be able to list the defect types found during, say, the last hour of its operation, the number of each defect type and possibly a small number of suggestions for corrective action that might be appropriate.

While speech synthesis is very versatile, the quality of the output is low and irritates some users, particularly in situations where a relatively small number of utterances are repeated often. In such cases, it may be better to replay pre-recorded spoken messages. MacProlog can achieve this in a straightforward way, although the details will not be given here.

4.1.11 On-line HELP

The provision of on-line HELP facilities is, of course, a major feature of much of the better commercial software available today. Effective interaction is fast! Hence, good HELP facilities are of special importance for an interactive image processing system, which is driven by a command-line interpreter and has a large repertoire of image processing operators. It is important, for example, that the user should be able to find what parameters are required by a given predicate, with a minimum of effort and delay. The significance of each parameter, its allowed range of variation and default values are all needed. In addition, the HELP facilities should provide a detailed description of what each command does and the type of image on which it operates. For these reasons, Prolog+ has been provided with two different HELP facilities:

(i) When the user depresses the '⌘' and '=' keys simultaneously, a dialogue box is displayed inviting him to type the name of a Prolog+ predicate. When he does so, the listing of that predicate is presented in the default output window.

(ii) When the user depresses the ⌘ / H keys simultaneously, a dialogue box is displayed inviting him to specify the name of an image processing command. The effect is to display a full description of the command, its use (i.e. suitable types of images, the effect, range and default values for each its parameters). (Figure 4.9) The user is also invited to choose a suitable stored test image, so that he can see the effect of that command for himself. Alternatively, he may elect to apply the operator to the current image. He is able to use either default values for the operator's parameters, or values specified by him. In this way, a close connection is maintained between the HELP facility and interactive image processing.

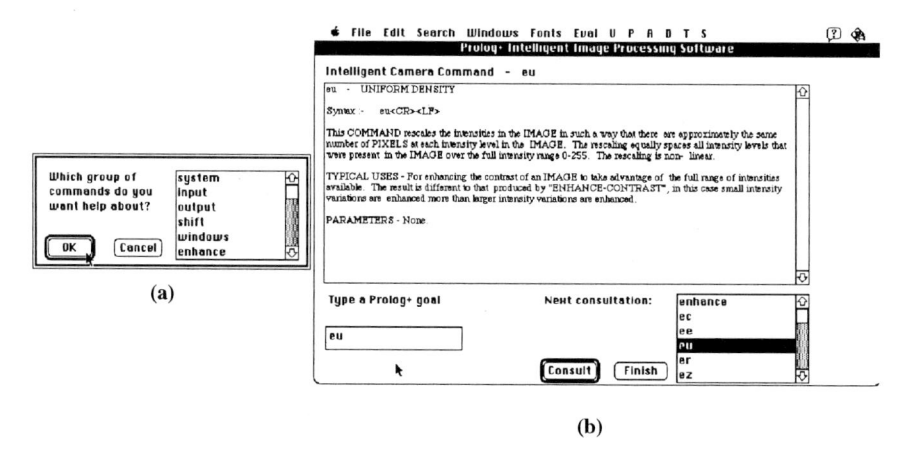

Figure 4.9 User HELP window (a) User initiated the HELP facility using ⌘/H. The window shown here invites the user to select which group of commands he wishes to peruse. (b) User selected the *eu* command, within the *enhance* group of commands. The text shown here relates to the Intelligent Camera implementation of Prolog+.

4.1.12 Cursor

It is important that the user should be able to investigate the intensity values in certain parts of the image he is viewing. Of course, a mouse, or track-ball, is especially convenient for this. The Prolog+ cursor function is provided by the predicate *cur(X,Y,Z)*. Initially, all three of its arguments are uninstantiated. As the user moves the cursor over the picture being investigated, the X- and Y-co-ordinates are displayed in the output window, together with the intensity at that point. To exit the cursor, the user simply double-clicks the mouse button. The effect is to instantiate X, Y and Z, which can then be used in a Prolog+ program. To illustrate this, we present a short program which allows the user to select a bright point in the Hough Transform of a given image, reconstruct the straight line whose slope and intercept are defined by the co-ordinates of that point and then superimpose the line on the original image. This is the inverse Hough Transform method mentioned in Section 2.6.1.

```
interactive_hough_analysis :-
      wri,                    % Save the input image (binary)
      hough,                  % Hough transform
      message(['Use cursor to select a peak in Hough transform
      image, visible now']), % Message for the user
      cur(A,B,_),             % Cursor
      reconstruct_line(A,B),  % Draw a line with slope A &
                              % intercept B
      rea,                    % Recover input image
      max.                    % Superimpose onto original figure
```

This predicate is particularly useful for interactively investigating the significance of the various peaks in the Hough transform image, which is often quite complicated in form. Objects in a binary image can be selected easily using the cursor. To do this, we need to use a Prolog+ predicate which shades blobs in some way. For example, we may use an operator (*shade_blobs* in the program below), which shades each blob according to its size. The blob with the greatest area is the brightest, the second biggest blob is the second brightest, etc.

```
% The following predicate isolates the blob ranked A out of a total
% of B blobs.
isolate_blob(A,B) :-
      count_blobs(B),     % Count the white blobs in the image
      shade_blobs,        % Shade blobs (e.g. according to area)
      cur(_,_,A),         % User chooses one of the blobs
      thr(A,A).           % Blob is isolated
```

The cursor is also useful for such tasks as drawing polygons around objects of interest in a grey-scale image. These objects can then be isolated, by simple image processing. Here is the program for drawing a convex polygon around a number of points defined by the user:

```
% The first clause is used simply to set up the results image
draw_polygon(A) :-
      keep,               % Save input image for use later
      zer,                % Draw black image
      wri,                % Save it
      draw_polygon([],A). % This bit does the real work

draw_polygon(A,B) :-
      fetch,              % Recover input image
      cur(X,Y,_),         % Interactive cursor
      rea,                % Get results image
      vpl(X,Y,X,Y,255),
                          % Add the point found to the results image
```

```
        chu,            % Draw the convex hull
        blb,            % Fill it (make it a solid white figure)
        wri,            % Save it again
        bed,            % Binary edge detector
        fetch,          % Get original image again
        max,            % Superimpose polygon onto original figure
        yesno(['Do you want to add any more points?']),
        draw_polygon([[X,Y]|A]. % Repeat operation.

draw_polygon(A,A).  % Everything done.
```

Of course, it is a straightforward matter to write predicates which draw an "open" polygonal curve, or a "closed" hollow polygon.

4.1.13 Automatic Script Generation and Optimisation

We have taken considerable pains in the earlier pages to explain that the use of interactive image processing is invaluable as a step towards discovering / developing a prototype algorithm for a given inspection task. A command recorder has been developed which allows the user to keep track of what he has typed during an interactive session and thereby develop programs with a minimum of effort. The program requires only a minor change to *interactive_mode* and hence need not be discussed in detail.

Quite a simple Prolog program is able to prune an algorithm search tree, to eliminate "dead" branches. This program is based upon the idea of *data inheritance*. To understand this term, recall that each image processing command transforms the current, alternate and stored images in a predictable way. For example, the command *neg* generates a new current image from the previous current image. Meanwhile, the new alternate image is also derived from (i.e. is identical to) the original current image. Other commands behave in a different way. Figure 4.10 illustrates, in graphical form, some of the different types of data inheritance that the Prolog+ image processing primitives possess. (The situation is simplified slightly, for the sake of clarity, but these details need not concern us here.) Figure 4.10(c) shows how data inheritance flows in a command sequence generated during a sample interactive image processing session. Notice the "dead branch" which does not alter the final outcome in any way and can safely be removed by pruning. As we indicated earlier, it is possible to write a simple Prolog+ program to prune a given command sequence, derived from the interactive image processor. The program relies upon the similar ideas to those written into the *ancestor* relationship used in analysing family trees and hence requires no further explanation here.

4.1.14 Linking to Other Programs

Thus far, we have emphasised the use of Prolog as the host language for image processing, although we have taken care to avoid suggesting that we should manipulate individual pixels at this level. There are, of course, occasions when some language, or application software package, other than Prolog is more appropriate. For example, we might want to use a spreadsheet, or a user-generated C program to implement certain functions that Prolog cannot do easily.

It is possible to interface LPA MacProlog to both C and Pascal programs and this facility is well documented elsewhere. [MCG-94] Interfacing MacProlog to these two languages is fairly straightforward, using the System 7 operating system software of the Macintosh computer. One possible interface for inter-application communication on the Macintosh range of computers uses *AppleEvents*. An AppleEvent is a message sent by one application to another, enabling data and / or commands to be transmitted. In theory, it is possible to communicate between any two applications which support AppleEvents, including MacProlog, Pascal, C, HyperCard, Excel, Lisp and LabView. In order to illustrate the general principles, we shall describe how a HyperCard stack can be used to communicate with MacProlog.

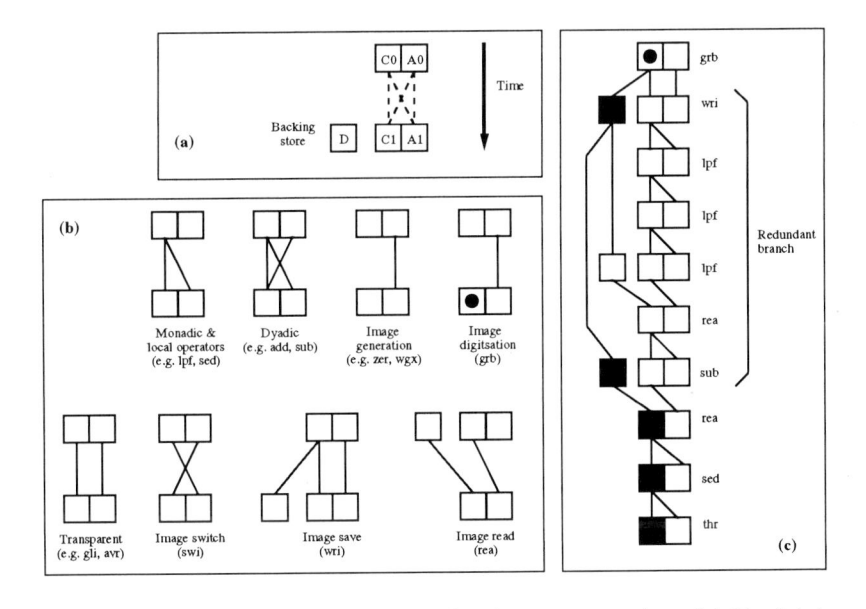

Figure 4.10 Data flow during interactive image processing. (a) Explaining the terminology. The squares labelled C0 and A0 represent the current and alternate images respectively, just prior to the execution of an image processing command. C1 and A1 represent the same images after execution. D represents an image stored either on disc or in "scratch-pad" (RAM) memory. (b) Data flow models for different kinds of image processing functions. (c) Data flow diagram for a short image processing sequence. Operations which affect the final result are indicated by shaded squares. Notice that the sub-sequence [*lpf, lpf, lpf, rea, sub*] can safely be deleted, since it does not influence the final result.

Hypercard Controller for a Flexible Inspection Cell

As we have already explained, the *Flexible Inspection Cell* consists of an array of twelve different lighting units (including a pattern projector and a laser), an (X,Y,θ)-table and a pick-and-place arm. (see Figure 4.8) The FIC at Cardiff currently has four different cameras. Setting up the cell is difficult for a command-based system, so a HyperCard application (called a *stack*) was developed for this purpose. A very convenient software mechanism for interfacing MacProlog and Hypercard has been developed by Stephen Cooper (formally of Uppsala University) and is used as the basis of the controller for the FIC. This interface software uses AppleEvents for signalling between Prolog and Hypercard and has allowed the authors and their colleagues to develop an integrated software environment for setting up and controlling the FIC.

Figure 4.11 shows a series of typical HyperCard screens for controlling the cell. (Also see Figure 4.8) Notice that the user is able to instruct the FIC to perform certain functions, simply by clicking on hot boxes: *buttons*. The FIC hardware can receive its control signals in one of four ways, (a) directly from Prolog, (b) from Prolog, via HyperCard, (c) from HyperCard, via Prolog and (d) directly from HyperCard. Eventually, the first option was selected as being the most convenient and reliable.

More will be said about this HyperCard stack in the following chapter. Let it suffice for our present discussion to note that one of the (HyperCard) cards is able to issue commands (i.e. goals) to Prolog+. (See Figure 4.11(b)) Clicking on the HyperCard button labelled *"cgr"* has exactly the same effect as typing the Prolog+ query *cgr(X,Y)*. When the user clicks on this button, the following sequence of events occurs:

(i) HyperCard sends an AppleEvent specifying the goal *"grb"* to MacProlog.
(ii) The goal *cgr(X,Y)* is satisfied in the usual way, by Prolog. This instantiates X and Y.
(iii) The values of the newly instantiated variables are returned to the application which originated the AppleEvent query (i.e. HyperCard).
(iv) The values of X and Y are available within HyperCard, for display or performing further calculations.

It is possible to specify numeric values using a "slider". (See, for example, the button labelled *"Par."* in Figure 4.11(b).) Among its other functions, the slider enables threshold parameter values to be specified. In addition, it can be used to determine the amount by which a picture is to be shifted (Prolog+ *psh* command), or the increment to the intensity values. (Prolog command *acn*.) The screen shown in Figure 4.11(b) was provided as a utility, enabling a person to set up the FIC and then perform a small amount of image processing, without needing to think in terms of Prolog+. The user can switch easily between the HyperCard control stack and Prolog+.

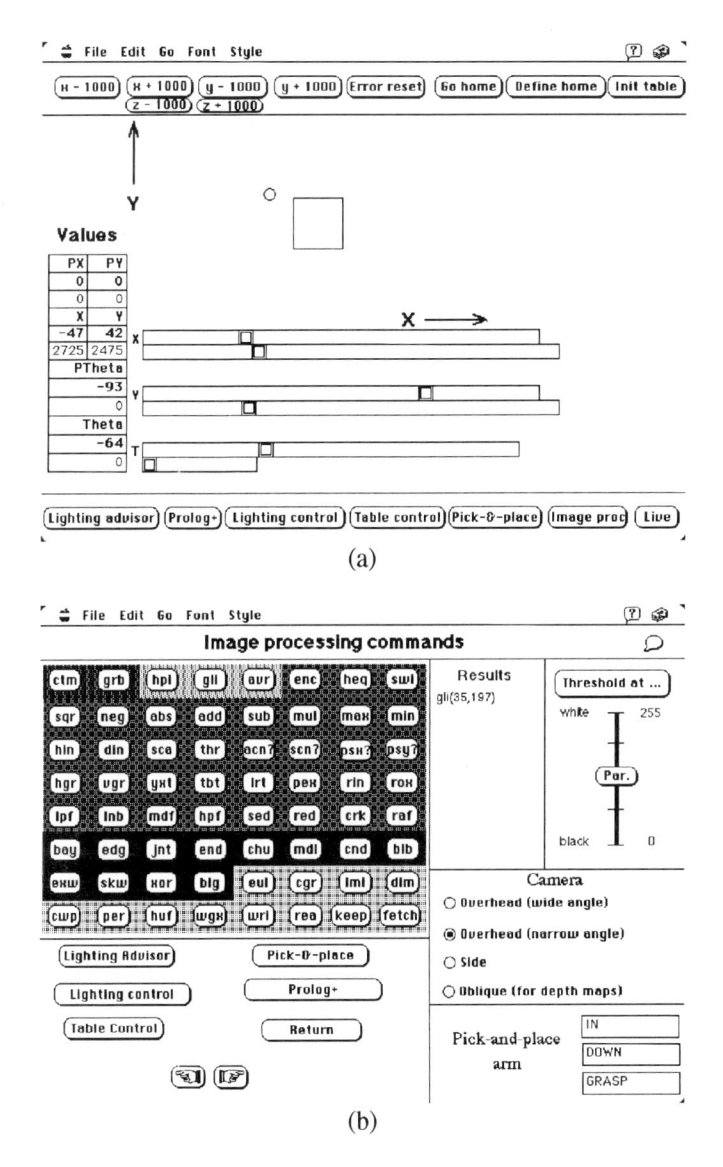

Figure 4.11 HyperCard controller for the Flexible Inspection Cell. (a) Controlling the (X,Y,θ)-table. The user has access to both coarse and fine controls (upper and lower bars) for each axis, labelled X,Y and T. The table movements are specified in both absolute units (millimetres) and as percentages of total travel (Px, Py and PTheta). (b) Controlling the image processor. The user has access to a range of image processing functions, simply by clicking on the appropriate button. Where the button label is followed by a question mark (e.g. *psy?*) the image processing function requires a numeric parameter, which is determined by the position of the slider on the right.

A button (in HyperCard) can be provided which automatically loads the MacProlog / Prolog+ software, or bring Prolog+ to the foreground. Similar facilities have been provided within Prolog+, for automatic loading of HyperCard, or switching to that application. The intention when designing this facility was to make the user feel that he is dealing with an integrated, "seamless" package. Thus, the HyperCard FIC control software is effectively part of the Prolog+ operating environment, as far as the user is able to judge.

4.2 Understanding Simple Spoken Instructions

Speech input has several attractions as a means of controlling industrial systems. It is faster, potentially more accurate and is certainly more natural for the user than a keyboard. In addition, speech input keeps both hands free and the equipment is less prone to damage by swarf, dust, dirt and splashes, all of which abound in a typical factory environment. Of course, safeguards are needed to ensure that noise does not disturb the speech recogniser. Indeed, it may not be possible to use it at all in a very noisy environment. Nevertheless, there are many situations where speech control could be advantageous, if used sensibly. It is not our intention to explore these possibilities here, merely to investigate the technology.

Speech input and natural language understanding form part of the broad environment in which Prolog+ operates and hence discussion of this topic could have taken place within the previous section. We have separated these issues, since our discussion of speech input involves a good deal of technical detail. The reader who wishes to skip this section, can do so in the knowledge that the Prolog+ language, which was, of course, developed initially as a tool for image processing, is also able to cope with the requirements of acquiring data using speech / natural language input. The user interface which we shall describe in this section, is fairly modest in its level of sophistication, being typified by the dialogues needed for controlling the position of an (X,Y,θ)-table, or moving the pieces around on a chess board. Compared to modern techniques for understanding Natural Language, our programs may seem naive but are certainly feasible in practice. The authors have deliberately aimed at producing a *practical*, realistic method for receiving and acting upon spoken commands.

4.2.1 Speech Recognition

There are several speech recognition systems available for users of modern desk-top computers. The remarks in this section relate to just one of these: the Voice Navigator II system (VN), manufactured by Articulate Systems, Inc. (99 Erie Street, Cambridge, MA 02139, USA). This is a hardware-software system which is connected to a Macintosh computer, via its SCSI port. A software version is also available for use with the more modern "AV" family of Macintosh computers, which have a speech input facility.

It is important to emphasise that the VN system is only able to recognise words that are spoken in isolation; it cannot accommodate continuous speech, although some more modern systems do boast this facility. Being restricted to speaking in isolated words is, of course, unnatural and slower than speaking normally. While the VN system has a vocabulary limited to 200 words, it is possible to perform "context switching", using a spoken command to select and load a new dictionary file, appropriate for a different subject (same speaker), or a different speaker (same / different subject). Thus, one user (suppose his name is *Bruce*) can work with a vocabulary of 200 words, one of which is *Paul*. When Bruce says *"Paul"*, the system loads Paul's dictionary file. Paul now takes over the control of the system, and has a working vocabulary of 200 words, one of which is *"Bruce"*. When Paul says *"Bruce"*, the original dictionary is reloaded and Bruce is able to use the system again[2]. An alternative scheme of operation is for a single user to have different dictionaries, corresponding to different subjects of discourse. For example, a single user might have *four* separate dictionaries for controlling the lighting, camera, $(X,Y-\theta)$-table and image processor. While these perform entirely separate functions, it is possible to switch from, say, the lighting dictionary to the dictionary for controlling the $(X,Y-\theta)$-table, simply by saying *"table"*. The user would be able to switch back again, by saying *"lights"*, or move on to operate the image processor, by saying *"image"*.

It is easy to train the Voice Navigator II System to respond to a speech command, so that it will have the same the effect as choosing an item appearing under one of the pull-down menus. Thus, the VN system provides an alternative to normal mouse-operated selection of menu items. (Other modes of operation are possible but are not relevant to our present discussion.) The VN system can operate with any pull-down menus, including system menus, menus created by application software (e.g. MacProlog), and menus programmed by a Prolog+ user, with the *"Extend menu"* option described in Section 4.1.7.

Suppose that the user wishes to order the Prolog+ system to calculate the convex hull of a blob in a binary image and that a pull-down menu has been created, which contains the term *"Convex Hull"*. When training the VN system, the speaker is invited to say the phrase *"Convex Hull"*. This process is repeated a total of three times, allowing the system to estimate the range of variation in the speaker's voice. Now, the user might choose to say *"Convex Hull"*, or to utter some other phrase, such as *"Rubber Band"*, or *"Smallest Polygon"*. As long as he is consistent, the VN system is able to trigger the action associated with the pull-down menu item *"Convex Hull"*, whenever that phrase is spoken in the future. The authors have trained the VN system to respond to a range of spoken commands, such as those listed in Table 4.3.

---

[2] If a speaker that is not known to the VN system tries to use it, the results will be very disappointing. There is no effective short-cut which avoids training the system properly.

a	an	and
anticlock_wise	by	clock_wise
degrees	down	down_wards
eight	eighteen	eighty
eleven	empty_list	fifteen
fifty	five	forty
four	fourteen	hundred
i	inches	left
left-wards	mm	move
nine	nineteen	ninety
one	pixels	platform
please	reposition	right
right_wards	rotate	seven
seventeen	seventy	shift
six	sixteen	sixty
stage	table	ten
that	the	thirteen
thirty	three	to
transfer	translate	turn
twelve	twenty	two
up	up_wards	will
xy-table	you	zero

Table 4.3 Vocabulary used to by the speech recognition system to control an (X,Y,θ)-table. These are the terminal symbols used in the grammar defined in Section 4.2. The control terms ("Begin", "End", "Cancel", "Kill") are not included in this table.

It is possible to perform a limited set of image processing functions, without touching the keyboard. Greater sophistication is needed, if numeric parameters are to be specified for such commands as thresholding, shifting and rotating images.[3] More will be said later in this section about using speech recognition in conjunction with a Prolog+ program, to control the position of the (X,Y,θ)-table in an FIC. Several features of the Voice Navigator II system make it awkward to use in practice:

(a) Homophonous words, such as *"to"*, *"two"* (numeral 2) and *"too"* have to be identified before the grammar rules are written, since they are indistinguishable to the speech recognition system.

---

[3] A few minutes' thought will show that the way that the integers 0 - 1,000 are expressed in English requires the definition of a non-trivial grammar. The analysis of phrases representing the integers is well within the capabilities of the Definite Clause Grammar (DCG) notation of Prolog, as we shall see later.

(b) There is no facility for using synonyms. Thus, *"rubber band"*, *"convex polygon"* and *"smallest polygon"* must all be included explicitly in the pull-down menus, if we wish to have the freedom to utter any one of these three terms. Synonyms are especially important in view of point (c).

(c) There is no record / replay facility, to remind the user what spoken phrase he used when training the VN system. For example, it is very easy for the user to forget that he said *"rubber band"*, rather than *"convex polygon"* or *"smallest polygon"*, since they are conceptually associated with each other.

(d) It can only respond properly, if the utterances it receives consist of well-pronounced isolated words.

4.2.2 Natural Language Understanding

Let us discuss how speech recognition can be used in conjunction with a Prolog program that is able to understand a simple form of Natural Language (English). Consider the limited task of controlling an (X,Y,θ)-table which moves in response to spoken commands. The vocabulary needed to accommodate all user utterances that could reasonably be expected as commands for moving the table consists of less than 200 words. Table 4.3 lists some of the terms that we might encounter. (This list contains 66 items and was compiled from a set of Definite Clause Grammar (DCG) rules defining acceptable commands to an (X,Y,θ)-table. These grammar rules will be discussed in detail later.) First however, we shall describe how the vocabulary defined by a set of grammar rules can be found. This is an essential step, before we can discuss how sentences are developed using the speech recogniser.

4.2.3 Automatically Building a Pull-down Menu

The following Prolog+ program automatically generates a pull-down menu (called the *"S"* menu), containing all of the *terminal symbols* used in the definition of a given grammar. Terminal symbols are those terms enclosed between square brackets in a set of grammar rules and collectively define the *vocabulary* used in conjunction with the grammar rules.

```
% Top level predicate for building the "S" menu
speech_terms :- menu_builder(0,[]).
                        % Do not confuse with "build_menus"

% Searching the DCG window for terms between square brackets: e.g.
% [term]
menu_builder(N,A) :-
     wsearch('DCGs','[',N,X,Y),
                  % Search for string beginning with '['
     wsearch('DCGs',']',Y,Z,_),   % End of string, denoted by ']'
     X1 is X + 1,
     wsltxt('DCGs',X1,Z,Q),
                  % Select text between '[' and ']'
     ((Q = '', R = empty_list); R = Q),   % Ignore empty lists
```

```
        !,
        menu_builder(Z,[R|A]).
                    % Repeat until window search complete

% Add control terms. See note immediately following the end of this
% clause
menu_builder(_,L) :-
        append([begin,end,cancel,'X'],L,A),
                            % Adding control terms to menu
        sort(A,B),          % Sort menu items in alphabetical order
        kill_menu('S'),     % Delete any previous offerings
        install_menu('S',B). % Install menu S using terms in list B
```

The second clause of *menu_builder* adds the following control terms to the *"S"* menu.

begin   end   cancel   kill   'X'

Thus, far, the *"S"* menu does nothing. In order to correct this, we must add the following items to our program:

```
'S'(begin) :- remember(sentence,[]),writenl('Please speak to me').

/* Notice that we could use the speech synthesiser here, to tell
the user what to do. In this case, the following clause should be
used instead of that immediately above.
'S'(begin) :- remember(sentence,[]),speak('Speak to me slowly and
clearly'). */

% What to do when the term "end" is selected from the menu.
'S'(end) :-
        recall(sentence, Z),              % Recall sentence
        writeseq(['The sentence given was:~M',Z]),
        phrase(sentence,Z),               % Apply the parser
        writenl('and was parsed successfully and the instruction
        obeyed').

/* What to do when the input sentence was not understood and cannot
be obeyed. Again, we could use speech synthesis here to good
effect. */
'S'(end) :- writenl('but was NOT understood').

/* What to do when the term "cancel" is selected from the menu.
Notice that this deletes the latest word that was added to the
partially developed sentence. */
'S'(cancel) :-
        recall(sentence, Z),       % Recall partial sentence
        reverse(Z,Z1),             % Reverse list
        Z1 = [_|Z2],               % Omit head
        reverse(Z2,Z3),            % Reverse list again
        writeseqnl(['Sentence has been reduced to:',Z3]),
                                   % Tell the user
        remember(sentence,Z3).     % Remember rest of the sentence

/* What to do when 'X' is selected from the "S" menu. (The user
might, for example, want to enter a number or an unpronounceable
data item (e.g. a product code), via the keyboard.) */
'S'('X') :-
        prompt_read(['Enter a valid Prolog term, please'],X),
        recall(sentence,Y),        % Recall partial sentence
```

```
          append(Y,[X],Z),            % Add new term to sentence
          remember(sentence,Z).       % Remember enlarged sentence

% What to do when anything else is selected from the "S" menu
'S'(X) :-
          recall(sentence,Y),         % Recall partial sentence
          append(Y,[X],Z),            % Add new item to the menu
          remember(sentence,Z).       % Remember enlarged sentence
```

It is possible to select items from the pull-down menu in the normal manner, using the mouse, or by speech control, via the Voice Navigator hardware. The *"S"* menu is intended for use with speech recognition hardware but, if this is not available, the ideas can be tested, almost as effectively, using the mouse. It is possible to build up sentences by selecting terms from the *"S"* menu, starting with *begin* and finishing with *end*. The sequence of operations is as follows:

(i) The user selects *begin*. This has the effect of clearing the input buffer, as a prelude to receiving further data.
(ii) The user selects one of the terms in the *"S"* menu. As each item is selected, the term is added to the buffer. In this way, a sentence is gradually built up, word by word.
(iii) At any time, the user can select the term *cancel* to delete the term added most recently.
(iv) Selecting *kill* causes the program to terminate and thereby ends the dialogue; the partially completed sentence is discarded.
(v) The user selects *end*. This has the effect of passing the sentence just entered to the parser. This checks that it conforms to the grammar.
(vi) Assuming that the sentence entered by the user is accepted by the parser, a second program extracts the relevant meaning. In the next section, we shall illustrate how this can be done for the task of operating an (X,Y,θ)-table.

4.2.4 Understanding NL Commands for an (X,Y,θ)-table

In a little while, we shall present set of grammar rule that was devised specifically for controlling an (X,Y,θ)-table. However, before we do so, we shall briefly explain an important feature of MacProlog that we will be needed to understand the meaning of a command. Consider the following grammar rule:

```
          subject1 --> article, adjective, noun.
```

The states that *phrase* may be replaced by *article*, followed by *adjective*, followed by *noun*. (We need not bother here with the precise definitions of the terms *article*, *adjective*, and *noun*, whose general meanings are obvious.) To verify that a given word sequence S conforms to this grammar, we simply apply the Prolog+ parser, *phrase*:

```
          phrase(subject1,S,_)        % We can safely ignore the
                                      % third argument here
```

This is a standard Prolog goal and either succeeds or fails, in the usual way. Let us assume that *phrase(subject1,S,_)* succeeds. Now, consider the modified definition, in which we have added a term enclosed within brackets, {......}.

```
subject2 --> article, {goal}, adjective, noun
```

The term in brackets is evaluated as a Prolog goal *during the parsing* of *subject2*; the goal *phrase(subject2,S,_)* will only succeed if *goal* succeeds *and subject2* conforms to the given grammar. By embedding portions of Prolog code in the middle of language-rule definitions, we are able to extract meaning, as we are about to see. Here is the promised grammar for commands which operate an (X,Y,θ)-table:

```
% Grammar rules for the (X,Y,Theta)-table
sentence --> table_command.
sentence --> lighting_command.
                    % Possible extension - not defined here
sentence --> arm_command.
                    % Possible extension - not defined here
table_command -->
      {remember(parser_output,[])},
                    % Prolog goal. Store [] in property
                    % "parser_output"
      courtesy, motion, table1, direction, amount.
                    % Basic grammar rule

% It is always best to respond to politeness!
courtesy -->
        [] |                            % No courtesy at all
        [please] |                      % Simple courtesy: "please"
        [will] , [you], [please] |      % A bit more elaborate
        [will],[you] |
        [please], [will], [you] |
        [i], [X], [you], [to] |         % Example: "I want you to"
        [i], [X], [that], [you].        % Example: "I demand that you"

% Verbs: Defining motion
motion --> [shift] | [move] | [rotate] | [turn] | [reposition] |
[transfer] | [translate].

% Noun phrase
table1 --> [] | article, table2.

% Synonyms for "table"
table2 --> [table] | [platform] | [xy-table] | [stage].

% Articles
article --> [] | [a] | [an] | [the].

% Directions
direction --> preposition, [left], {save_parser_output(left)}.
direction -->[left-wards], {save_parser_output(left)}.
direction -->preposition, [right], {save_parser_output(right)}.
direction --> [right_wards], {save_parser_output(right)}.
direction --> [up], {save_parser_output(up)}.
direction --> [up_wards], {save_parser_output(up)}.
direction --> [down], {save_parser_output(down)}.
```

```
direction --> [down_wards], {save_parser_output(down)}.
direction --> [clock_wise], {save_parser_output(clock_wise)}.
direction --> [anticlock_wise], save_parser_output(anticlock_wise)}.

% Saving the output from the parser
save_parser_output(Z) :-
      recall(parser_output,X),     % Find partial sentence
      append([Z],X,Y),             % Add a bit more to it
      remember(parser_output,Y).   % Save it for later

% Prepositions. Note the problem of homophonous terms such as "to"
% and "two".
preposition --> [] | [to] | [to], [the].
preposition -->
      [ two],{writenl('Do not worry - I will cope with this
apparent error')}.
preposition -->
      [two], [the],{writenl('Do not worry - I will cope with this
apparent error')}.

% How far do we move?
amount --> numeral, dimension.
amount --> [by], amount.

% Coping with a common speech recogniser error: confusion of "I"
% and "by"
amount -  .  [: :, amount.

% Dimensions
dimension --> [mm], {save_parser_output(mm)}.
dimension --> [inches], {save_parser_output(inches)}.
dimension --> [pixels], {save_parser_output(pixels)}.
dimension --> [degrees], {degrees}.

% Rule for recognising numbers such as "six hundred and fifty two"
numeral -->
      [X], [hundred], [and], [Y], [Z],
      ( phrase(units,[X]),   % Note: Parser used inside Prolog
      phrase(tens,[Y]),      % Parser inside Prolog to check "tens"
      phrase(units,[Z]),     % Parser applied to check "units"
      words_to_digits(X,X1),
      words_to_digits(Y,Y1),
      words_to_digits(Z,Z1),
      concat([X1,Y1, Z1],W1),
                             % Create 3-digit numeral, in form "652"
      pname(W3,W2),
      save_parser_output(W3) }.

/* Rules are required for recognising several other word-sequences
that are sometime used when naming the integers, in the range 0 -
999. The following list contains one example of each type:
      [six, hundred, and, fifty, two]
      [six, hundred, fifty, two]
      [six, fifty, two]
      [six, five, two]
      [six, hundred, and, eleven]
      [six, hundred, eleven]
      [six, eleven]
      [six, hundred, and, fifty]
      [six, hundred, fifty]
      [six, hundred, and, two]
      [six, hundred, two]
      [six, hundred]
```

```
        [fifty, two]
        [fifty]
        [eleven]
        [seven] */

% Basic definitions
units --> [zero] | ['O'] | [a] | [one] | …… [nine].
tens  --> [ten]  | [twenty] | …… [ninety].
teens --> [ten]  | [eleven] | …… [nineteen].

% Converting words to numerals
words_to_digits(no,0).
words_to_digits(none,0).
words_to_digits('O',0).       % People often say (letter) 'O' rather
                              % than 'zero'
words_to_digits(zero,0).
words_to_digits(a,1).
words_to_digits(one,1).
words_to_digits(two,2).
……
words_to_digits(nineteen,19).
words_to_digits(twenty,2).
……
words_to_digits(ninety,9).
```

4.2.5 Sample Sentences

The following sentences all conform to the grammar defined above.

```
[please, move, up_wards, fifty, two, mm]
[i, want, you, to, shift, up_wards, fifty, two, mm]
[i, demand, that, you, shift, to, the, right, by, four, inches]
[i, want, you, to, shift, to, the, right, by, one, hundred, and,
sixty, two, mm]
[rotate, clock_wise, by, four, hundred, and, two, degrees]
[rotate, clock_wise, by, twenty, two, degrees]
[please, turn, clock_wise, by, seventy, two, degrees]
```

4.2.6 Interpreting the Parser Output

We have defined the grammar rules in such a way that, when the parser is applied to any valid sentence, the result is a 3-element list, stored in the MacProlog property *parser_output*. To appreciate this, consider the following compound goal:

```
remember(parser_output,[]),          % Initialise parser output list
X = [please, turn, clock_wise, by, seventy, two, degrees],
phrase(sentence, X,_),               % Apply the parser
recall(parser_output,Y)              % Find the parser output
```

The effect is to instantiate Y to *[degrees, 72, clock_wise]*. This enables us to understand how the "meaning" of a command can be extracted from a sentence. In a little while, we shall present a program which interprets the parser output and actually moves an (X,Y,θ)-table. This program uses a predicate called *move(X,Y,Z)*. Let it suffice for the moment to say that its arguments have the following functions:

First argument: Move along the X-axis, units are specified in millimetres.
Second argument: Move along the Y-axis (millimetres).
Third argument: Rotate by an angle specified in degrees.

Here is the program to interpret the parser output.

```
interpret :-
      recall(parser_output,X),      % Get the parser output, (a 3-
                                    % element list)
      interpret1(X).                % Now, interpret X

% How to interpret [clock_wise, degrees, numeral] (The 3 elements
% occur in any order)
interpret1(A) :-
      member(clock_wise,A), member(degrees,A),
      member(X,A), number(X), move(0,0,X).
                                  % Rotate the table by X degrees

% How to interpret [anticlock_wise, degrees, numeral]
interpret1(A) :-
      member(anticlock_wise,A), member(degrees,A),
      member(X,A), number(X), X1 is -X, move(0,0,X1).

% How to interpret [left, mm, numeral]
interpret1(A) :-
      member(left,A), member(mm,A), member(X,A),
      number(X), X1 is -X, move(X1,0,0).

% How to interpret [right, mm, numeral]
interpret1(A) :-
      member(right,A), member(mm,A),
      member(X,A), number(X), move(X,0,0).

% How to interpret [up, mm, numeral]
interpret1(A) :-
      member(up,A), member(mm,A), member(X,A),
      number(X), move(0,X,0).

% How to interpret [left, mm, numeral]
interpret1(A) :-
      member(down,A), member(mm,A),
      member(X,A), number(X), X1 is - X, move(0,X1,0).

% What to do if we cannot move the table according to the given
% command
interpret1(A) :- message([A,'was not understood']).
```

4.2.7 Review

The following points should be noted:

(i) The Voice Navigator II speech recognition system is able to select any item from any pull-down menu.

(ii) It is possible to construct a pull-down menu (called the *"S"* menu) from a set of grammar rules, using *menu_builder*. The *"S"* menu contains a small number of control terms (*begin, end, cancel, kill, 'X'*), in addition to a list of the terminal symbols defined in the grammar.

(iii) Items from the *"S"* menu can be selected using either the mouse or the speech recogniser.

(iv) A set of grammar rules, defining commands for controlling an (X,Y,θ)-table, has been defined.

(v) When *menu_builder* is applied to these rules, an *"S"* menu is built and contains the items listed in Table 4.3.

(vi) To enter a command for the (X,Y,θ)-table, the user can enter a sequence such as: [*begin, please, move, up_wards, fifty, two, mm, end ....*].

(vii) We have explained how the key-words in such a sentence can be extracted. The result is a list, such as [*up_wards,mm, 52*].

(viii) A set of rules for moving the (X,Y,θ)-table has been defined (predicate *interpret*). These rule use the 3-element key-list as "input".

In addition to the table controller, the speech recogniser has been used to operate the lights in the Flexible Inspection Cell and to perform standard image processing functions. It is a notable experience to observe the (X,Y,θ)-table move, or the lamps turn ON / OFF, in response to spoken commands, since the user feels as though he has great power. While these ideas are still under development, they do illustrate the fact that speech recognition, linked to simple grammar-rule analysis is both practical and highly attractive, even to computer literate people. For these reasons, the authors are convinced that many uses will be found for this combination of technologies, in the future.

4.3 Aids for Designing Vision Systems

Designing a machine vision system is more complicated than simply choosing an image processing algorithm. Selecting an appropriate lighting and viewing configuration is of vital importance, before any image processing is ever contemplated. The Lighting Advisor described below is one of several design tools that have been devised, or are still under development, as part of the general drive towards providing assistance for the vision engineers. (Another web version of this design tool is also available [WWW-1].) Expert systems, hypermedia programs and deductive systems are also being developed for such tasks as choosing the appropriate camera, selecting a suitable lens, preparing samples for inspection, etc. The one outstanding requirement is for a system that can give advice about which are the most appropriate image processing algorithms to consider for a given application. While this is still a long way off, one system for doing this, within the narrow confines of inspecting objects made by a single multi-product manufacturing line has been developed. [CHA-95]

4.3.1 Lighting Advisor

The Lighting Advisor is a hypermedia catalogue, describing about 150 different lighting and viewing techniques, suitable for industrial machine vision systems. The Lighting Advisor is in the form of two interconnected HyperCard stacks. For each lighting and viewing technique, there are two cards: one provides notes in the form of plain text, while the other shows the optical layout diagram. The Lighting Advisor can also be linked to other HyperCard stacks, such as the controller for the Flexible Inspection Cell described in Section 5.5. Additional stacks are currently being developed and will eventually include details about cameras, references to the technical literature and preparing a sample for visual inspection.

Lighting and viewing are now widely regarded as being of critical importance for the successful development of machine vision systems, whether they are being used for such tasks as inspection, measurement, grading, sorting, monitoring or control for industrial applications. [BAT-85, BIE-91] In the early years of the development of interest in Automated Visual Inspection, it was often argued that an experimental approach to lighting design was essential. [BAT-80] While this point of view is still valid, there is undoubtedly a need for more formal methods and suitable computer-based design aids. [BAT-91b, BIE-91, BAT-92] There has been a concerted effort recently by several groups to develop software tools by which a vision engineer can obtain advice about which lighting and viewing techniques are appropriate for a given machine vision task. Advisors for lighting and viewing have been developed by Ball Corporation Inc., [PEN-88] Industrial Technology Institute, Dearborn, MI [ITI-89] and one of the present authors [BAT-89]. In recent years, very flexible lighting systems have been developed, most notably ALIS 600 [DAU-92] and Micro-ALIS prototyping systems. [ALIS] Other notable work in this area has been described by Ahlers, who devised feedback control systems for industrial lighting units [AHL-91]. The Flexible Inspection Cell (FIC) has many of the advantages of the automated ALIS 600 system and contains a versatile manipulator. [BAT-94b] While the control of the FIC will be described in detail in the next chapter, we merely need to note here that systems such as these emphasise the need for a program which can assist in choosing an appropriate lighting / viewing arrangement. The Lighting Advisor should be viewed as being part of a much larger prototyping system (see Figures 4.12 to 4.19), which will eventually incorporate databases on:

(a) Lighting and viewing techniques.
(b) Cameras.
(c) Lenses.
(d) Sample preparation for easier viewing.
(e) Image processing techniques.
(f) Technical literature on machine vision.
(g) Addresses of suppliers of hardware and software, research institutes.

In the future, it is anticipated that the Lighting Advisor software will also be used to control the Flexible Inspection Cell. It is already interfaced to the Prolog+ software. It is of interest to note that the possibility of using HyperCard for a Lighting Advisor was first considered in 1988 but this approach was soon abandoned in favour of one based upon a Prolog program. [BAT-89] The decision to revert to HyperCard was made in 1993, following preliminary work by a Cardiff student. [WIL-93]. The reason for this was the improved facilities offered by HyperCard, most notably the ability to show high-quality graphics, on a large-screen display unit.

Stack Structure

The Lighting Advisor consists of three interconnected HyperCard stacks called *"Lighting Advisor", "Lighting Diagrams"* and *"Lighting Pictures"*. Apart from a short preamble, the *"Lighting Advisor"* stack consists entirely of so-called *Methods cards*, having the form shown in Figure 4.12. The buttons along the bottom of the *Methods* cards have the following functions:

Find	Find text. Dialogue box appears.
Index	Go to "Index" Card (Figure 4.13).
Layout	Go to corresponding card in "Lighting Diagrams" stack.
Print	Print the present card
Begin	Go to the first card in the "Lighting Advisor" stack.
Prev.	Go to the previous card seen
Exit	Go to HyperCard "Home" stack

The text fields in the *Methods* cards are obvious from Figure 4.12. At the time of writing, there are over 150 *Methods* cards, relating to different lighting and viewing techniques. This compares with 63 different lighting and viewing methods described in a catalogue of such techniques, published 1985. (See [BAT-85] and [BAT-94b].) Notice that each Method card in the *"Lighting Advisor"* stack is uniquely associated with one card in the *"Lighting Diagrams"* stack and another in the *"Lighting Pictures"* stack. Each card in the *"Lighting Diagrams"* stack shows the layout of an optical system, while the *"Lighting Pictures"* stack shows sample images obtained (see Figure 4.14).

Search Mechanisms

Apart from the buttons just described, there are several other ways to navigate, in the Lighting Advisor: (a) *"Index"* card (Figure 4.13); (b) *"Map"* card (Figure 4.16); (c) First card in the *"Lighting Advisor"* stack (Figure 4.15); (d) Automatic text search.

Each of these will now be described in turn.

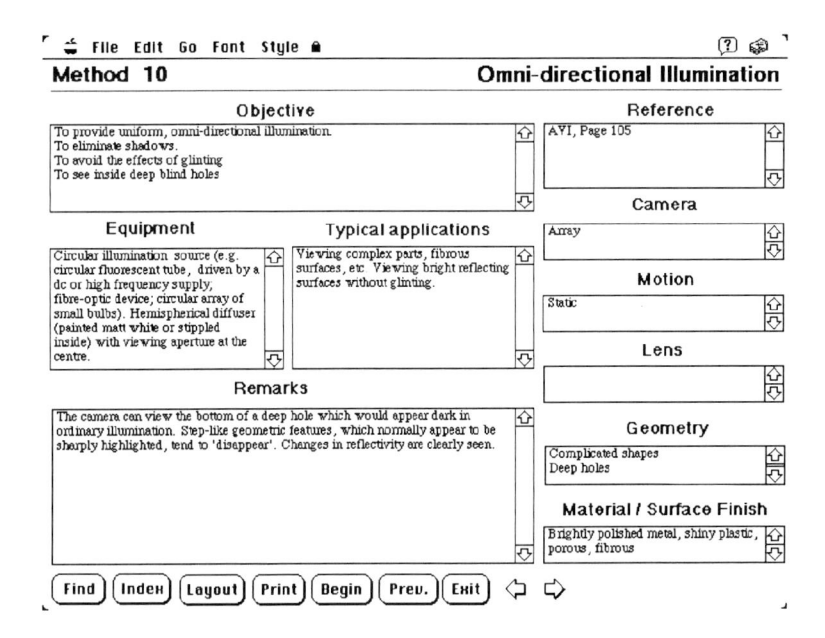

Figure 4.12 Typical *"Method "* card from the Lighting Advisor Stack.

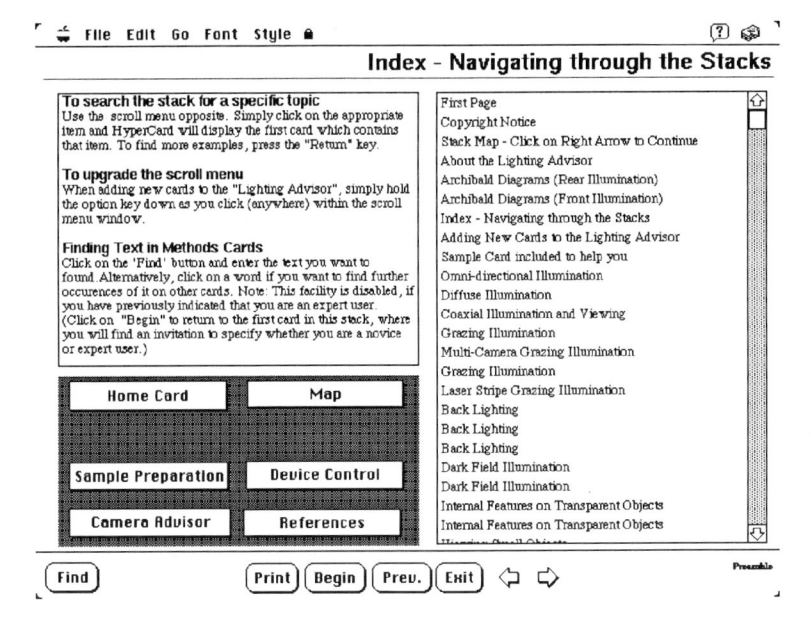

Figure 4.13 *"Index"* card. *Top left:* Information window. *Bottom left:* Buttons for navigating through this stack or moving to other stacks. *Right*: scroll menu. Clicking on an item in this scroll menu, causes Hypercard to move to the corresponding card in the *"Lighting Advisor"* stack.

138

(a)

(b)

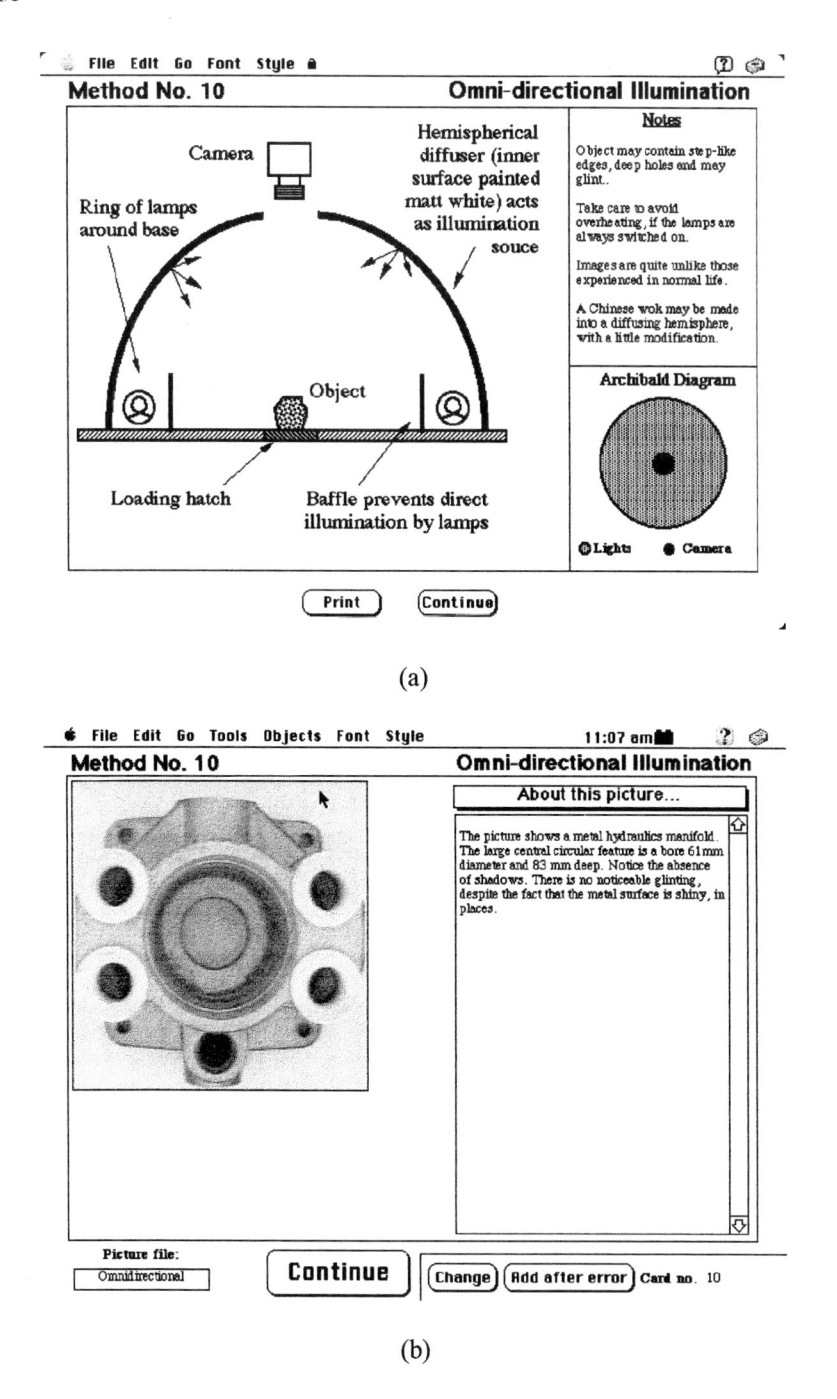

Figure 4.14 Secondary cards, linked to the card shown in Figure 4.12. (a) Card from the *"Lighting Diagrams"* stack. (b) Corresponding card from the *"Lighting Pictures"* stack.

"Index" card:

The *"Index"* card is shown in Figure 4.13. In addition to the buttons described above for cards in the *"Lighting Advisor"* stack, there are seven buttons, labelled:

Sample Preparation	Go to the "Sample preparation" stack
Camera Advisor	Go to the "Camera Advisor" stack
Map	Go to the "Map" card
References	Go to the "References" stack
Device Control	Go to the "Device Control" stack
Home Card	Go to the Hypercard "Home" stack.

In addition, there is a scroll menu, whose entries correspond to the text field at the top-right-hand corner of each of the *"Methods"* cards in the *"Lighting Advisor"* stack. Clicking on an item in this scroll menu, causes Hypercard to move to the corresponding card.

"Map" card:

See Figure 4.16 It is recommended that the stack structure of any complex HyperCard system be modelled, on a card known as a *"Stack map"*. The map card follows this practice and shows how the *"Lighting Advisor"* and *"Lighting Diagrams"* are inter-linked. Clicking on a card, or stack, causes HyperCard to move as appropriate.

First card in the "Lighting Advisor" stack:

Buttons can conveniently be added to the this card, whenever other stacks are added to the system. Templates for the "Camera", "References" and "Sample Preparation" stacks have all been devised. From the first card of the "Lighting Advisor" stack, the user can elect to move to a detailed description of the stack (via the "Preamble" button), or skip to the Methods cards, describing the various lighting / viewing techniques. Of special note here is the invitation to the user to indicate his level of experience. While an experienced user can add / modify cards in the "Lighting Advisor" and "Lighting Diagrams" stacks, a novice user can navigate through the "Lighting Advisor" stack using the automatic text search facility.

Automatic text search:

The automatic text search facility provides a very simple way for a novice user to search for items of interest. Consider Figure 4.12. Suppose that the user wishes to find more information relating to the word *"glinting"*. To find another card which contains this term, the user simply clicks once on that term, in one of the Methods cards. By repeating this exercise, it is possible to find all cards which share this term in a very short time. Browsing in this way is, of course, one of the strengths of HyperCard.

140

Figure 4.15 The first card in the "Lighting Advisor" stack. Notice the presence of several buttons to initiate navigation through this stacks. The functions of these buttons are explained in the text. The user can indicate his level of his expertise, by clicking on the button on the lower right. (The button label toggles between "Expert" and "Novice".)

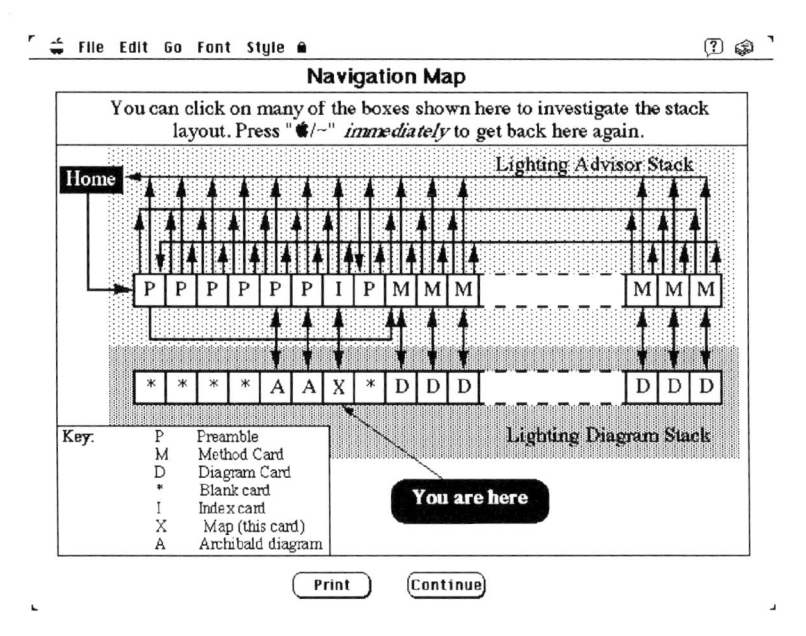

Figure 4.16 "Map" card, showing how the *"Lighting Advisor"* and *"Lighting Diagrams"* stacks are interconnected.

Remarks About the Lighting Advisor

Throughout this book, we repeatedly emphasise that machine vision is concerned with much more than image processing, which is the particular forté of Prolog+, as it was originally described in Chapter 3. One of the primary areas that requires detailed consideration in any application is that of designing the image acquisition sub-system (i.e. lighting, optics, viewing angle and camera). Since 1985, when the original catalogue of lighting and viewing methods was published, a large number of other techniques have been used for machine vision. Rather than publish another catalogue in book / paper form, it was decided to write a much more comprehensive lighting advisor program using a modern version of HyperCard and link this to Prolog+.

At the time of writing, there are 150 different lighting / viewing methods described in the Lighting Advisor. However, it is not intended to be a static entity; as new lighting and viewing methods are devised, its database will grow. Even a moderately experienced Hypercard user can do this with ease. There are several obvious steps for the future development of the Lighting Advisor:

(a) Restore the inferential power of Prolog, which was present in the earlier Lighting Advisor [BAT-89], but which is lacking in the present implementation. This would provide a "smart" dialogue to elicit information from the user about the application. A Prolog program can, for example, make useful logical inferences: that oily or wet objects glint; that glinting is also called specular reflection and that glinting can be cured by using crossed linear polarisers.
(b) Control the Flexible Inspection Cell. Of course, this facility will only be possible for some of the methods represented in the "Lighting Advisor" stack.
(c) Add descriptions of further lighting and viewing methods to the stack.
(d) Add further material, to provide a tutorial introduction to lighting and viewing methods for industrial applications.

The Lighting Advisor can, of course, be linked to the HyperCard stack which controls the FIC and hence forms part of the general Prolog+, environment. Thus, a person using Prolog+ can consult the Lighting Advisor, set up the recommended lighting configuration and return to Prolog+ simply by clicking the Macintosh mouse. The procedure for navigating between these utilities is as follows:

In Prolog+	Select from the "Utility" menu
In HyperCard (FIC control)	Click on button marked "Lighting Advisor"
In HyperCard (Lighting Advsor)	Click on button marked "FIC Control"
To return to Prolog+	Click on button marked "Prolog+"

It is the authors ultimate ambition to eventually add several other advisory programs to the extended Prolog+ system, including a *Camera Selection Advisor* and a *Sample Preparation Guide*. To date, only the templates for these additional facilities have been devised (Figures 4.17-4.19). Collecting the relevant information from equipment manufacturers, then restructuring it to form a comprehensive, up-to-date, database, requires greater man-power resources than the authors can muster at the moment.

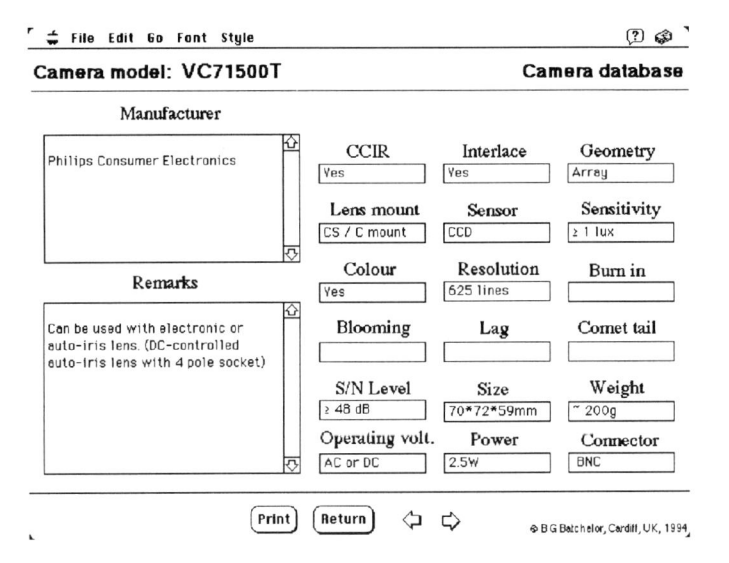

Figure 4.17 Template for the *Camera Advisor*.

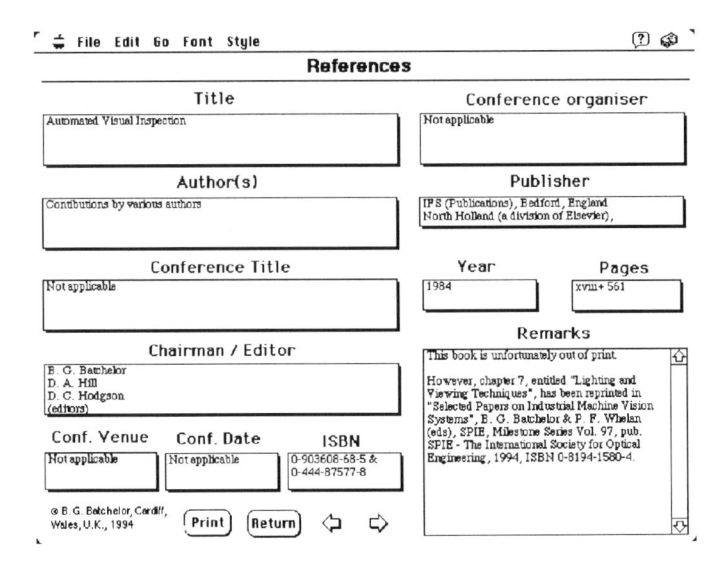

Figure 4.18 Template for the *References* database. This will eventually form part of a comprehensive hypertext data search facility, for machine vision.

```
File  Edit  Go  Font  Style                                    ?
```

Sample Preparation
To Improve Image Contrast / Make Image Processing Easier

Name of Technique
Dye Penetrant Indicator Method

Objective
To highlight cracks in non-ferrous components (e.g. forgings).

Description
1. Immerse the component in a vividly coloured dye, which has a low viscosity and high surface tension
2. Gently blow / shake to remove excess dye
3. Coat the component in an absorbent white powder, such as talc.
4. Any dye which has penetrated the cracks is drawn out by capilliary action

[Print] [Return]

Figure 4.19 Template for the *Sample Preparation* database.

4.3.2 Other Design Aids for Machine Vision

The authors long-term research goal is to develop a comprehensive, fully integrated, CAD facility for machine vision systems. There is no doubt that, we are very rapidly approaching a bottle-neck on further development of the subject, caused by an acute shortage of trained personnel. For this reason, the authors believe that the development of good design tools is essential, if the enormous potential of machine vision technology is ever to be realised in full. While considerable progress has been made already in certain areas, some gaps remain. The following is a list of notable achievements by other workers in this general field, but which cannot be integrated easily with Prolog+, and outstanding "problem" areas requiring further work:

(i) *Opto*Sense®* is a comprehensive database of commercial systems suppliers [WHI-94]. The software runs under the MS-DOS operating system and hence is not compatible with the Prolog+ software.

(ii) A specialised calculator for lens selection has been distributed by the Machine Vision Association of the Society of Manufacturing Engineers. [MVA] The *Machine Vision Lens Selector* is in the form of a slide rule, which allows the user to find the focal length of a lens, given the object distance and field of view. Alternatively, the user can find the f-number from the depth of field and desired resolution.

(iii) A Macintosh desk accessory for lens selection has been devised. [SNY-92] Since this runs on the Macintosh computer, it can "co-exist" with the Prolog+ / HyperCard software but does not interface directly to it. The software is in the form of a specialised screen-based calculator that can be called by selecting the appropriate item from the Apple menu. This program performs the same calculations as the slide rule mentioned in (ii). It would not be a major task to rewrite this program using Prolog or Hypercard.

(iv) At least one sophisticated ray-tracing program for the Macintosh computer is known to exist [KID]. This program can, of course, "co-exist" with the Prolog+ software, but does not interface directly to it. The program is able to calculate all of the necessary design parameters for a given optical set-up.

(v) The basic frame-work for a program which can give advice about which camera to use have been developed by two post-graduate students working at Cardiff [PET-90, WIL-92]

(vi) A "general purpose" program which is able to give advice about suitable image processing operations for a given application is not yet available. Some progress has been made in this area, for just one specific application area: inspecting cakes made in the form of a continuous ribbon. [CHA-95]

4.4 Multi-camera Systems

The embodiment of Prolog+ described in Chapter 3 allows the use of only one video camera. However, there are many instances when it is useful to be able to combine two or more views of an object. (Table 4.4 and Figures 4.20 and 4.21.) In this, the last part of this chapter, we shall discuss some of the ways in which we can build an *intelligent* vision system that is able to digitise and process data from several cameras.

4.4.1 Multiplexed-video Systems

Consider Figure 4.22(a), which shows perhaps the simplest and potentially most flexible multi-camera system. We shall refer to this as a *Video-Multiplexed (V-M)* system. Notice the presence of the video multiplexor on the input of the frame-store, and that the multiplexor is controlled by the image processing computer. Many low-cost image processing systems are organised in this way and can typically accept up to four cameras. In a V-M system, the images from the various cameras can be superimposed, if desired. This facility is not possible with Figure 4.22(b) and some of the other schemes described below. It is a simple matter to add a video multiplexor on the input to *any* image processor, provided it can generate the necessary control signals. (This will be discussed again in Chapter 5.) The high flexibility of a V-M system is achieved by performing operations in sequence, but of course this reduces its operating speed, compared to the concurrent systems about to be described.

Application	Remarks
Packing/depletion e.g. cutting leather for shoes, bags, etc.	One camera is used for inspecting leather surface, prior to cutting. A "flaw map" is fed to packing program. Cutting templates and hide are viewed by different cameras.
Car wash	Several cameras are needed to inspect the whole body surface before and after the wash. Other cameras are needed to monitor the washing process.
Monitoring bakery, or similar continuous-flow manufacturing plant	Several cameras are needed for inspecting the mixing and feeding of raw materials, the unbaked dough (formed by the extruder), and the product after baking, after decoration, before and after cutting, during and after packing.
Brick laying robot	Separate views are needed of both sides of wall being built, placing of mortar on wall and bricks, cutting of bricks.
"Golden sample" inspection of populated printed circuit boards	At least, one camera is needed for each board. (May be more, if both high- and low-resolution viewing is needed.)
Pruning plants	At least three cameras are needed to obtain all-round view. (May be more, if both high- and low-resolution viewing is needed.)
Multi-lane production processes	One camera cannot provide sufficiently high resolution.
Power-press, or similar machine	Separate cameras are needed to monitor material feed, inspect dies and to inspect the finished product.
Painting large complicated structures using a robot	Several cameras are needed, for navigation and for inspecting the work-piece before during and after spraying. Cameras may be mounted on robot arm, to examine every nook and crevice.
3D inspection of foreign bodies in packaged food.	Several containers may be filled simultaneously on a modern, high-speed production line. Hence jars can never be viewed in isolation from one another. (For example, arrays of 2*3 jars of tomato sauce may be filled at the same time. Two orthogonal views are needed to locate a foreign body.)
Flexible Manufacturing System	FMS is typically left unattended for long periods of time. Vision is good means of monitoring progress.
Monitoring factory (for safety hazards)	Many cameras needed to make sure gangways are clear, doors are closed, floor is clean, pallets are correctly loaded, stacked boxes are stable, no leaks from complex pipe-work installations, etc.
Robotics	Monitoring robot work area for human intruders, obstacles which could get in way of the robot.

Table 4.4 Some practical applications of intelligent multi-camera vision systems.

data it receives from the Slave units operating below it and *occasionally* sends them data to control / modify their actions. The arrangement shown in Figure 4.24 is even more powerful. Each of the Slaves can control as many as eight image processors (for example, Intelligent Cameras), through the use of an RS232[4] multiplexor. Each of the image processors can have several remote image sensing heads. (Up to 4 for the Intelligent Camera.) As many as 32 Macintosh computers can be connected to a single AppleTalk network. Hence, we realise that data from up to 1024 cameras can be processed by the system.

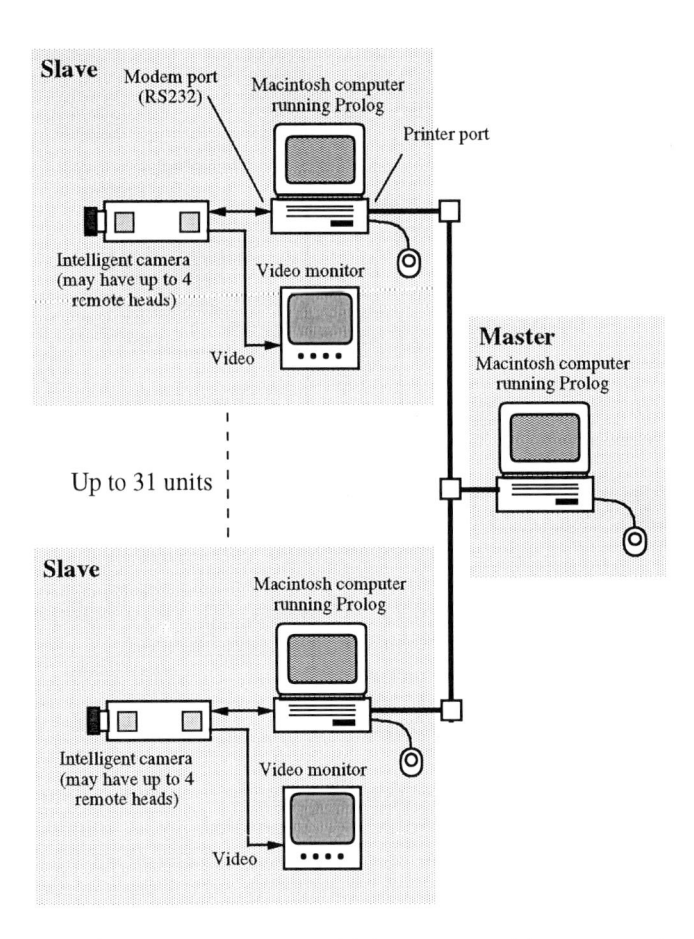

Figure 4.23 Organisation of a multiple camera vision system in which several Prolog+ sub-systems are controlled by a single processor.

---

[4] Although the Macintosh computer uses the RS422 protocol, it is possible to operate RS232 devices.

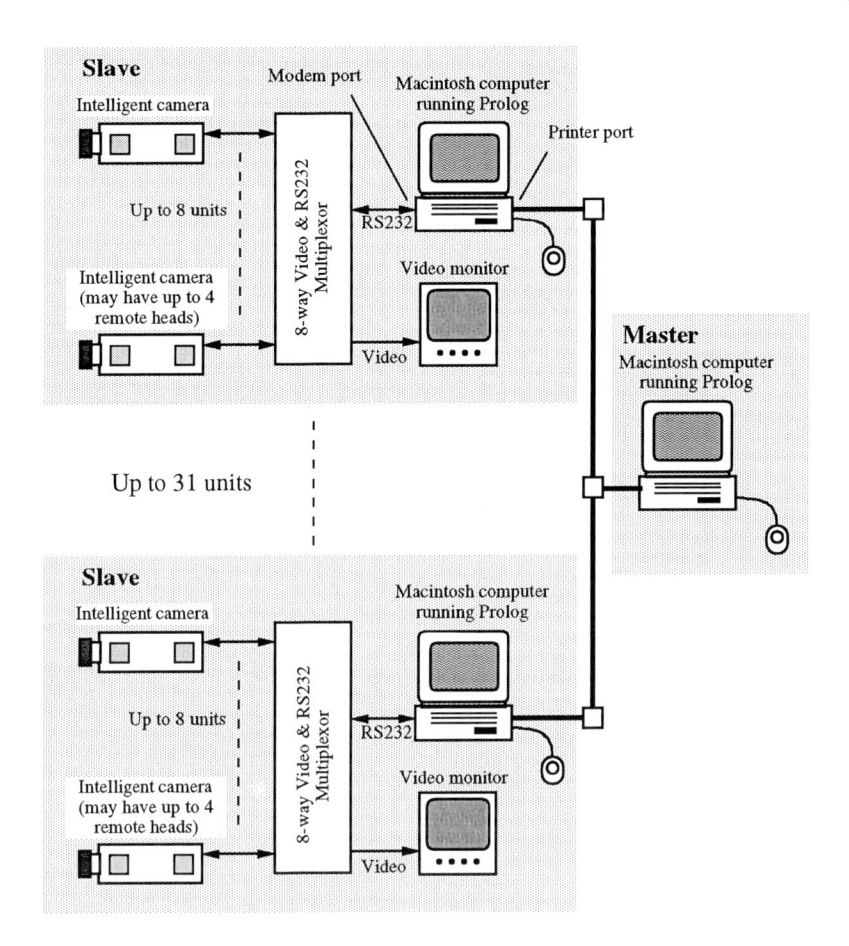

Figure 4.24 Enhanced networked vision system. Each of the slaves is a Video-Multiplexed (V-M) system and hence can control up to 8 image processors, via a multiplexor.

The role of the Master requires some explanation. Since it is not likely to be able to respond individually to each signal sent to it by the Slaves, it is right to question what this arrangement actually achieves. Decisions can be made locally and rapidly by the Slaves, which are themselves intelligent Prolog+ systems. Typically, the Slaves would generate performance statistics and report special conditions that require the attention of the Master. They merely report *in broad, symbolic terms* what they have found, without giving the Master too much detail.

4.4.3 Master-Slave System Organisation

The system illustrated in Figure 4.24 is co-ordinated using AppleEvents, which we have encountered already. Using AppleEvents, a Prolog+ program (P1) can pass a goal to another Prolog+ program (P2), running on either the same computer or another remote machine. This mode of operation places certain limitations upon the structure of the software:

(a) The most important point to note is that P1 can pass a query to P2, but only if P2 is not already running, i.e. trying to satisfy a goal. If P2 is already trying to satisfy a goal G when the query from P1 arrives, the latter will be held in obeyance until G has been evaluated. The incoming goal specified by P1 will be evaluated after G.
(b) It is possible for P1 to assert and retract relations in P2's database, in a slightly indirect way. P2 must contain a predicate which does the assertion / retraction on P1's behalf.
(c) When the goal defined by P1 has been evaluated, P2 can pass results back to P1. If these results from P2 do not arrive within a defined time limit, written into program P1, the goal in P1 fails. This is a very useful facility, since it allows recovery from a *deadly embrace*.

With these points in mind, we see that it is desirable to employ a *passive* Prolog program which acts as a buffer between programs P1 and P2, so that they communicate indirectly with each other. This will be called the *Blackboard*. By the term *passive*, we mean that the Blackboard does not satisfy goals on its own account, only when ordered to do so, by receiving a remote query from another Prolog program. The Blackboard contains definitions for only a very small number of relations, so that it can assert / retract on behalf of P1 and P2.

We shall consider a network in which the Master and each Slave is resident on a different computer. Each Slave host runs *two* copies of the MacProlog software: one with the Slave program and the other with a copy of the Blackboard program. Thus, each Slave has a (passive) Blackboard program dedicated to it. (See Figure 4.25(a)) We could also use a common Blackboard to good effect, enabling the Master to broadcast information to all Slaves. (See Figure 4.25(b).) However, we shall not consider this option any further, since it provides few advantages compared to the alternative arrangement and is a potential bottle-neck, limiting the speed of the system.

Figure 4.26 shows the communication paths that we need to consider in a Master - multi-Blackboard/Slave system. Before we can discuss this in detail, we need to consider how one Prolog application (the *local process*) can send data to another (the *remote process*).

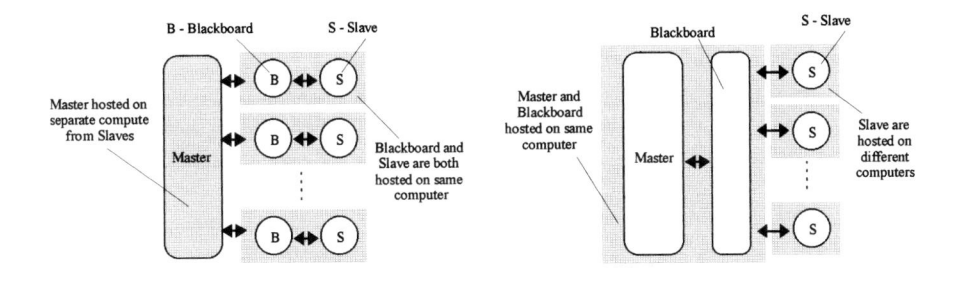

Figure 4.25 Blackboard organisation (a) Blackboard dedicated to the Slave. (b) Common Blackboard for all Slaves.

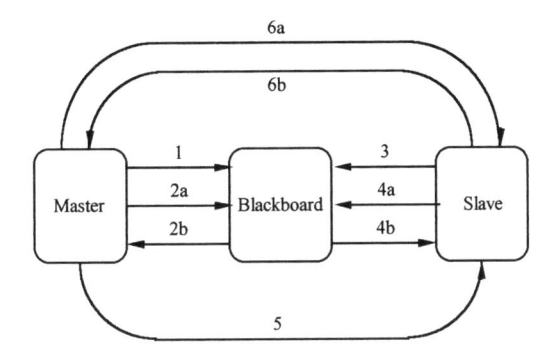

Figure 4.26 Communication paths in a Master - multi-Blackboard/Slave system. *1.* Master uses △ *remote_asserta(A)* to place message for Slave on Blackboard. *remote_asserta(A)* is sent from Master to Blackboard. *2a.* Master receives message from Slave by sending remote query to Blackboard. (Example: △ *info(X)*) *2b.* Blackboard satisfes remote query (*info(X)*) received from Master and instantiates X; *info(value1)* is returned to the Master. *3.* Slave uses △ *remote_asserta(B)* to place message for Master on Blackboard. *remote_asserta(A)* is sent from Slave to Blackboard. *4a.* Slave receives message from Master by sending remote query to Blackboard. (Example: △ *info(Y)*). *4b.* Blackboard satisfes remote query (*info(Y)*) received from Slave and instantiates Y; *info(value2)* is returned to the Slave. *5.* Master satisfies △ *remote_goal* to start the Slave running. *remote_goal* is sent from Master to Slave. *6a.* Master tries to satisfy △ *true*, to test whether the Slave is still active. *6b.* Slave satisfies *true*, if it is *not* running another program and signals "goal succeeds" to the Master. If the Slave is already running, the timer in *remote_query* (in the Master program) causes *remote_machine_running* to fail.

4.4.4 Remote Queries

The MacProlog built-in predicate *remote_query* allows the user to pose queries via one copy of Prolog (local process) but which are satisfied by another, possibly running on a different computer (remote process). This facility is discussed in detail in the MacProlog manual. [MAC] *remote_query* uses the following arguments:

    A - Query to be sent to the remote process.
    B - Identification of the remote process. (Derived from the MacProlog built-in predicate *ppc_browse*. See *remote_reset*, defined below.)
    C - Maximum time delay allowed for the remote process to respond to the query.
    D - List of values returned from the remote process.

The following program segments are useful for running remote queries.

Getting started. Finding the ID number of the target process where remote queries will be sent.

```
remote_reset(A) :-
    ppc_browse('Select the MacProlog application that you want to
    use',_,B),                  % Get ID for the target process.
    retractall(index_number(A,_)),
                                % Forget previous ID number
    assert(index_number(A,B)).
                                % Remember current process ID number
```

This predicate allows the user to locate any application running on the network interactively. The target application is identified by the user, with an interactive dialogue generated by *ppc_browse*. The goal *remote_reset(A)* associates the label A with the ID number of the target application. After satisfying *remote_reset(A)*, the Prolog database contains information needed (in *index_number*) to despatch AppleEvents to that application whose "address" is A. It is possible for a user working at the Master terminal to locate each of the Slave and Blackboard applications on the network, using *remote_reset*. It is not necessary for the user to visit each of the Slaves. Exactly how this can be avoided will be explained later.

Useful operator (△). This provides a simple syntactic form for remote queries. These may be of the form △ *goal(defined_arguments)* or △ *goal(Results)*.

```
△ A :-
    A =.. [_|B],       % Analyse arguments
    B = [C|_],         % Get first argument
    var(C),            % Checking that argument is a variable
    run(A,[B]),        % Run remote query A. Results go into B
    !.                 % Do not allow backtracking to next clause
```

```
△ A :- run(A,_).       % What to do if there are no variable
                       % arguments

% Handling remote queries at simplest level for user. A is query.
% B contains results.
run(A,B) :-
        index_number(D),       % Consult database for ID of target
                               % process
        remote_query(A,D,600,C),
                               % Arguments are described above
        prolog_solns(C,B),     % Decode solutions found. Not defined
        !.                     % Avoid resatisfaction on backtracking.
```

Interactive Operation of the Remote Process

```
transparent :-
        prompt_read(['Please type a goal for the REMOTE process to
        satisfy'],X),
        not(X = end),  % Terminate interactive session
        △ X,           % Do it
        writeseqnl(['Remote...',X]),
                       % Keep log of what happened during interaction
        !,             % Make recursion more efficient
        transparent.   % Do it all again

transparent.           % Finish with success
```

Examples of remote queries :

Local goal	Remote goal	Action
△ grb	grb	Digitise an image
△ thr(123)	thr(123)	Threshold image
△ hil(12,34,56)	hil(12,34,56)	Highlight intensities
△ avr(X)	avr(P)	Average intensity
△ cgr(X,Y)	cgr(P,Q)	Centroid
△ lmi(X,Y,Z)	lmi(P,Q,R)	Least MOI

Starting an infinite remote process We might wish to start the following remote program which performs *process* in an endless loop.

```
remote_goal :-
        process,       % Remote machine has definition of "process"
        !, remote_goal .
```

The *local* goal △ *remote_goal* will start *remote_goal* running on the *remote* machine. After a period of time defined by the programmer of the local machine, △ *remote_goal* succeeds and *remote_goal* continues running on the remote machine. The two machines are now running independently.

Is the remote machine running? Clearly, the local machine needs to know whether or not it can communicate with the remote machine again. Here is a program which the local machine can run, in order to find out whether or not the remote machine is actively trying to satisfy some (unknown) goal or is passively waiting for some further instructions from the local machine.

```
remote_machine_running :-
        !,              % Avoid resatisfaction on back-tracking
        not(△ true).    % Satisfied if remote machine is NOT running
remote_machine_running.
                        % Remote machine running. Local goal succeeds
```

The first clause fails if the remote machine does not satisfy △ *true* within a time limit defined by *remote_query*.

Reassign query not understood locally It is possible to reassign any query that is not understood by the local machine to the remote machine. Error number 2 in MacProlog signals that a goal has no definition. However, by redefining the error handler in the following way, it is possible to divert a query to the remote machine.

```
'<ERROR>'(2,A) :- △ A.      % If error 2, try goal A on remote
                            % machine

'<ERROR>'(2,A) :- abort.    % That failed, so we really do have to
                            % give up
```

Thus far, we have described the mechanism for forcing a remote copy of MacProlog to answer a query posed to it by the local process. We now need to consider how these ideas can be put to use in organising a Master-Slave message signalling system. First, we need to consider the structure and use of the Blackboard, since this is central to the proper co-ordination of the Master-Slave system.

4.4.5 Blackboard

The Blackboard is a passive Prolog program which can receive data from the Master or from any Slave, either of which can also consult it at any time. The organisation of communication in a Master-Blackboard-Slave system is illustrated in Figure 4.26. The Blackboard contains only a few items:

(a) *remote_asserta* which performs an *asserta* on behalf of the Master or a Slave, neither of which are able to perform this operation directly. (It is not possible for the Master or a Slave to issue the remote goal △ *asserta*.) Similarly, *remote_assertz* performs *assertz* remotely.

(b) *remote_clear_and_assert(A)* clears any exisiting relations with the same name and arity as A and then asserts A into the Blackboard database.

(c) *remote_retractall(A)* retracts all clauses in the Blackboard database that have the same arity as A.

Here are the definitions for these predicates.

```
% Perform "asserta", on behalf of the Master or a Slave
remote_asserta(A) :-
     attach_data(A,'Data window'),
                    % Defines window where "assert" will be done
     asserta(A).    % Assertion on behalf of Master or Slave.
% Perform "assertz", on behalf of the Master or a Slave
remote_assertz(A) :-
     attach_data(A,'Data window'),
                    % Defines window where "assert" will be done
     assertz(A).    % Assertion on behalf of Master or Slave.
% Perform "assert" on behalf of the Master or a Slave
% The following predicate clears any existing relations first.
remote_clear_and_assert(A) :-
     A =.. [B|C],   % Separate functor and arguments of A
     length(C,D),   % Find how many arguments in A
     abolish(B,D),  % Remove all clauses of arity D with name B
     attach_data(B,'Data window'),
                    % Defines window where "assert" will be done
     asserta(A),    % Assertion done on behalf of Master or Slave
     !.             % Avoid backtracking
/* Remote version of "abolish". The following predicate uses number
of arguments in A to define the arity. */
remote_abolish(A) :-
     A =.. [B|C],   % Separate functor and arguments of A
     length(C,D),   % D is number of arguments in A
     abolish(B,D).  % Remove all clauses of arity D with name B
```

In addition to the definitions given above, the Blackboard may contain a large number of messages, which are in the process of passing from Master to Slave, or *vice versa*. Examples of these messages will be given later.

Master and Slave Program Elements

To avoid confusion, the same predicate names should be used in the Master, Slave and Blackboard programs. Thus, both Master and all Slaves should contain the following definitions:

```
remote_asserta(A)         :- △ remote_asserta(A).
remote_assertz(A)         :- △ remote_assertz(A).
remote_clear_and_assert(A) :- △ remote_clear_and_assert(A).
remote_abolish(A)         :- △ remote_abolish(A).
```

4.4.6 Controlling the Master-Slave System

We are now in a position to understand how the Master-Slave system can be controlled using the Blackboard. We shall discuss the operation of the system step by step.

Starting the System

In order to use the Blackboard properly, both the Master and Slave must be running Prolog programs. If a Slave is not already running a query, it can be started directly by the Master. Either *run* or the $\triangle$ operator can be used to do this:

```
△ remote_goal       % Remote query (remote_goal) is run by Slave
```

Notice that, prior to satisfying $\triangle$ *remote_goal*, we must make sure that AppleEvents will be directed to the appropriate Prolog application. This is achieved using *remote_reset*. In order to start a Slave via its Blackboard, we need to invoke the $\triangle$ operator *twice*:

```
△ (△ slave_program)
```

(Since the $\triangle$ operator is right associative, the brackets can be removed without altering the meaning.) Of course, for this to work, the $\triangle$ operator and the predicate *slave_program* must be defined in the Blackboard. Either the Master or the Blackboard can invoke *remote_machine_running* at any time, to test whether the Slave is still running.

Stopping a Slave

The Master can order a given Slave to terminate its activity by placing the message *stop_slave* in the Blackboard's database. In order to do this, the Master should satisfy the local goal $\triangle$ *remote_clear_and_assert(stop_slave)*.

The following predicate must be defined in the Slave.

```
stop :-
        △ stop_slave,   % Consult the Blackboard database
        remote_abolish(stop_slave),
                        % Clear the stop message from the Blackboard
        abort.          % Stop whatever we are doing
stop(_).                % No stop message, so carry on running slave
```

It is also necessary for the programmer to place the goal *stop* somewhere in his program. This is a minor nuisance, but it does allow the programmer of the Slave program to terminate the program at the most convenient point in its execution cycle. Of course, if he does not do so, the Slave will continue running for ever!

Passing a Message to the Slave

In a similar way to that just described, the Master can pass messages and other information, via the Blackboard, to the Slave, while the latter is running. This permits the Slave to modify its actions, under the command of the Master, while both Master and Slave applications are running. For example, the following Slave program cycles endlessly (*idling mode*), until the Master places an appropriate "command" in the Blackboard. This command is then "executed' by the Slave.

```
slave_program :-
     message_from_master(A),
                         % Slave consults Blackboard, similar to "stop"
     call(A),            % Perform operation defined in Blackboard
     !,                  % Included for greater efficiency of recursion
     slave_program.      % Do it all again

slave_program :-        % Program defined by Master failed. Keep
                        % idling
     !,                 % Included for greater efficiency of recursion
     slave_program.     % Idle until Blackboard contains a command
```

The Master can pass messages to the Slave via its Blackboard, at any time, using the $\triangle$ operator.

Receiving Data from a Slave

It is possible for the Master to receive data directly from a Slave, or indirectly via the Blackboard. The latter process is very similar to that just described for passing messages in the opposite direction, from the Master to the Slave. There follow three short programs showing how data derived from images can be passed from the Slave to the Master, via the Blackboard.

Slave Program

```
Slave_program :-
     beep(3),           % Audible signal for the user
     grb,               % Digitise image
     gli(A,B),          % Get upper & lower intensity limits
     remote_clear_and_assert(parameters([A,B])),
                        % Place data into Blackboard
     stop,              % Stop if ordered to do so by Master
     !,                 % Included for more efficient recursion
     Slave_program      % Do it all again
```

Blackboard (Snapshot of Database, Changing Constantly)

```
parameters([31, 251])

master_program :-
     repeat,            % Begin processing cycle
     run(parameters(Z),Q), % Consult Blackboard
     Q = [W],           % Decode results (quirk of AppleEvent format)
     writenl(W),        % Write results list for the user to see
     fail.              % Force backtracking (to "repeat")
```

An important point to note here is that the Slave and Master programs are unsynchronised. The periods of the program cycles may be similar, or completely different. The Slave may be much faster than the Master, or *vice versa*. However, a number representing the Slave program cycle number may be included in the list of parameters passed to the Master. We can use the built-in MacProlog predicates *time*, or *gensym* to provide a "time stamp".

Direct communication from a Slave to the Master can be achieved in an obvious way. The difference between this and indirect communication is that, in the latter case, the Master can do nothing but wait, until the Slave sends the data. Notice that the Master program can make use of the timer built into *remote_query* to escape from a deadly embrace. In this event, some of the data from the Slave will be lost.

4.4.7 Crash Recovery

The network described above experiences some difficulty, if a Slave goes into a tight recursive loop, containing no *stop* sub-goal. The Master cannot then regain control whatever AppleEvent message it sends, either directly to the Slave or via its Blackboard. There is a solution to this problem but it requires the construction of a small piece of electronic hardware. Recall that the default option for (software) interrupts to MacProlog allows goal satisfaction to be terminated by the user pressing "⌘." (COMMAND / period) on the keyboard. A small, low-cost hardware unit can be connected to the Apple Desktop Bus (ADB) so that, on receipt of an appropriate signal from the Master, it generates a signal sequence equivalent to "⌘." being pressed on the Slave keyboard. Thus, to regain control, the Master sends an interrupt signal to the Slave ADB, via the hardware unit. (In the next chapter we shall describe how a whole range of hardware devices can be controlled from a Prolog program.) The action of MacProlog following the receipt of an interrupt signal can be programmed using the '*<INTERRUPT>*' predicate [BAT-91b]. The following clause in the Slave database forces the Slave to stop whatever it is doing, when the interrupt from the Master arrives:

```
'<INTERRUPT>'(_) :- abort.
```

Programming the Slave from the Master

It is important to note that the Master can write complete programs in the Slave's database. This is straightforward if the Slave already contains a definition of *remote_asserta(A)*. In this event, the Master can write (and delete) programs in the Slave; the master simply satisfies a clause of the form △ *new_slave_clause*.

4.5 Comments

In this chapter, we have touched on a number of topics, all of which are connected by our desire to make vision systems both more powerful and easier to use. The choice of Prolog as the basis for intelligent image processing was made as long ago as 1985. The original arguments in support of this choice, rather than, say, Lisp, have long been superseded by the experience that we have subsequently gained. In other words, we have been able to achieve far more than was ever thought possible in those early days of the development of Prolog+. It has been possible to develop a flexible environment around MacProlog, to assist vision system design engineers. This includes such features as pull-down menus, replaying recorded speech, speech recognition, speech synthesis, automatic generation and pruning of macros, a rich program library, the ability to link Prolog+ to programs written in other languages and commercial packages (e.g. HyperCard). Prolog+ can also be extended to control a number of image processors in a Master-Slave network.

However, there are several important aspects of Prolog+ that we have not yet been able to develop. It is important to note that MacProlog can itself be extended, by linking it to:

(a) *Flex* an Expert System tool-kit [LPA].
(b) *Prolog++*, an Object Oriented Programming package [LPA].
(c) *Parlog*, a parallel declarative language.

There are, in addition, a number of notable programs, written in Prolog, which could be advantageous to our work on machine vision. Natural language processing, via the medium of Prolog, has been developed to a high level of refinement. [GAZ-89]

5

Controlling External Devices

5.1 Devices and Signals

This chapter is based upon the axiom that every industrial vision system must be capable of communicating with a range of devices that exist in the world outside itself. To be more specific, each machine vision system must be able to perform at least three functions:

(i) Sense the electro-magnetic radiation (visible light, IR or UV) emanating from the object or scene being examined.
(ii) Analyse data derived from the camera.
(iii) Decide what action its associated effector mechanisms should perform.

The authors assert that (iii) is just as important as (i) and (ii), even though few books and articles on machine vision, or image processing, even bother to mention the topic of controlling external devices. There is a fourth action which some, but not all vision systems perform: receiving and analysing data from various (non-imaging) sensors, a process which helps to synchronise the vision system with other machines.

In this chapter, we shall discuss various aspects of the general subject of interfacing to / from vision systems and will introduce more specific issues, such as communications requirements, interfacing standards, designing a "general purpose" interface unit, creating user-friendly control software, aligning and calibrating a robot vision system. It is impossible, within the small space available here, to consider any of these topics in great depth. As a result, we shall merely highlight the more important issues involved. In addition, we shall describe one particular physical arrangement of lamps, cameras, and mechanical handling equipment, which was devised to form a general purpose test bed for studying inspection and robot guidance techniques. This is the *Flexible Inspection Cell (FIC)* referred to earlier [BAT-85b]. The FIC will provide the focus for illustrating many of the more important general design considerations.

Referring to Figure 5.1, we see how an industrial machine vision system is expected to interact with the external devices surrounding it in a factory. A system inspecting objects on a production-line conveyor belt must be able to synchronise its activities to a number of machines, located both up- and downstream from its camera. Whatever the inspection task, it is important that image data be captured from the camera at the most appropriate moment in the manufacturing cycle. System synchronisation might typically be achieved using a light-beam and photo-detector to sense the arrival of parts, travelling along the conveyor. In this archetypal system, the accept/reject mechanism can be operated by a simple 2-level signal. The accept/reject mechanism might be a simple air-jet "blower", or a swinging arm, operated by either a solenoid or pneumatic cylinder. In other applications, it may be necessary to use a fixed-function pick-and-place manipulator, or even a multi-axis robot, operating under the control of the vision system. It might also be necessary to flash a stroboscope lamp, or to acquire and then combine two or more images, derived from the same camera but using different lighting conditions.

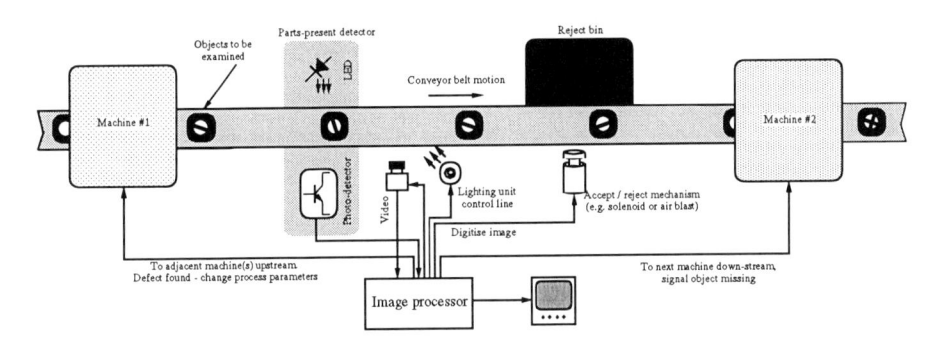

Figure 5.1 Archetypal machine vision system for inspecting objects as they travel along on a conveyor belt. Notice that the vision system is expected to interact with several other machines: a photo-optical (or infra-red) sensor which detects when an object is correctly located for examination; the lighting control unit; the accept/reject gate; process machines located upstream and down-stream

5.2 Protocols and Signals

In the past, many different types of device have actually been interfaced, or proposed for connection, to vision systems. By inspecting Table 5.1, it is not difficult to understand why such a wide range of standardised and *ad hoc* data-communications and control protocols have been used or proposed. The list includes but is not restricted to the following:

(a) *Serial digital data* (RS232 / RS 422) used to transfer numeric and symbolic information (e.g. name of defect type, position and orientation of the component).

(b) *Parallel digital bus.* This may be a simple "unstructured" array of signal lines, with / without timing signals, or conform to a well-defined bus standard, such as the IEEE 488 protocol. Both TTL (5 volt) and industrial 24 volt standards are popular for parallel I/O.

(c) *Computer bus* (parallel). Of particular importance is the bus associated with the IBM-compatible PC (such as the PCI bus), while other vision systems use the *Nubus* (Macintosh family), and VME bus.

(d) *Analogue.* (Bandwidth, impedance and voltage levels may all vary over a wide range. Analogue I/O ports may be balanced or non-balanced.)

(e) *Digital data network.* (e.g. Ethernet, AppleTalk, Novell, etc.) Few vision systems provide connections to these networks, at the moment. However, it is clear that this is an area for future developments. Indeed, we investigate this possibility in Chapter 4.

(f) *Video.* As far as we are concerned, there are two important video standards: RS170 (60Hz field scan, used throughout USA) and CCIR (50Hz field scan; widely used in Europe). In addition, there are various high definition television standards in use. Notice however that high-definition laser scanners and solid-state line-scan imaging systems are usually designed individually, without reference to internationally agreed standards.

5.2.1 Interfacing to Commercial Systems

In this section, we shall briefly review the types of interfaces available on existing machine vision systems. A typical example, is provided by the Intelligent Camera [INT], which has the following facilities:

(a) *Two RS232 ports* (connections are made via 9-way D-type connectors). One of these allows the host processor to control / program the Intelligent Camera and the other to an ancillary device, such as a printer or (X,Y,θ)-table. In addition, both D-type connectors provide DC power for external devices.

(b) *One 8-bit parallel I/O port*, with opto-isolators (provided via a 25-way D-type connector). These signal lines permit the camera to drive a stroboscope lamp unit. It is also possible to delay image capture and processing until the arrival of a "parts present" signal, generated by a proximity sensor. In addition, the Intelligent Camera can drive an *accept / reject* mechanism (e.g. a solenoid, air blast, or pick-and-place arm). It is also possible to operate a device, such as a warning bell or flashing lamp. Once again, power is made available via the D-type connector for driving external devices.

(c) *Video output* (BNC connector). This allows the Intelligent Camera to be connected to a video monitor, so that a person can watch the image processing taking place.

Sensor	Interface	Comments
Proximity sensor	Single TTL	May use micro-switch, optical, magnetic, or capacitance sensing.
Pressure	Single TTL, serial (RS232), or parallel	Device may indicate when certain pressure limit is passed, or provide a string of measurement values.
Temperature	Single, TTL or serial (RS232)	Device may indicate when certain temperature limit is passed, or provide a string of measurement values.
Force	Single, TTL or serial (RS232)	Device may indicate when certain force limit is passed, or provide a string of measurement values.
Gauges, micrometer, etc	Serial	Device may indicate when diameter limits are violated, or provide a measurement value.
Bar-code reader	Serial (RS232)	Used for input of product type / identity so that vision system can perform appropriate inspection function.
Range	Serial (RS232)	Range measurement may use radar, lidar, IR, or optics for sensing. Device may indicate when certain range limit is passed, or provide a string of measurement values. Particularly useful for controlling auto-zoom.

Cameras / Lenses	Interface	Comments
Zoom	Serial (RS232)	Zoom motor requires intelligent controller.
Focus	Serial (RS232)	Focus motor requires intelligent controller.
Aperture	Serial (RS232)	Aperture control motor requires intelligent controller.
Pan-and-tilt	Serial (RS232)	Pan-and-tilt control motors require intelligent controller.
Synchronisation trigger	TTL	Some types of camera can initiate their scanning cycle on command, rather than waiting for synchronising to the video "fly wheel".
Scan mode	Parallel TTL serial (RS232)	Camera controller may have serial or parallel interface.
Filter wheel	Serial (RS232)	Controller has serial interface.

Table 5.1 The tables list some of the devices that have been interfaced to vision systems. Notice that display and input devices needed to support the human-computer interface are not included here. The table is divided into several parts, indicating families of devices that have been interfaced, or proposed for connection, to industrial vision systems.

Lighting Device	Interface	Comments
Stroboscope lamp unit	Single TTL	Used to "freeze" rapidly moving objects. Lamp must be synchronised to video camera and image capture.
Lamps (ON / OFF)	Mains power	Solid-state switch suitable for switching lamps ON / OFF.
Pattern projector	Mains power	Solid-state switch suitable for switching lamps ON / OFF.
Laser	5 - 12V	Used for generating light stripe for structured lighting. Simple electronic amplifier may be needed to drive laser from TTL.
Multi-LED lighting	Multiple TTL	Used to provide light-weight illumination unit for mounting on robot arm.
Lamps (variable)	Serial (RS232)	Serial line may be used to provide multi-level control for an array of lamps.

Mechanical Actuators	Interface	Comments
Solenoid	Single line, 24V, 240V, or 415V	Often used to operate mechanism for deflecting reject parts from conveyor belt.
Pneumatic	Single line, 24V, 240V, or 415V	Often used to operate mechanism for deflecting reject parts from conveyor belt.
(X,Y,Z,θ)-table	Serial (RS232)	(X,Y,Z,θ)-table may be part of machining centre e.g. drilling, milling machine, lathe, water-jet, or laser cutter, etc. Table may be servo-controlled, or stepper motor.
Multi-axis robot	Serial (RS232)	Robot may be pneumatic, hydraulic or electrical and may form part of a complex manufacturing cell, performing a range of functions, such as cutting, turning, welding, paint spraying, soldering, etc.

General Purpose Devices	Interface	Comments
Remote computer	Parallel, Serial, SCSI	Computer may provide interface to LAN, WAN, Ethernet or Internet. Data rate may be very high for transferring video images.
Programmable logic controller (PLC)	TTL, 24V	PLC may synchronise several other machines, as well as the vision system.

Table 5.1 (Cont'd)

Miscellaneous Devices	Interface	Comments
Specialist image processing equipment	Video, SCSI, Parallel	High-speed processing of images in external hardware augments software-based processing.
Video recorder / player	Parallel, TTL or serial RS232	Vision system might operate player for frame-by frame processing of video signal, or for controlling time-lapse recording.
Video monitor	TTL, serial (RS232)	Scan size, positive/negative image display, pseudo colour, superimposition of text can all be controlled.
Alarm bell, buzzer	TTL	Warning for general alarm.
Automatic test equipment	Serial (RS232), or Parallel	Vision system might adjust brightness / scan-size of picture when calibrating television receiver.
Printer (paper)	Serial (RS232)	Product statistics.
Ink-jet / bar-code printer	Serial (RS232)	Used for printing product type, grade, dimensions, etc. directly on the product. Also used for marking defect regions of web materials, etc.
Plant emergency stop	TTL or 24V	Stops all activity in the manufacturing plant if certain dangerous conditions have been detected.

Table 5.1 (Cont'd)

Other vision systems have been built which provide direct interfacing to a Programmable Logic Controller, or to one of the standard industry interfaces, such as IEEE 488, VME, SCSI and Ethernet. Interfacing standards are, of course, intended to make the task of the industrial design engineer easier. However, many standards interfacing protocols have been defined and much of the equipment which fits certain standard interfaces is expensive. It is inevitable that, whichever standard has been adopted, some important facility is not provided, in a convenient form. It is often cheaper to buy exactly those electro-mechanical devices that are needed for a given application and then interface them directly to the vision system. While this takes greater effort, researchers, university staff and others working in an environment where funds are severely limited may find this "do-it-yourself" solution preferable. The authors are in this category and, in their research, are attempting to develop "general solutions" to a range of inspection and other industrial vision applications. Faced with such a varied range of interfacing requirements, connecting vision systems to industrial sensors and effectors can be problematical. However, as we shall show, there is an alternative approach, which the authors have found to be effective. In Section 5.4 we describe a "general purpose" interface module which was designed to allow even the simplest computer-based vision system (i.e. one with only a single RS232 port) to control a range of electro-mechanical, electrical, pneumatic, illumination and video devices. First however, we shall explain why the Programmable Logic Controller (PLC) does not fit our requirements exactly.

5.3 Programmable Logic Controller

A Programmable Logic Controller (PLC) is a versatile digital electronics device, intended for industrial control; a PLC is typically used to co-ordinate the operations of a group of machines used in manufacturing. For example, a robot might be used to load metal blanks into a power press. At the end of the metal-forming process are then transferred from the press to either an "accept" or "reject" bin, depending upon the decision reached by an automated visual inspection system. (Figure 5.2) Even a simple production work-cell like this is most likely to contain three machines made by different companies and have completely different operation cycles. A PLC is often used in a situation like this, to synchronise the three machines, so that they perform their various functions at the appropriate times.

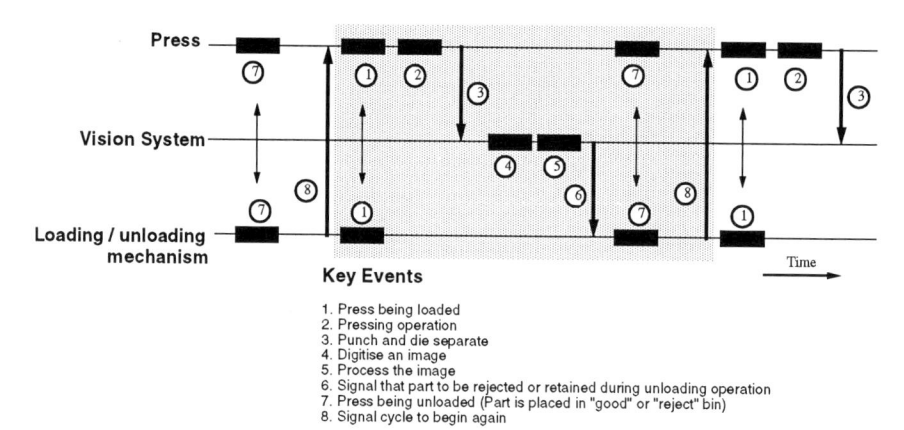

Key Events
1. Press being loaded
2. Pressing operation
3. Punch and die separate
4. Digitise an image
5. Process the image
6. Signal that part to be rejected or retained during unloading operation
7. Press being unloaded (Part is placed in "good" or "reject" bin)
8. Signal cycle to begin again

Figure 5.2 Timing diagram for an hypothetical production cell, in which a robot is used to load raw material into a power press. Following the pressing operation, the finished parts are examined by a vision system. Defective products are transferred by the robot to a "reject" bin. On the other hand, "good" products are taken to the "accept" bin by the robot. Notice the need to synchronise the three machines, so that appropriate operations are performed at each stage in the processing cycle. Rather than interconnecting machines like these directly to each other, it is common practice to use a PLC to sense the status of each of the machines and to issue appropriate control signals to them. Thus, the interconnection diagram resembles a (3-pointed) star, with the PLC at the centre.

In order to explain in semi-formal terms what a PLC is, we may, in the first instance, think of it as being a "black box", which has a set of N binary inputs $(X_1, ..., X_N)$ and M binary outputs $(Y_1,...Y_M)$ ($M \neq N$). Each of the X_i is derived from an ON/OFF sensor attached to one of the machines that the PLC is intended to control and synchronise. Typically, most, if not all, of the Y_j outputs operate a set of control actuators. The output Y_j is calculated from the N inputs: $Y_j = Y_j(X_1, ..., X_N), j \in [1,..., M]$

The output vector of the PLC varies with time; the output variables $(Y_1, ..., Y_M)$ change at discrete time intervals. Suppose that the time is represented in the form of a P-bit Boolean vector $(Z_1, ...,Z_P)$. Then, each of the Y_i ($i = 1, ...M$) is a Boolean function of an (N+P)-bit vector, $(X_1, ..., X_N, Z_1, ...,Z_P)$. While a PLC can be implemented easily using a random access memory of size M*(N+P) elements, programming it at a low level (i.e. in terms of 0/1 bits) would be very tedious and difficult. The task is made very much easier by the use of a computer, which allows the input and output signal lines to be assigned symbolic names (e.g. *start, reset, doors_open, tool_broken, temperature_high* etc.). Of course, the computer helps the programmer in many other ways (e.g. by providing facilities for cut-and-paste, storing programs, programming with subroutines, using FOR-loops and IF-THEN program statements).

However, PLCs are more complex than this; a modern device is likely to contain a set of timers. In addition, a PLC is often expected to:

(a) *accept analogue inputs* (The analogue inputs are compared to pre-defined limits; if an analogue input is out of some pre-defined range, an internal variable, equivalent to one of the X_i inputs described above, is set to logic level 1.)

(b) *accept input sequences of alphanumeric characters* (RS232 signal line) (When a pre-defined sequence of characters is detected on the input, an internal variable, equivalent to one of the X_i inputs, is set to logic level 1.)

(c) *provide outputs in the form of character sequences.* (The PLC contains a character-sequence generator which is able to form a string of alpha-numeric characters when a certain output variable, equivalent to one of the Y_j, is set.)

The PLC is of particular importance to designers of vision systems, because it is frequently used for co-ordinating and synchronising machines in a complex manufacturing plant. One (or more) of the machines connected to the PLC might well be a vision system. In the simplest case, there are just two 2-level signal paths between the vision system and the PLC. (See Figure 5.3) One of these transmits a signal from the PLC to trigger the vision system, prompting it to digitise an image and begin its processing cycle. The other sends a *pass/fail* signal from the vision system to the PLC, which in turn, operates the *accept/reject* mechanism. This last device may be located some way down-stream from the camera and must be triggered at the appropriate moment, at the instant when the object to be rejected is passing it. (Figure 5.4) In a typical factory application, the PLC may also be required to co-ordinate the actions of a large number of very varied machines (e.g. robot, metal cutting, moulding, assembly, spraying, coating, packing, printing, etc.), using data derived from a large range of sensors, in addition to the vision system. This means that the PLC, not the vision system, may be responsible for deciding *what to do,* in the event of a defective item being found.

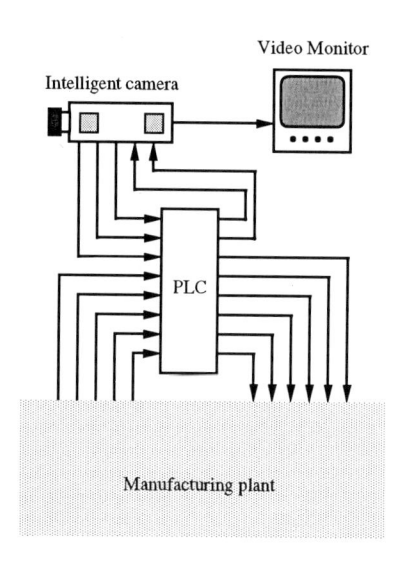

Figure 5.3 Connecting a Programmable Logic Controller (PLC) to a vision system such as the Intelligent Camera. Notice that the PLC lies at the heart of the control system. For this reason, it is difficult to add "intelligence" to such a system.

This is in marked contrast to our approach in this book, where we assume that the computational processes needed to interpret observations from the sensors are integrated with the intelligent interpretation of images, within the vision system. The approach implicit in Figure 5.3 is limited in that it does not permit the operation of the vision system to be modified easily, by taking note of the state of the manufacturing plant. On the other hand, the approach that we advocate in this chapter allows various sensor signals to be interpreted within the same symbolic reasoning (Prolog+) program as we are use to analyse image data. While the primary task of this program, measured purely in terms of computing effort, is analysing image data from the camera(s), its other functions (e.g. analysing data from the input ports and operating the output effectors) are of equal importance for the harmonious operation of the manufacturing. The main advantage of this approach is that the full reasoning power of Prolog is available to interpret the signals representing the state of the external world. If necessary, the vision system can re-evaluate its decisions iteratively. If there is a conflict in the data arising from different sensors, this can be particularly valuable. Using a PLC does not permit any intelligent reasoning to take place.

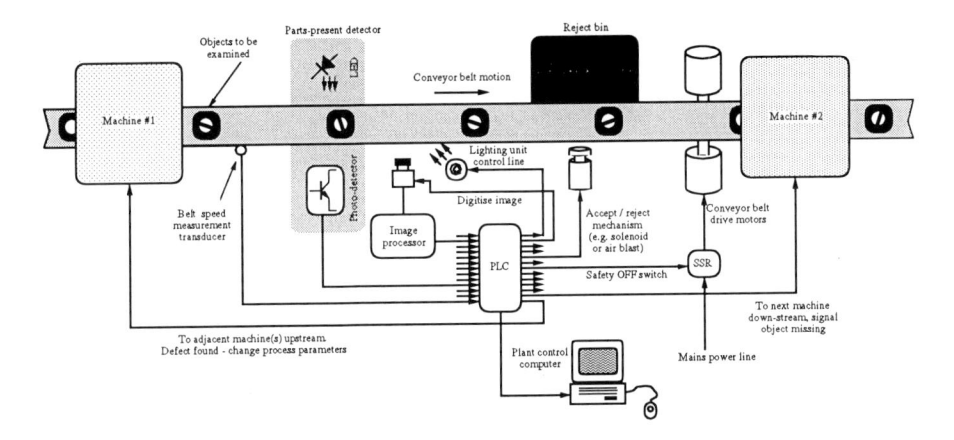

Figure 5.4 Using a PLC to control an inspection system operating on a conveyor belt. Many of the signals entering and leaving the PLC are binary, including those on the lines connecting it to the lighting unit, the *accept / reject* mechanism, the safety switch, the machine located down-stream and the parts-present transducer. The signal from the belt-speed transducer produces a numeric output, while the signal sent to the upstream process machines is likely to be symbolic in nature (i.e. conveyed as a string of ASCII characters).

5.4 General Purpose Interface Unit

In this section, we describe a simple, low-cost interfacing device (Figure 5.5), which was developed as an aid for prototyping industrial machine vision systems. [BAT-94] The device, which we shall refer to as the MMB unit, provides facilities for controlling the following devices, via a single serial (RS232) port.

(a) Ten mains-operated devices, such as lamps, relays, solenoids, or other ON / OFF devices. Each individual channel has a current rating of 2.5A (at 240V), while the *total* current rating for all ten lines is 10A. Ten flush-mounted 3-pin mains sockets are fitted to the rear of the MMB unit.
(b) Four 3-port (i.e. simple ON / OFF) pneumatic valves. The connectors for these air lines are fitted to the MMB unit's front panel, while the air input line is fitted at the rear.
(c) Two 5-port pneumatic valves. Each of these is capable of operating a PUSH / PULL cylinder. Again, the connectors for these air lines are fitted to the MMB unit's front panel.
(d) One 8-way video multiplexor. Nine BNC connectors (8 inputs and one output) are fitted to the front of the MMB module.

(e) Six programmable-speed serial (RS232) communication ports. The connectors for these are 9-way D-type connectors, fitted to the MMB front panel.

(f) Six opto-isolated, 8-way, parallel I/O ports. (Again, the connectors for these are 9-way D-type connectors, fitted to the MMB front panel.)

(g) A laser stripe generator (used for structured lighting).

(h) A pattern projector. (This is a ruggedised slide projector and is able to project patterns from a standard 50mm slide.)

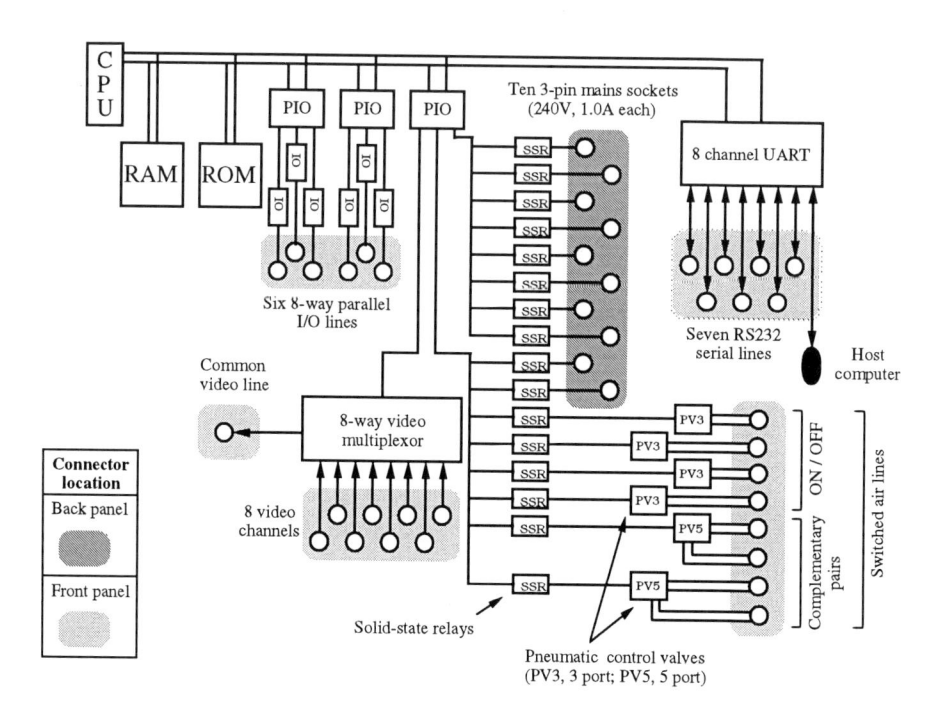

Figure 5.5 Internal hardware organisation of the MMB module. This is a conventional microprocessor system with a rich set of interfacing chips connected to its bus. Some of these, in turn, drive the video multiplexor, solid-state relays and pneumatic air valves. All of the parallel I/O lines are opto-isolated, for safety. Key: *CPU*, Central processing unit (Intel 80C86); *RAM*, Random Access Memory (64 Kbytes); *ROM*, Read Only Memory (128 Kbytes); *PIO*, Parallel I/O (Intel 8251); *UART*, Universal Asynchronous Receiver Transmitter (Philips Octal UART SCC 2698B); *OI*, Opto-Isolator (Quad type, NEC 2502-4); *SSR*, Solid-state relay (RS 348431, rated at 2.5A) *PV*, Pneumatic valve, (Manufacturer, SMC; Model, 3-port valve: VT307; Model for 5-port valve: V25120.) (Design by MW Daley.)

In addition, the MMB module provides regulated power at +12V and +5V. These power supply lines are useful for driving cameras, lamps and various electronic devices connected to it. The MMB module's control port is connected to a host computer, running Prolog+. Hence, all I/O traffic flows via a single RS232 line; the MMB module may be regarded as a *data funnel*, which allows the control port to collect / distribute data to a large number of other data pathways. Using the MMB unit, it is possible for Prolog+ or other programs, which contain only the most rudimentary I/O facilities, to operate a range of electro-mechanical devices. For example, a HyperCard program can switch lamps and pneumatic air lines ON / OFF, control the movements of an (X,Y,θ)-table and select different video cameras. The electro-mechanical devices just mentioned form part of the *Flexible Inspection Cell* (FIC), which is described in detail below. Let it suffice for the moment to say that the FIC contains a variety of illumination devices, several cameras, an (X,Y,θ)-table and a pneumatic pick-and-place arm, all controlled from by Prolog+ program. Neither Hypercard nor MacProlog were originally intended to control external devices like these.

The MMB interface module has also been used in a multi-camera vision system in which an image processing module (Intelligent Camera) controls both gantry-type and SCARA[1] robots. (See Section 5.7, Figure 5.11.) A Prolog+ system, interfaced to a gantry robot and a simple conveyor belt, has been programmed to play dominoes, against a human opponent. Given certain *minor* changes to the program, the FIC could also play dominoes. The FIC was used in the experiments on the automated packing of arbitrary shapes reported in Chapter 7.

5.4.1 Motivation for the Design

The cost of designing industrial vision systems remains high, despite recent advances in software and hardware. The motivation for the design of the MMB interface module, originated with our concern for easy interfacing between intelligent image processing software on the one hand and a variety of hardware modules on the other. As we have seen, the latter range from simple ON / OFF devices, such as lamps, pneumatic valves and solenoids, to sophisticated multi-axis robotic manipulators. The MMB module was designed to act as the interface controller for Prolog+, which has only rudimentary I/O facilities. (In its present implementation, Prolog+ is only able to operate a single serial, RS232, port.) An important aspect of the design is that the MMB module should provide a unified high-level software interface for a variety of devices. Of course, operating speed is reduced by the data funneling, but in the context of research, training, education and prototype development, this is of little significance, compared to ease of use and flexibility. Realising this led to the simple design which we shall now describe in more detail.

---

[1] SCARA is the acronym for *Selectively Compliant Articulated Robot Arm.*

5.4.2 Hardware Organisation

Figure 5.5 shows the internal organisation of the MMB hardware module. During the design, keeping the cost of the MMB at a minimum was regarded as being of prime importance. The objective was to achieve minimum cost, while maintaining a high level of flexibility. At the heart of the MMB hardware is an Intel 8086 micro-processor, connected to a series of serial and parallel interfacing chips, including: one 8-channel UART; three 3-port PIO chips[2] which is connected to one 8-way video multiplexor and sixteen ON / OFF solid-state relays capable of switching mains power. Six of the solid-state relays are dedicated to operating pneumatic control valves, while another ten are intended to switch mains operated devices, such as lamps, solenoids, relays, motors, etc. An important safety feature of the MMB unit is the fact that the parallel I/O lines are all optically isolated. Figures 5.6 and 5.7 show the layouts of the front and rear panels of the MMB unit. Notice that the pneumatic air lines use simple 6mm plug-in connectors, which make the task of fitting and removing air pipes particularly easy.

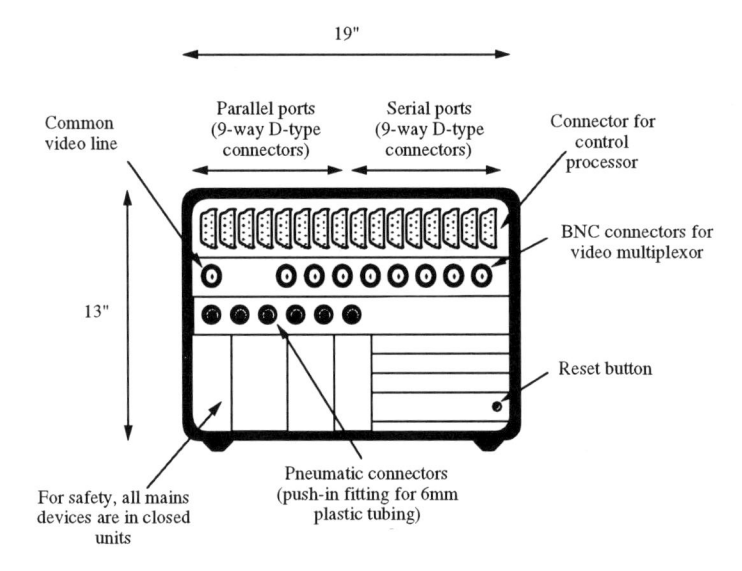

Figure 5.6 Front panel layout of the MMB module. The diagram shows that there is adequate space for all of the connectors on the front panel. Indeed, the number of connectors could be doubled without undue crowding.

---

[2] We have recently realised that these three chips could be replaced by a single device. This would simplify the board design and programing, whilst reducing the cost.

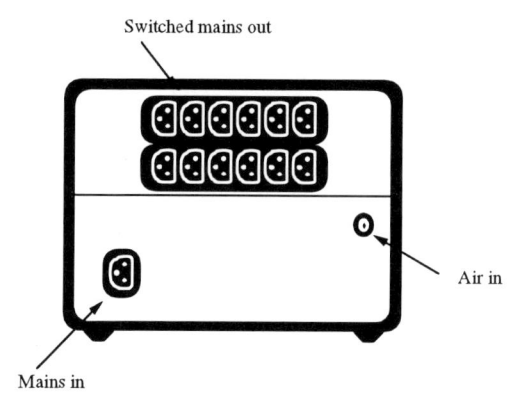

Figure 5.7 Rear panel layout of the MMB module. Standard recessed instrument-panel mains sockets were used. Once again, it is clear that there is adequate space for more connectors if needed.

5.4.3 Programs

Programs for the MMB can be stored, either in battery-backed random access memory (RAM), or in read-only memory (ROM). Of course, the latter can be programmed using a conventional ROM-programmer, connected to a standard personal computer. Of particular note here is that it is possible to program the MMB module so that it can operate in a semi-autonomous manner, thereby eliminating the need to control certain useful but tedious tasks using Prolog+. For example, a program has been written which automatically initialises the Intelligent Camera, on start-up; by pressing the RESET button; or at the command of the host processor. Another MMB program initialises the (X,Y,θ)-table controller, while a third operates the FIC's pneumatic pick-and-place arm. The MMB is currently programmed in 8086 assembly code. The cross-assembler runs on a personal computer but this is removed when the MMB unit is in use, controlling the FIC.

5.4.4 Digression on Lighting

Of course, lighting control is well advanced in the theatre, where complex lighting patterns and sequences are pre-programmed using sophisticated units, capable of switching many kilowatts of power. These are almost invariably ON / OFF devices. When we investigated theatre lighting devices, we did not find any that satisfied our seemingly more modest needs. One of the principal objections to using theatre lighting was its high cost, which simply reflected the fact that these units have facilities which we do not need. More importantly, we did not discover a unit that provided good facilities for controlling the lighting from a remote device, such as a vision system.

The MMB module provides only ON / OFF control of lighting. For finer control than this, a multi-lamp control unit has been designed. It is capable of operating 16 filament lamps, with each lamp having any one of 16 different brightness levels. The principle of operation is shown in Figure 5.8.

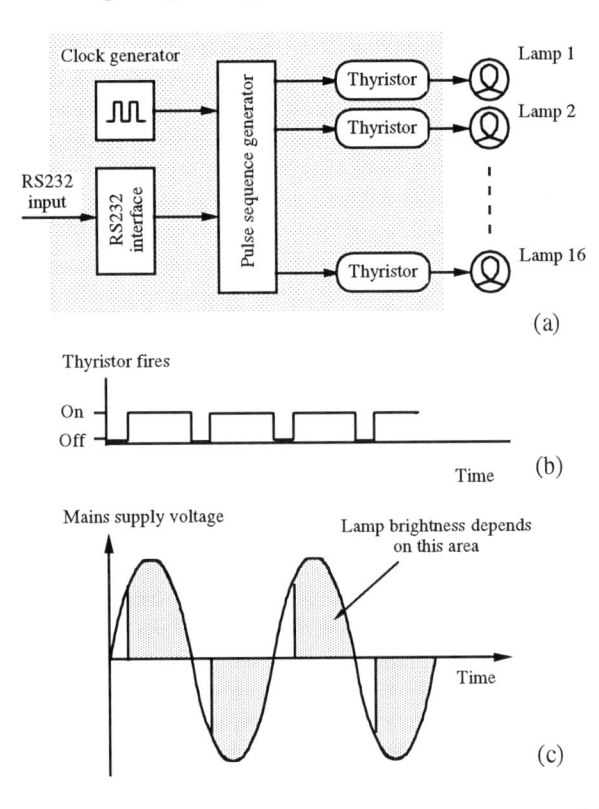

Figure 5.8 Lighting system, providing almost continuously variable control of the intensities of 16 filament lamps. (a) Block diagram. The pulse sequence generator determines when each of the thyristors fires during the mains power cycle; the timing of the output pulses fixes the instant of firing. (b) Thyristor firing pattern. (c) Each thyristor is switched on for only part of the mains power cycle. The longer it is switched on, the brighter the lamp connected to it will be.

Each lamp is controlled via a thyristor switch, which is fired by a pulse that is carefully timed to arrive at a certain instant in the mains power cycle. By making this pulse arrive early in the cycle, the thyristor will be switched on for longer and the lamp will be brighter. It is a simple matter to arrange for an N-bit microprocessor to generate suitable pulses for an N thyristor-lamp combinations. By multiplexing the microprocessor outputs, an array of M*N lamps can be controlled, where M is a small integer. The value of M that can be used in practice is limited by the speed of the microprocessor, the cost of the hardware and the tolerable power dissipation. In the design that was executed in Cardiff, 16 lamps were controlled by an 8-bit microprocessor. (N = 8; M = 2) A more modern 16-bit microprocessor, together with other low-cost electronic hardware, could easily generate the switching control signals for 64 lamps (M = 4), although the power dissipation might well limit this number.

It is unlikely that many applications would require the use of as many as 64 high-energy lamps (> 10W each). However, the development in recent years of super-bright LED's has opened up a new possibility. Moderately priced multi-LED lighting units are now available and are, of course, often seen on information display boards. Special purpose lighting units, based upon the same principles of operation, are now available commercially. These may contain several hundred LED's, arranged in some convenient pattern, such as an annulus. (The camera lens can view through the central hole.) Since the power requirement for each LED is modest, they can be driven directly from standard, cheap electronic switching devices. The result is that it is easy to design control circuits for multi-LED illumination heads, in which each LED can be individually controlled. While each LED is an ON / OFF device, an array may contain so many of them that the overall effect can be to produce nearly continuous brightness variations as more LED's are switched on. One limitation of using LED's is the limited number of colours available. (Red, yellow and green LED's are currently sold.)

Various companies have developed lighting control units which monitor the light output from a source and used this information in a feedback control system to maintain very nearly constant light levels. R. J. Ahlers designed such a system for providing in the primary light source for fibre optic illuminators. [RJA] A trivial extension of this idea allows the light level to be determined by an external signal, perhaps arising from a vision system. However, these units are not yet cheap enough to allow us use a large number in an FIC.

An alternative to controlling the lighting is provided by a camera whose gain can be changed under the control of a signal from an RS232 input port. One such camera [Philips VC7105T] achieves this effect by altering the light integration time in its CCD image sensing chip. The dynamic range achievable by this technique is currently 1000:1. (This ignores the possibility of achieving an even greater dynamic range by using an auto-iris lens as well.) Now, it is obvious that, by altering the camera gain, the effect is the same as changing the brightness of all lamps simultaneously, by the same factor. It is important to notice that adjusting the camera gain does not permit the relative brightness of different parts of a scene to be modified. However, there is one very useful ploy: digitise several images, each obtained using a single lamp switched on, and then add them in the appropriate proportions, within an image processor. To do this properly, we need to have an accurately calibrated camera. Thus, by sacrificing time, we can achieve the same overall effect as would be obtained with many lights being switched on simultaneously.

176

Finally, we must mention the ALIS lighting systems, made by Dolan Jenner, Inc. [ALIS] The ALIS 600 system consists of a light-proof cabinet containing a range of lighting devices, of different kinds and arranged at various points around the object mounting stage. The whole ALIS 600 system is capable of being controlled by computer and hence could, in theory, be operated automatically using an image processor. The object mounting stage can be moved using an (X,Y,θ)-table. The result is a lighting system that is very similar to the FIC, except that ALIS is far more precise (and expensive!) than the authors "home-made" unit. The same company also sells a set of interconnecting fibre optic and other mutually compatible lighting units. They are intended for manual rather than automatic adjustment and are collectively known as Micro-ALIS. They are intended for use when investigating lighting system design experimentally.

5.4.5 Languages for Robotics

A number of specialised languages [MCL, PAL, RAIL, RPL, VAL, CURL, AML, JARS] for controlling a robot have been developed, by extending standard computer languages, such as PASCAL, Concurrent PASCAL, FORTRAN, LISP, BASIC, PL/1 and APL. It is not difficult to envisage how languages such as these can be enhanced, by providing a subroutine library for robot control. Apart from the obvious functions of moving the robot to a defined pose, robot control languages frequently contains facilities for performing appropriate co-ordinate transformations. This is important, because it allows the user to control the movement of the robot using parameters expressed in terms of world co-ordinates (X,Y,Z), arm-joint angles, or a simple pixel address, derived from the vision system. The software must also provide facilities for initialising and calibrating the system. The reader is referred elsewhere for further details of specialised robot-control, languages. [FU-87] The important point to note here is that Prolog+ is in good company, having been extended in a similar way. The fundamental differences between the imperative languages on the one hand and the declarative languages, exemplified by Prolog, is preserved.

The decision to use Prolog in intelligent machine vision systems was not a once-and-for-all choice, forever forsaking the more conventional languages, for the simple reason that it is possible to link code written in C and Pascal into MacProlog programs. [MAC] Hence any "awkward" functions could be programmed using either of these popular languages. So far, this has not been necessary. (There is only one notable exception to this: the implementation of Prolog+ in software, where C code is used to achieve high efficiency on image processing tasks. See Appendix D.) Many of the device control functions that might seem, at first, to suggest the use of Pascal or C routines embedded in Prolog can be performed instead using the MMB unit. For example, initialising the (X,Y,θ)-table controller was accomplished satisfactorily in this way. This is hardly surprising when we examine the nature of the task: a small amount of data (e.g. a command to a low level device controller) initiates a self-contained task.

In effect, the assembly code software in the MMB module is able to perform the "awkward" operations, such as device control, which we might have assigned to a C or Pascal program running on the same computer as Prolog.

5.5 Flexible Inspection Cell, Design Issues

Figure 5.9 shows a sketch of the physical layout of the Flexible Inspection Cell. The main portion of the frame consists of a cube constructed using 12.5mm black steel tubes, 1m long. The lamps are domestic "spot" lamps, rated at 60W. The light-stripe generator consists of a 3mW solid-state diode laser, fitted with a cylindrical lens beam expander. The pattern projector is a rugged 50mm slide projector and can be fitted with various transparencies, including intensity wedge or staircase, single or multiple stripes, "rainbow" patterns, or dot-matrix patterns. The (X,Y,θ)-table is driven by stepper motors and has a travel of about 300*375mm, with a resolution of 2.5µm / step. A pneumatic pick-and-place arm, not shown here, allows objects to be repositioned on the table. The same arm can also load / unload the table. At the time of writing, the cameras are four remote-head units, connected to an Intelligent Camera. Notice that there are two overhead cameras, providing wide-field and narrow-field views of the top of the (X,Y,θ)-table. The (X,Y,θ)-table is driven by three Shinkoh stepper motors, controlled by a drive unit supplied by Time and Precision Ltd. The horizontal (IN/OUT) movement of the pneumatic pick-and-place arm used in the FIC is provided by a rod-less pneumatic cylinder (stroke 375 mm, manufacturer SMC, Model MY1M256-400). A standard pneumatic cylinder with a 60mm stroke provides the UP/DOWN movement. The gripper fitted at the time of writing is a simple vacuum device, with a circular suction cup, 12mm in diameter.

5.5.1 Lighting Arrangement

To reduce reflections, the FIC has a matt-black, rigid steel frame, holding the lighting units. The latter consists of the following devices:

(a) Four 60 W filament lamps, around and just above the work-table, to provide glancing illumination.
(b) Four 60 W filament lamps, around the table provide illumination at about 45° to the vertical axis.
(c) Two lamps directly above the table and very close to the overhead camera provide nearly coaxial illumination and viewing. (An alternative would be to use a ring illuminator, located around the lens of the overhead camera.)
(d) Two 60 W filament lamps, connected in parallel, which cannot shine directly onto the table but provide high intensity illumination for a grey sheet of

material. This enables the *side camera*[3] to view an object on the table with either a bright or dark background.

(e) A rugged slide-projector, which enables the side camera to obtain range information. The projector is normally fitted with a 50mm slide, which contains a series of broad parallel black-white stripes, although this can be changed manually. Among the other interesting possibilities is a slide consisting of a set of coloured stripes.

(f) A solid-state laser light-stripe generator, fitted immediately above the table. This is normally used in conjunction with the oblique camera to generate depth maps. The laser can be connected directly to one of the parallel I/O lines.

(g) Infra-red and ultra-violet lamps can also be fitted to the FIC.

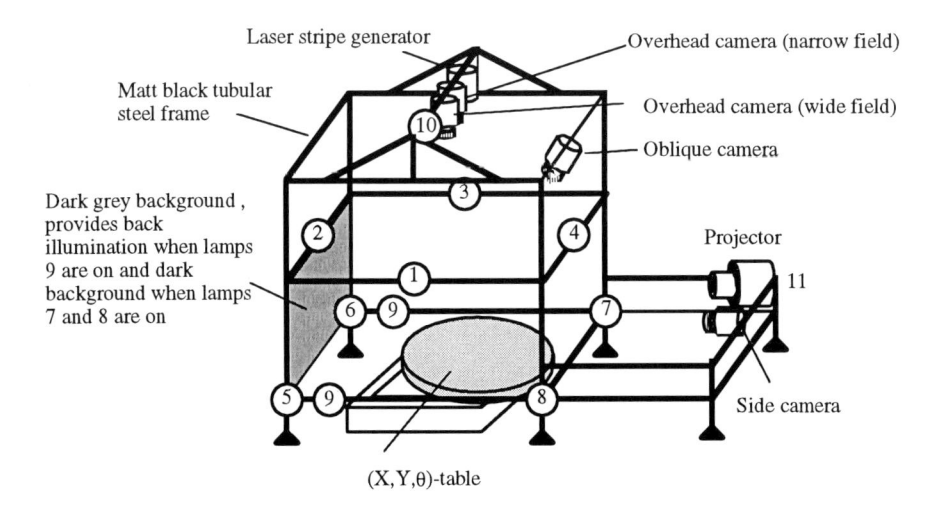

Figure 5.9 Flexible Inspection Cell (FIC).

5.5.2 Mechanical Handling

The FIC contains a high-precision (X,Y,θ)-table, driven by standard stepper motors and which provides movement control as follows:

(i) X-axis, travel 375 mm, step resolution $\pm$ 2.5µm.
(ii) Y-axis, travel 300 mm, step resolution $\pm$ 2.5µm.
(iii) θ-axis (rotation), 360° movement, step resolution 0.1°.

The interface card for the (X,Y,θ)-table controller (Digiplan IF5 card) has a standard serial (RS232) input port, and has a straightforward and well-defined

---

[3] That is one whose optical axis is horizontal and views objects on the table from the side.

command language. (Table 5.2) The FIC has been fitted with a pneumatic pick-and-place arm, arranged so that an object can be place on the table and then removed, or simply lifted up, while the table is shifted beneath it. This arm is connected to the MMB module, simply by plugging 6mm plastic piping into both units. The pick-and-place arm requires the use of the two 5-port switched air lines mentioned earlier (One of these operates the *in-out* movement and the other for the *up-down* movement of the end-effector.) One of the MMB unit's ON/OFF air lines is used to operate a suction gripper.

Command	Function
< 10	Define start / stop speed to be 10 steps s^{-1}
^ 15	Define acceleration / deceleration rate to be 15 steps s^{-2}
X 1000 @ 200 $	Shift 1000 steps in +X direction, at peak rate of 200 steps s^{-1} ($ initiates the movement)
Y-3000 $	Shift 3000 in -X direction
Z 250 @ 50 $	Rotate by 250 steps at peak speed of 50 steps s^{-1}
X+ Y- $	Shift in +X and -Y directions previously-programmed distance
X @ 500 G $	Travel continuously in +X direction at constant speed
Z @ 1000 G $	Rotate continuously in +Z direction at constant angular speed
#	Cancel previous instruction (continuous or indexed movement)
B - 20	Store backlash distance of 20 steps (See next two commands)
X 100 B $	Move in +X direction for 120 units then reverse for 20 units
X-100 B $	Move in -X direction. No backlash correction necessary when travelling in this direction

Table 5.2 Control language for the Digiplan stepper motor controller. Only the most basic operations are described here. More advanced facilities, such as defining and storing move sequences, delaying movement until a trigger signal arrives on a parallel input port and generating output control signals are not included, since they are not required in the FIC.

5.5.3 Cameras and Lenses

To be effective, an FIC requires a minimum of three cameras:

(a) *Overhead camera*, looking vertically downwards, used to view the top surface of object(s) on the table.

(b) *Side camera*, providing a horizontal view. This camera is also able to generate depth maps, when used in conjunction with the pattern projector, fitted with a suitable slide.

(c) *Oblique camera*, used in conjunction with the laser stripe generator. (In the FIC, this camera is used to acquire depth maps of the top surface of the object on the table.)

However, there are good reasons for using more than three cameras in an FIC. The principal advantage of adding extra cameras is that it is possible to fit them with different lenses, in order to achieve different levels of magnification. In many applications, it is necessary to locate an object using a camera fitted with a wide-angle lens and then take more detailed observations using another camera with a narrow-angle lens. A standard lens, fitted to the overhead camera in the FIC can view the whole of the table-top, when it is in its "home" position (i.e. at the centre of its range of travel). A second overhead camera, located very close to this one but fitted with a narrow-angle lens, may be used to obtain a higher magnification and thus to see small detail, which would not be clearly visible to the wide-field overhead camera. (Figure 5.10) Notice the offset caused by the fact that these cameras are not in the same place. This is an important point and it has a significant influence upon the calibration of the system. When a robot vision system is first set up, the relative positions of the cameras and mechanical handling sub-system are unknown and have to be found by taking measurements experimentally. When there is more than (overhead) camera, this process has to be extended, in order to relate their fields of view together.

5.5.4 MMB-Host Interface Protocol

The interface between the host computer and the MMB module follows the protocol described below. (Also see Table 5.3)

(i) All output from the host computer is sent directly to the currently selected serial port, with the exception that the control sequences defined in (iii) are trapped by the MMB module and are not transmitted onwards. Notice that the serial ports connecting the MMB to satellite devices may operate at a different speed from the control port, connecting the MMB to the host processor.

(ii) All signals received on the currently selected serial port are sent to the host computer. Signals received by the MMB module on its other serial ports are simply ignored.

(iii) Control sequences are of the form: ¶ *a b {c}*, where *a* is an integer in the range [1,8]; *b* is an integer and c is an optional integer. (See Table 5.3.) All characters between "¶" and carriage return are ignored when sending data to the serial ports.

A range of special functions (¶ 8 ... commands) are available for operating the FIC. These special functions are not essential but make the task of writing software for the host processor easier. For example, '¶ 8 ...' commands have been written for initialising the Intelligent Camera, initialising the (X,Y,θ)-table and operating the pick-and-place arm from software resident in the MMB module.

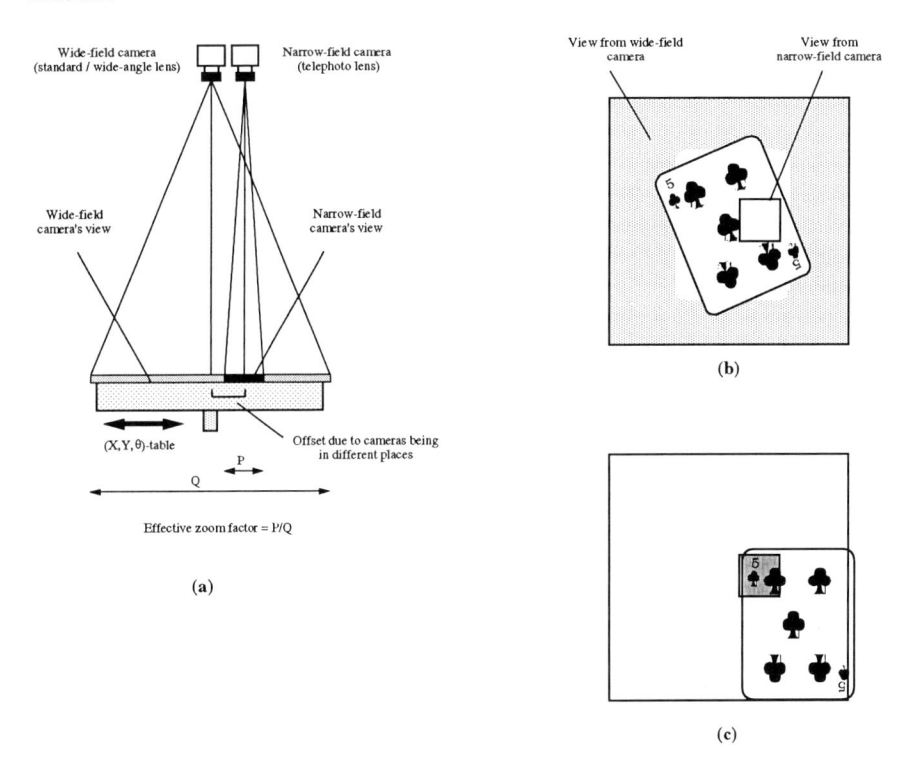

Figure 5.10 Two cameras with different lenses can provide different magnifications. (a) Optical arrangement. (b) The wide-field camera views the whole object to be examined, in low magnification. From this image, the appropriate table-movement parameters are calculated. (c) After the table and the object on it have been shifted, the narrow-field camera can examine a selected part of the object in high magnification.

5.5.5 Additional Remarks

It must be emphasised that the MMB module was designed to fulfil a range of functions; it was not designed specifically to control the present FIC. The module has also been used to operate two commercial multi-axis robots: one a SCARA-type robot, and the other a gantry robot. (Both robots are made by Cybernetic Applications Ltd. and are intended for educational use.) The latter has been used recently in a student exercise, building a machine that is able to play dominoes against a human opponent. Other tasks that the MMB module has performed include starting / stopping a conveyor belt and controlling a pneumatic "blower" for diverting defective items from a conveyor belt inspection system. The MMB module can also be used for such tasks as operating relays, solenoids, warning lamps, bells and sensing the arrival of components as they pass the camera during inspection.

Function	First	Second	Third	Example
Prolog signals to MMB that it wants to receive data	0	Data port required (Integer, 1-6)	None	¶ 0 3
Divert output to a different serial port	1	Port no. (1-7)	Baud rate (e.g. 9600)	¶ 1 4 9600
Switch video multiplexor	2	Port no. (1-8)	None	¶ 2 3
Switch mains power lines on / off	3	Lamp state vector (Hex integer, 0 - FF)	None	¶ 3 B4
Switch pneumatic air line on / off	4	Pneumatics state vector (Hex integer, 0 - F)	None	¶ 4 9
Superbright LED's (reserved for future)	5	Lamp state vector (Hex integer, 0 - FF)	None	¶ 5 7
Send data to parallel port	6	Port no.	Hex integer	¶ 6 4 137
Receive data from parallel port	7	Port no.	None	¶ 7 5
Initialise all ports	8	0	None	¶ 8 0
Initialise Intelligent Camera	8	1	None	¶ 8 1
Initialise (X,Y,θ)-table	8	2	None	¶ 8 2
Switch laser stripe generator on / off	8	3	0 / 1	¶ 8 3 0
Switch pattern projector on/ off	8	4	0 / 1	¶ 8 4 1

(Argument columns: First, Second, Third)

Table 5.3 MMB control signals.

Many software packages have limited I/O capabilities. This statement includes both MacProlog and HyperCard, which is discussed in a little while. The MMB unit provides a means of interfacing these and similar software packages to a range of external devices, which would not otherwise be possible. In situations in which operating speed is of less importance than ease of use and versatility, the present design provides a low cost solution to a wide range of interfacing problems. Although it is conceptually naive, the absence of such a unit, in the past, has hindered our research work in machine vision. The component cost of the present unit is about £1000 (US$1500). Much of this cost lies in the case, connectors, solid-state relays and pneumatic valves. The central electronics unit costs under £100 (US$150). Hence, a lower cost unit could easily be made, with a subsequent reduction in its functionality.

5.5.6 HyperCard Control Software for the FIC

In Chapter 4, we described how the FIC can be controlled using a HyperCard program ("stack"). HyperCard provides an excellent range of set of tools for constructing easy-to-use and highly effective human-computer interfaces. The HyperCard controller has been developed in such a way that it simplifies the task of setting up the (X,Y,θ)-table, lighting, etc. in the FIC, without hindering access to all of the features of Prolog+. The HyperCard controller co-exists and co-operates with Prolog+. For example, the user can specify Prolog+ goals in Hypercard; results are returned to HyperCard for viewing by the user. Moreover, the user can switch quickly from one program to the other, with a minimum of effort.

As we have seen in Chapter 4, Hypercard can control the FIC in two possible ways: by direct control of the I/O port, and via Prolog+. The first of these makes use of Hypertext's (i.e. HyperCard's programming language) ability to address the Macintosh computer's serial (RS232) ports. The second option uses so-called *AppleEvents*. An AppleEvent is a message that is passed from one Macintosh application (e.g. HyperCard) to another (e.g. Prolog). The message may be a request for the receiver application to operate in some particular way. For example, Hypercard may send MacProlog an AppleEvent which specifics a goal (e.g. *goal(A,B,C)*) that is to be satisfied. If *goal(A,B,C)* does succeed, the variables A, B and C will be instantiated (say A = 1, B = 2, C = 3). The message *goal(1,2,3)* is then despatched by MacProlog, back to HyperCard, which can then perform further calculations on the returned values (1,2,3). AppleEvents provide a very useful medium for inter-application communication on the Macintosh computer. Similar facilities exist, of course, on other computers. The authors have written programs to control the FIC using both direct control of the I/O ports by Hypercard and by using AppleEvents to specify Prolog+ goals, which in turn operate the FIC.

5.6 Prolog+ Predicates for Device Control

Standard Prolog contains relatively unsophisticated facilities for controlling external devices, compared to those found in many other languages. The inventors and early pioneers working with Prolog apparently never envisaged it being used to operate devices such as lamps, lasers, (X,Y,θ)-tables, robots, electro-plating baths, milling machines, injection moulders, etc. Of necessity, the original language did contain facilities for operating a printer and the "output" predicates listed in this section follow the lead this provides.

Consider the Prolog built-in predicate *write(X)*. This always succeeds but is never resatisfied on backtracking. Strictly speaking, *write(X)* is a test, not a command; as a side effect of trying to prove that *write(X)* is true, the printer operates and types whatever value X currently has. [CLO-87]. In the same way, the predicate *lamp(A,B)*, which forms part of the extension of Prolog+, always succeeds but is not resatisfied on backtracking. As a side effect, lamp number *A* is switched to the state indicated by B. (B may be *on, off, half-on,* or some integer, indicating the desired brightness level.) Other predicates controlling the "output" operate in the same way. To see how this is achieved in practice, here is the Prolog code for *lamp(A,B)*, which controls a (two-level) lamp:

```
lamp(A,B) :-
     lamp_state_vector(C),
                  % Consult database for lamp state vector, C.
     list_element_set(A,B,C,D),
                  % Set A-th element of list C to value B. D is
                  % the new lighting state vector.
     retract(lamp_state_vector(C)),
                  % Forget the previous lighting state vector
     assert(lamp_state_vector(D)),
                  % Remember the new lighting state vector
     writeseqnl(modem, ['¶3 ',D]),
                  % Command to MMB. See Table 5.3.
     !.           % Prevent resatisfaction on backtracking
```

Now, let us consider the "input" predicates. Standard Prolog is provided with the built-in predicates *read* and *get* and, at first sight, it seems that it would be a simple matter to use these to obtain information about the outside world, via the MMB module. However, there is an important point to note: Prolog relies upon programmed I/O. Hence, all requests for data to be transferred from the MMB module must be initiated by Prolog. Here is a suitable data-transfer protocol needed for input, expressed in Prolog:

```
get_data(A,B) :-
       ticks(C),       % Integer P indicates time. Units: 1/60th sec.
       D is C + 60,    % Q is latest time allowed for MMB response
       writeseqnl(modem, ['¶0 ',A),
                       % '¶0' is test sequence used by Prolog to
                       % sense whether MMB module is ready to receive
                       % data. A indicates what data is requested.
       mmb_response(D),  % Succeeds only if MMB is ready to send data
       read(modem,B).  % Get data from MMB unit.
get_data(_,_) :-
       beep,           % Audible warning
       writenl('The MMB Module was not available to send data'),
       fail.           % Forced fail

% Clause 1 fails, if the time limit has been passed.
mmb_response(A) :-
       ticks(C),       % C is the time now
       C > A,          % Is C greater than given time limit, A?
       !, fail.        % Force failure

% Clause 2 succeeds, if the character '$' has been detected
mmb_response(A) :-
       serstatus(modem,in,B),
                       % B = number of characters in input buffer
```

```
            B ≥ 1,              % There are some characters to be analysed
            get(modem,C),       % So, let's see one
            C = '$',            % Is the character '$'?
            !.                  % Yes, so the goal succeeds
% Time limit has not been reached & character '$' has not been seen
% yet - Try again
mmb_response(A) :-
            !,      % Improve efficiency of recursion
            mmb_response(A).
                    % Repeat until '$' found or time limit is reached
```

Following these general principles, a wide range of I/O predicates have been defined for controlling the FIC via Prolog+.

5.7 System Calibration

Whenever a vision system is intended for use with a multi-axis manipulator (i.e. (X,Y,θ)-table, or multi-axis robot), it is vitally important that their co-ordinate systems be related to one another, before the system is used. The procedure for doing this will be referred to as *calibration*. To appreciate the importance of this calibration for robot vision systems in general, consider the FIC, which, of course, has an (X,Y,θ)-table and pick-and-place arm. The FIC is one example of a $2\frac{1}{2}$D robot and is functionally equivalent to a SCARA arm. (Figure 5.11(a).) Both of these devices are able to move parts around on a flat table and can conveniently be guided by a vision system, with a single overhead camera. When the FIC is first set up, the position of the camera relative to the (X,Y,θ)-table is unknown. In addition, the camera may be skewed relative to the (X,Y,θ)-table, so that their co-ordinate axes are not parallel. Furthermore, the camera's magnification is not known exactly. Finally, in the case of the FIC, the relative positions of the pick-and-place arm and the camera are not known exactly. It is obvious that a series of parameters must be calculated, so that a point determined in terms of the vision system's co-ordinates can be related to co-ordinates describing the position of the (X,Y,θ)-table. An experimental procedure for calibrating the FIC forms the subject of discussion in this section.

There are two possible methods for calibrating the overhead camera of the Flexible Inspection Cell.

(a) "Once-and-for-all" calibration, using careful measurement and dead reckoning.
(b) Automatic self-calibration, using some convenient marker to measure the movements of the table.

By the former, we mean to imply that the calibration is done only once. It is then assumed that the camera, its lens and the table (or robot) base are not moved at all. This is a particularly dangerous assumption in any environment, such as a factory, where unauthorised tampering is ubiquitous. Even in a research laboratory, the dangers of mischievous fingers must be taken into account when designing experiments. It is far safer to re-calibrate the machine vision system every time that it is used. In a factory, it is good practice to perform the calibration procedure at least once during every shift. While it may be possible to bolt the camera and table rigidly to the same frame-work, it is often necessary to adjust the lens, particularly during cleaning. (Of course, changing the camera zoom, focus and aperture controls can all be accomplished accidentally during cleaning.) Frequent re-calibration is needed, even in the best controlled environments. Manual calibration is tedious, slow, imprecise and prone to gross errors. On the other hand, automatic self-calibration is fast, precise and reliable, provided that sensible checks are built into the procedure. In the remainder of this section, we shall therefore concentrate upon the self-calibration procedure, restricting our discussion to the FIC. A very similar process can be devised for a SCARA robot, used in conjunction with an overhead camera. The reader will not be surprised to discover that a 3D robot, such as the gantry type (Figure 5.11(b)), requires a more complicated procedure. However, this is a straightforward extension of the ideas discuss here: we calibrate the system using the overhead and side cameras separately. Finally, we relate the pictures for the two cameras together, using some suitable target object, which both cameras can see, at the same time.

5.7.1 FIC Calibration Procedure (Overhead Camera)

The process consists of a number of distinct steps:

(a) The geometry of the work-space is shown in Figure 5.12. First, it is necessary to set up the wide-angle overhead camera manually, so that it is focused on the top of the (X,Y,θ)-table and can view the whole of the desired field of view. In order to make the best possible use of the available image resolution and to avoid over-complicating the image processing, it is suggested that the camera should not be able to see much detail outside the disc forming the top of the (X,Y,θ)-table.

(b) Next, a small high-contrast circular marker is placed near the centre of the table. This can be done either by hand, or preferably by the pick-and-place arm. Suppose that the table is dark, nominally matt black. Then, the marker should be nominally white and it should not glint unduly. The lighting should be arranged to come from above the table and be diffuse (multi-directional). This will avoid casting sharp shadows. The overhead camera is then used to view the table top. A straightforward Prolog+ program is then used to analyse the image of the marker and find the co-ordinates of its centroid. Let these co-ordinates be denoted by (X11,Y11). Notice that these parameters relate to *vision system co-ordinates*.

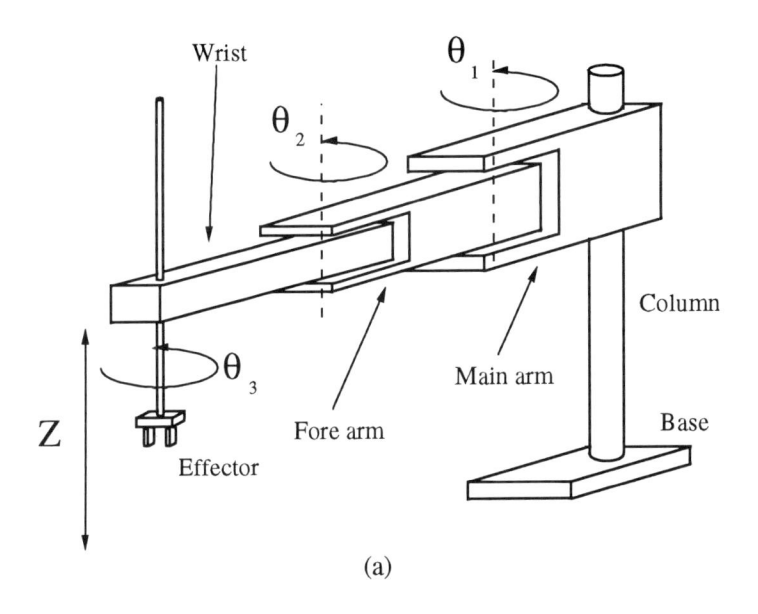

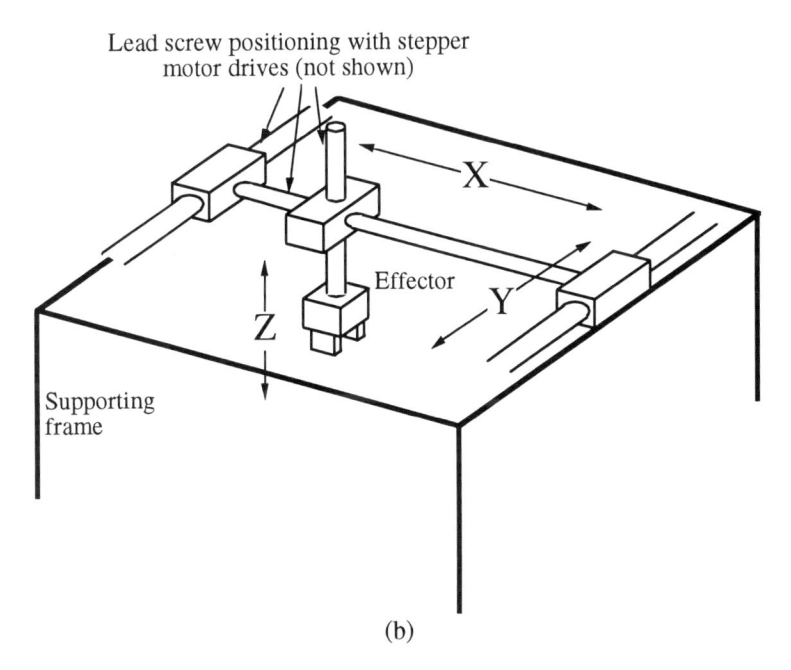

Figure 5.11 Two types of multi-joint robot (a) SCARA robot (DP Batchelor). (b) Gantry robot.

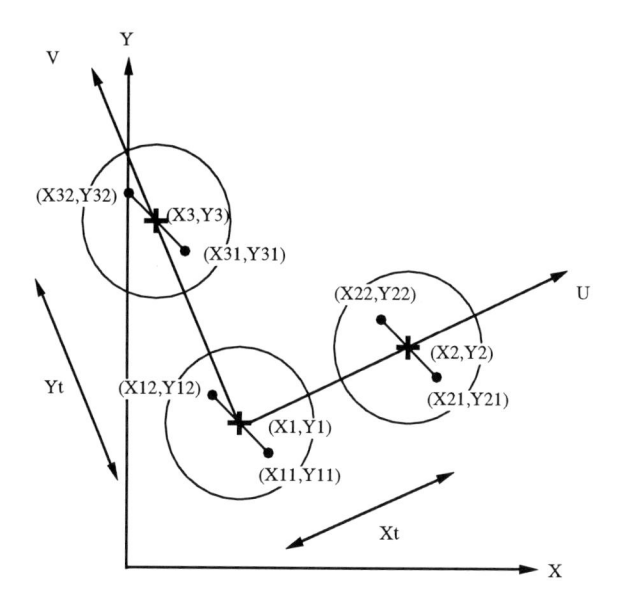

Figure 5.12 Geometry of the work space for calibrating the overhead camera in the FIC

(c) The (X,Y,θ)-table is then rotated by 180°. The centroid of the marker is recalculated. Suppose that the new co-ordinate values are (X12,Y12). Again, these numbers relate to *vision system co-ordinates*. The centre of the table is now at (X1, Y1), where

$$(X1, Y1) = ((X11 + X12)/2, (Y11 + Y12)/2) \qquad \ldots(5.1)$$

(d) The table is then moved by a distance Xt along its X axis. (Notice that this movement may not be parallel to the x-axis of the vision system.)
(e) Steps (b) and (c) are repeated. This process yields the new position for the centre of the table at (X2,Y2). (In fact, it is not necessary to rotate the table by 180°. We could, for example, derive all of the relevant parameters using just four points (i.e. (X11,Y11), (X12,Y12), (X21,Y21) and (X31,Y31)). However, it is conceptually simpler to explain the procedure, in terms of movements of the centre of the table.)
(f) The table is then returned to the original position, as at the end of step (a).
(g) A procedure akin to steps (b) - (f) is then performed, except that the table is now moved an amount Yt along its Y axis and the centre of the table is discovered to be at (X3,Y3).
(h) The axis-transformation equations are therefore

$$X = U.\left[\frac{X2 - X1}{Xt}\right] + V.\left[\frac{X3 - X1}{Yt}\right] + X1 \qquad \ldots(5.2)$$

$$Y = U.\left[\frac{Y2-Y1}{Xt}\right] + V.\left[\frac{Y3-Y1}{Yt}\right] + Y1 \qquad ...(5.3)$$

where the position of the centre of the table, according to the table's co-ordinate system, is (U,V), and its position according to the vision system is (X,Y). The origin of the table's co-ordinate system is the centre of the table in step (b). This is the point (X1,Y1), in the vision system's co-ordinate space. It is normally more useful to have the equations rearranged so that the robot co-ordinates, (U,V), can be calculated from the vision system co-ordinates, (X,Y).

$$U = Xt.\frac{\left[\left[\dfrac{Y-Y1}{Y3-Y1}\right]-\left[\dfrac{X-X1}{X3-X1}\right]\right]}{\left[\left[\dfrac{Y2-Y1}{Y3-Y1}\right]-\left[\dfrac{X2-X1}{X3-X1}\right]\right]} \qquad ...(5.4)$$

$$V = Yt.\frac{\left[\left[\dfrac{X-X1}{X2-X1}\right]-\left[\dfrac{Y-Y1}{Y2-Y1}\right]\right]}{\left[\left[\dfrac{X3-X1}{X2-X1}\right]-\left[\dfrac{Y3-Y1}{Y2-Y1}\right]\right]} \qquad ...(5.5)$$

Of course, if the table and vision system co-ordinate axes are parallel, these equations can be simplified to

$$U = Xt.\left[\frac{X-X1}{X2-X1}\right] \qquad ...(5.6)$$

$$V = Yt.\left[\frac{Y-Y1}{Y3-Y1}\right] \qquad ...(5.7)$$

The general axis-transformation equations are of the form

U = A.X' + B.Y' ...(5.8)
V = C.X' + D.Y' ...(5.9)

where

X' = X - X1 ...(5.10)
Y' = Y - Y1 ...(5.11)

$$A = -Xt \cdot \left[\frac{\left[\dfrac{1}{X3-X1}\right]}{\left[\dfrac{Y2-Y1}{Y3-Y1}\right] - \left[\dfrac{X2-X1}{X3-X1}\right]} \right] \qquad \ldots(5.12)$$

$$B = Xt \cdot \left[\frac{\left[\dfrac{1}{Y3-Y1}\right]}{\left[\dfrac{Y2-Y1}{Y3-Y1}\right] - \left[\dfrac{X2-X1}{X3-X1}\right]} \right] \qquad \ldots(5.13)$$

$$C = Yt \cdot \left[\frac{\left[\dfrac{1}{X2-X1}\right]}{\left[\dfrac{X3-X1}{X2-X1}\right] - \left[\dfrac{Y3-Y1}{Y2-Y1}\right]} \right] \qquad \ldots(5.14)$$

$$D = -Yt \cdot \left[\frac{-\left[\dfrac{1}{Y2-Y1}\right]}{\left[\dfrac{X3-X1}{X2-X1}\right] - \left[\dfrac{Y3-Y1}{Y2-Y1}\right]} \right] \qquad \ldots(5.15)$$

Hence, A, B, C and D can be calculated once, before we begin using the FIC in earnest. Thereafter, we simply apply Equations 5.8 - 5.11. Notice that the origin (i.e. the (0,0) position) for the (X,Y,θ)-table's co-ordinate axes is given by the position of the centre of the table at the beginning of the calibration procedure.

5.7.2 Calibration, SCARA and Gantry Robots (Overhead Camera)

The procedure just described can be modified slightly to calibrate a SCARA or gantry robot, used in conjunction with an overhead camera. The changes needed for this are fairly obvious and, for this reason, will be described in outline only.

(a) Pick up a white object, called the marker, in the robot's gripper.
(b) Use the robot to place the object on a flat, matt black table. The robot releases the marker when it is at its "home" position.

(c) Use the overhead camera to view the marker. (The robot arm is, of course, moved out of the sight of the camera.) Digitise an image and locate the centroid of the marker. (Co-ordinates: (X11,Y11))
(d) Pick up the marker. Rotate the gripper by 180° and then replace the marker on the table.
(e) Repeat step 3. (Marker co-ordinates: (X12,Y12))
(f) Compute the centre point, between (X11,Y11) and (X12,Y12) using Equation 5.1.
(g). Pick up the marker. Move the robot to position (Xt, 0) relative to its home position. Repeat steps (c) to (f). (Marker co-ordinates: (X2,Y2).)
(h) Pick up the marker. Move the robot to position (0,Yt) relative to its home position. Repeat steps (3) to (6). (Marker co-ordinates: (X3,Y3).)
(i) Use the equations given in the previous section to translate the position (X,Y), given in terms of vision system co-ordinates, into co-ordinates defining the robot's position.

This procedure is needed because an overhead camera cannot see an object while it is being held in the robot gripper.

5.7.3 Calibration Procedure (Overhead Narrow-view Camera)

The calibration procedures just described are suitable for use with either wide-angle or narrow-angle lenses. However, there is an additional requirement: relating the wide and narrow-field cameras to each other. To simplify the analysis, let us assume that both cameras

(a) see a circular white marker, resting on a black table;
(b) are aligned so that their co-ordinates axes are parallel;
(c) generate digital images with a resolution of $X_{max}*Y_{max}$ pixels;
(d) generate digital images which have a 1:1 aspect ratio when a square is being viewed.

Furthermore, we shall assume that the narrow field-of-view is entirely contained within the wide one. Suppose that by using the narrow-field camera, we discover that the marker is at (X_n, Y_n) and has a diameter of D_n, while the equivalent figures for the wide-field camera are (X_w, Y_w) and D_w. Then, the optical-magnification ratio of the two cameras is equal to R, where

$R = D_w/ D_n.$ ...(5.16)

The axis-translation equation is given by

$(X', Y') = ((X - X_w)/R + X_n, (Y - Y_w)/R + Y_n))$ ...(5.17)

where (X',Y') is the position of a point, given in terms of the narrow-field camera, and (X,Y) is its position related to the wide-field camera. (Figure 5.13) It is useful to locate the centre of the field of view of the narrow-field camera in terms of the wide-field camera co-ordinates.

This can be achieved by solving the equation

$$(X_{max}/2, Y_{max}/2) = ((Xc - Xw)/R + Xn), (Yc - Yw)/R + Yn)) \qquad ...(5.18)$$

for Xc and Yc. The resulting equation is

$$(Xc, Yc) = (R.(X_{max}/2 - Xn) + Xw), R.(Y_{max}/2 - Yn) + Yw) \qquad ...(5.19)$$

Suppose that we identify something of interest, using the wide-field camera, and then move it, using the (X,Y,θ)-table, so that we can take a more detailed look with the narrow-field camera. The procedure for doing this is straightforward. (Figures 5.10 and 5.13) Suppose that the wide-field camera has found the object at point (X_p, Y_p). We simply calculate the table movement by substituting ($(X_p - X_c), (Y_p - Y_c)$) for (X,Y) into Equations 5.4 and 5.5.

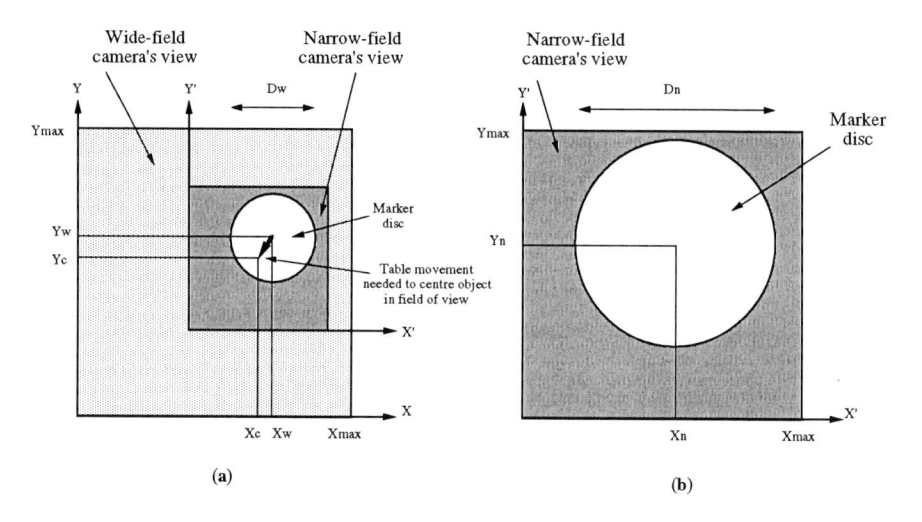

Figure 5.13 Relating the two overhead cameras together. (a) Wide-field camera's view (light shaded area). The narrow-field camera's field of view is also shown (dark shaded area.) Notice that the white marker disc is visible to both cameras. By moving the table by an amount given by the thick black arrow, the disc will be (approximately) centred on the narrow-field camera's optical axis. (b) Geometry of the narrow-field camera's view of the marker.

5.7.4 Calibration Procedure (Side Camera)

In one respect, the calibration procedure is simpler for the side camera, since objects resting on top of the (X,Y,θ)-table, or held in the gripper of a gantry or SCARA robot are visible. However, there are complications, due to the fact that the object can move along the optical axis (Y axis, Figure 5.14) of the side camera. This can lead to mis-focussing of the side camera, as well as causing the apparent size of the object to vary. These problems can be avoided altogether, if we use the overhead camera and (X,Y,θ)-table or robot to fix the object position along the Y-axis, before images from the side camera are ever digitised and processed. In the arrangement shown in Figure 5.14, the overhead camera locates the object in the (X,Y) plane. The table / robot is then moved, so that the feature of particular interest is placed at a known Y-position. The side camera sees object features projected onto the (X,Z) plane.

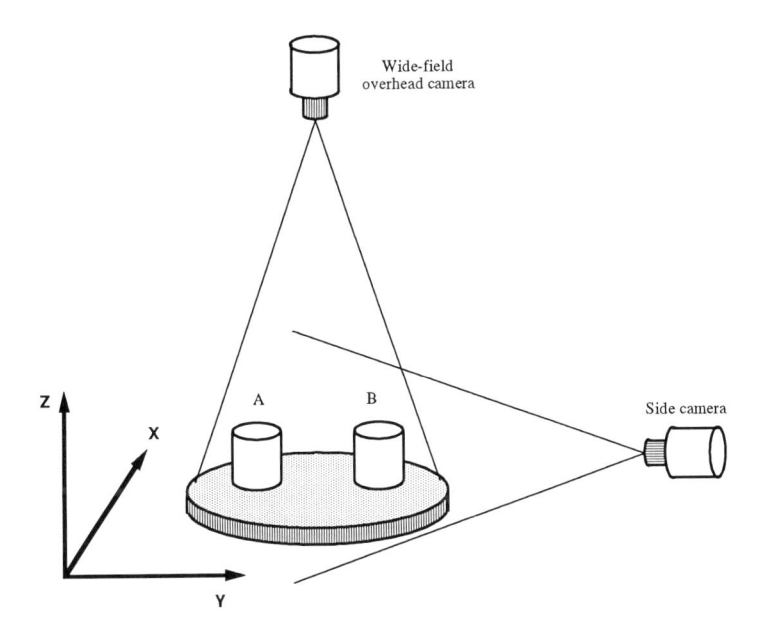

Figure 5.14 The side camera has an unknown magnification, if the position of the object along the Y-axis is not known. Thus, the object appears to get bigger, if it moves from A to B. The overhead camera can provide some help in determining the Y-position. However, this is not always possible, because suitable "landmark" features might not be visible from both cameras. The possible effects of mis-focusing are ignored here.

When the target has been placed in fixed Y-position, it is a trivial matter to calibrate the robot vision system. Note, however, that it is advisable to use a clearly recognisable target, since the view from the side camera may be far from simple. (The gripper of a robot may have brightly polished metal surfaces. The side-frame of the (X,Y,θ)-table, slide-way, carriage and θ stepper-motor may all be visible, below the table top and are likely to glint to some extent.) There are two obvious ways to simplify the image, by employing a target which generates an image with a very high contrast. (a) Use a calibration target with a small retroreflective patch stuck to it and arrange for lighting to (appear to) come from the side camera. (A small lamp, or LED, placed very close to the side camera will suffice.) (b) Use a target which has a bright beacon (fibre optic or LED "point" source") mounted on it.

The calibration procedure consists of simply moving the (X,Y,θ)-table along the X axis only and noting the movement, as observed by the side camera. When a robot that is capable of moving in the vertical (Z) direction is being calibrated, the target must be moved along both the X and Z axes. The equations relating the table / robot movement to the shifts in the image seen by the side camera are simple to derive.

5.8 Picking up a Randomly Placed Object (Overhead Camera)

The general manipulation task that we shall discuss in this section is that of picking up a thin, laminate object, placed on the table in unknown orientation and position. For simplicity in our discussion, the table will be assumed to be very dark and the object very bright, so that image segmentation can be performed easily, perhaps by thresholding, or some other simple process.

Consider Figure 5.15. The radius of the table is A units. The centre of the table can move anywhere within a square of side 2.A units. The camera *apparently* needs to be able to view a square of side 4.A units. This arrangement *guarantees* that the camera can see an object placed on the table, wherever the latter is located. It is assumed, of course, that the object lies wholly within the perimeter of the circular top of the turn-table. Suppose, however, that the turn-table is always moved to the centre of its range of travel, whenever an object upon it is to be viewed. Then, the camera need only view a square of side length 2.A units. Of course, this enables the camera to sense smaller features than the arrangement suggested in Figure 5.15(a). Notice too that the corners of the image can safely be ignored. This is important because we have no control over the appearance of the corners of the image, which show the slide-way and drive motors of the (X,Y,θ)-table mechanism. The slide-way might well have brightly polished metallic surfaces, which would cause problems due to glinting. The corners of the image should therefore be removed by masking, either optically, or by software. (Alternatively, a larger black disc, whose radius is at least $\sqrt{2}$.A units, could be fitted to the turn-table mechanism.) Of course, a well-designed program for controlling the FIC, should always be aware of the position of the (X,Y,θ)-table. In this case, it is not difficult to move the table to the centre of its range of travel, every time the object upon it is to be viewed. Notice, however, that this does increase the time spent moving the table, since two movements rather than one are now needed. Thus, we can make better use of the camera's finite resolution but at the cost of reducing the speed of the system, using a camera with the smaller field of view.

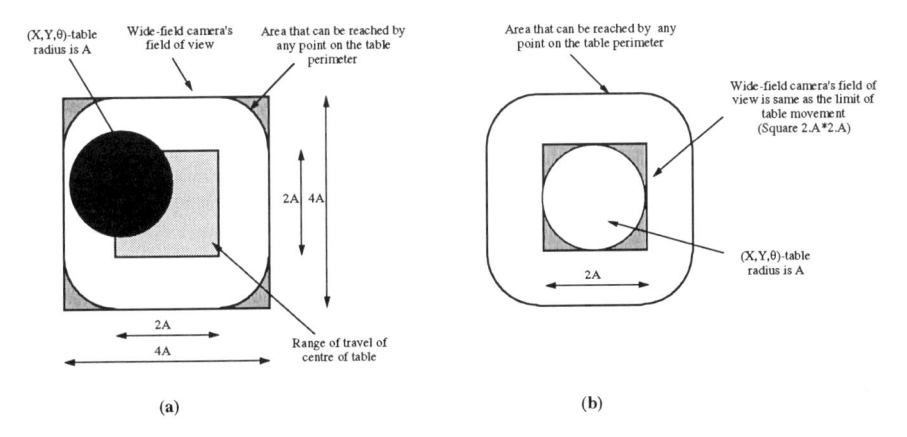

Figure 5.15 Picking up a randomly placed object, using the overhead camera to guide the (X,Y,θ)-table. (a) Viewing an object anywhere on the table, which can be located anywhere within its defined limits of travel. (b) Viewing an object when the table is always located at the centre of its range of travel. The magnification can be increased by a factor of two.

In order to normalise the position and orientation of an object positioned in an arbitrary way on the table, we may use the following procedure. (See Figure 5.16)

1. Use the vision system to determine the position and orientation of the object.
2. Calculate the robot position parameters using the formulae listed in Section 5.7.
3. Move the table to the position and orientation found in step 1.
4. Rotate the table by $-\theta°$.
5. *Optional: Use the vision system to verify that the object position and orientation have been correctly normalised. If not, return to step 1.*

The co-ordinates of the centroid are, of course, useful for locating the object. Let the centroid co-ordinates, as estimated by the vision system, be denoted by (X_0, Y_0). The orientation of the object, relative to some appropriate reference axis, will be represented by θ. The value of θ might be found by calculating the orientation of the axis of minimum second moment, or of the line joining the centroid of the object silhouette to the centroid of the largest "bay", or "lake". Figure 5.17(a) illustrates the most obvious way to determine the position of the object. This involves computing the co-ordinates of its centroid. Alternatively, the centre of that line which joins the centroids of the two holes could be used to determine the position. Of course, this method will not work if there do not exist at least two clearly distinguishable holes. The orientation can be found in a number of ways, of which four are illustrated in Figure 5.17(a). The axis of minimum second moment is one possibility. A second is provided by finding the orientation of the line joining the centroid to the furthest edge point. The line joining the centroid of the silhouette to the centroid of the largest hole provides the basis for the third method. The fourth method is similar to the third, except that the hole that is furthest from the centroid of the silhouette is used. Alternatively, several other techniques, based upon joints and limb ends of the skeleton of the silhouette, can be used. It is also possible to derive position. Figure 5.17(b) illustrates how the skeleton of a rigid blob-like object can provide a number of key features by which the position and orientation may be determined. For example the centre point of the line joining the skeleton limb-ends B and C, or the skeleton joint F would be suitable places to locate a gripper. The orientation of the object may be determined from the line BC. A good place to grip a flexible object might be to hold it near its limb ends. In this case, a "four-handed" lift, with the grippers placed close to limb-ends A, B, C and D would be appropriate. Since the use of four grippers might be difficult in practice, a "four-handed" lift might use gripping points placed close to A and D.

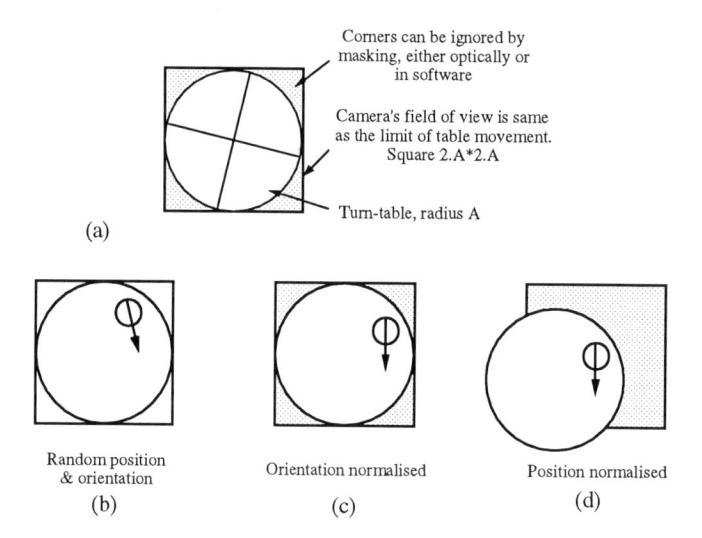

Figure 5.16 Procedure for picking up a randomly placed object, using the overhead camera to guide the (X,Y,θ)-table. (a) The wide-field camera's field of view is a square of side length 2.A. The radius of the circular top of the turn-table is A and the travel is 2A*2A. (b) Object is in random position and orientation. (c) First step: Digitise an image and then determine the orientation of the object. Second step: rotate the table to normalise the orientation. (d) Third step: Digitise another image and determine the position. Fourth step: Shift the table to normalise the position.

Axis of minimum
second moment

Centroid of hole
furthest from centroid

Edge point furthest
from centroid

Y centroid

Centroid of
largest hole

Axis of minimum
second moment   X centroid

(a)

(b)

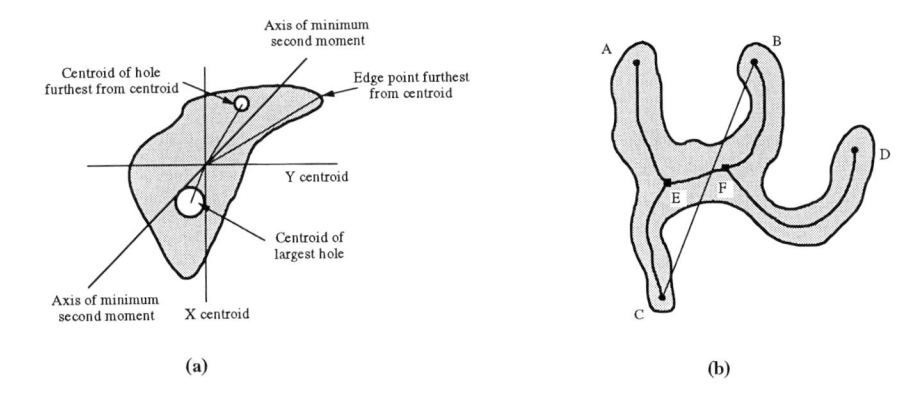

Figure 5.17 Some suggestions for calculating position and orientation parameters for a laminate object viewed in silhouette.

5.8.1 Program

The Prolog+ program listed below is able to normalise the position and orientation of the biggest white object among several others, distributed at random on top of the (X,Y,θ)-table, which is assumed to be black.

```
locate_and_pick :-
    home,           % Go to the table's home position
    grb,            % Digitise an image from the camera
    enc,            % Linear stretch of intensity scale
    thr,            % Threshold at mid grey
    biggest,        % Isolate the biggest blob
    lmi(_,_,Z),     % Find orientation
    Z1 is -Z*100,   % Rescale angle measurement
    writeseqnl(['Rotate table by',Z1,'degrees']),
                    % Tell user what is happening
    table(0,0,Z1),  % Normalise the orientation. Rotate the table
    where_is_table,
                    % Tell the user where the table is now
    grb,            % Digitise an image from the camera
    enc,            % Linear stretch of intensity scale to range
                    % [black, white]
    thr,            % Threshold at mid grey
    biggest,        % Isolate the biggest blob
    cgr(X,Y),       % Find centroid co-ordinates
    convert_table_axes(X,Y,Xt,Yt),
                    % Convert to table coordinates (Xt, Yt)
    table(Xt,Yt,Z1),
                    % Shift the table to normalise position.
    grb,            % Digitise another image
    vpl(128,1,128,256,255),  % Draw vert. line through centroid
    vpl(1,128,256,128,255),  % Draw horiz. line through centroid
    pick.           % Operate pick-and-place arm to pick up object
```

The operation of *locate_and_pick* is illustrated in Image 5.1. The reader may like to consider how this program can be modified to pick up all the objects on the table.

5.9 Grippers

We conclude this chapter with a brief discussion of grippers and how they relate to the selection of a suitable grasping point. We shall briefly consider three types of gripper: *Suction*, *Magnetic* and *Multi-finger*.

When discussing *Suction* and *Magnetic grippers*, we shall limit our attention to objects having flat top surfaces, while for *Multi-finger grippers* we shall assume that the object has sufficiently rough vertical sides to ensure that a friction-based gripper can hold the object securely.

5.9.1 Suction Gripper

The suction gripper is assumed to be a circular rubber cup. Suppose that its diameter is D and that it is being used to pick up smooth flat-topped laminate objects. An obvious place to position the gripper is at the centroid of the object silhouette. (Figure 5.18) If the gripper is small enough, its whole gripping surface will lie within the perimeter of the object. However, if the edge of the gripper overlaps that of the object, the grasp will not be secure, since air will leak into the suction cup. If the object has any holes, these must also be taken into account. It may, of course, be possible to use a small sucker but this does reduce the loads that can be lifted. Here is a very simple program to determine whether gripping at the centroid is safe. It is assumed that a binary image, representing the object silhouette, has already been computed and is stored in the current image.

```
safe_grasp :-
    cgr(U,V),         % Centroid is at (U,V)
    cwp(N),           % How many white pixels in silhouette
    wri,              % Save image for use later
    zer,              % Generate black image
    gripper_size(X),       % Consult database for size of gripper
    draw_disc(U,V, X, white),
                      % Draw white "guard" disc, at (U,V), radius X
    rea,              % Recover image of silhouette
    max,              % OR disc and silhouette images
    cwp(M),           % Count white points. Compare to earlier value
    M is N.           % Test for equality of pixel counts
```

safe_grasp fails if the disc representing the gripper is not entirely covered by the object silhouette.

5.9.2 Magnetic Gripper

A magnetic gripper is more tolerant of overlap than the suction type, just discussed. Hence, it is possible to draw a "guard disc" that is smaller than the gripper itself; provided that the guard disc is entirely covered by the object silhouette, the grasp will be secure. (Figure 5.19) Hence, we can use *safe_grasp* again, except that *gripper_size* yields the radius of the guard disc, not that of the gripper itself. Another approach simply counts the number of pixels which lie within the gripper "footprint" but do not lie within the object silhouette. The following program is a simple modification of *safe_grasp*:

```
safe_grasp1 :-
        cgr(U,V),       % Centroid is at (U,V)
        wri,            % Save image for use later
        zer,            % Generate black image
        gripper_size(X),
                        % Consult database for gripper size
        draw_disc(U,V, X, white),  % Draw white "guard" disc
        rea,            % Recover image of silhouette
        sub,            % OR disc and silhouette images
        thr(0,100),     % Isolate pixels in disc
        cwp(N),         % Count white pixels
        M ≤ 100.        % We can safely allow 100 pixels overlap
```

Note that in Image 5.2(d) the circular suction gripper could not grasp this object securely, due to air leakage in the small regions enclosed within this circle but not within the silhouette. A magnetic gripper placed here and of the same size might well be able to do so.

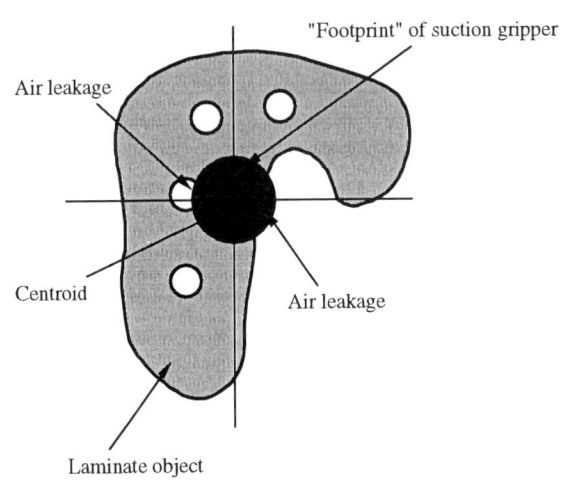

Figure 5.18 Using a suction gripper to hold a flat-topped object. The dark circle represents the suction gripper, which is so large that it overlaps the outer edge of the object and one of its holes. Since air will leak, via the overlap region, this does not provide a safe lifting point.

5.9.3 Multi-Finger Gripper

A multi-finger gripper lifts a laminate object by grasping its sides, not its top surface as both magnetic and suction grippers do. This means that when the gripper is lowered onto the table, the tips of the fingers are actually lower than the top of the object that it is about to lift. This presents the potential danger of the gripper colliding with the object to be lifted or another object placed nearby. However, this can be avoided in a simple and obvious way, by drawing the shadow of the gripper finger tips (or *"footprint"*) onto the image of the table top. In Image 5.2, the fingers are assumed to have rectangular tips. If these footprint rectangles do not overlap any white object in the image of the table top, it is safe to lower the gripper. Whether or not the grip will be secure depends upon friction, the gripping force and the extent of any deformation of either gripper or object to be lifted. (Such features cannot, of course, be understood by a vision system.) The same technique can be used if the fingers are circular, or there are more than two of them. (Also see Section 7.3.4.)

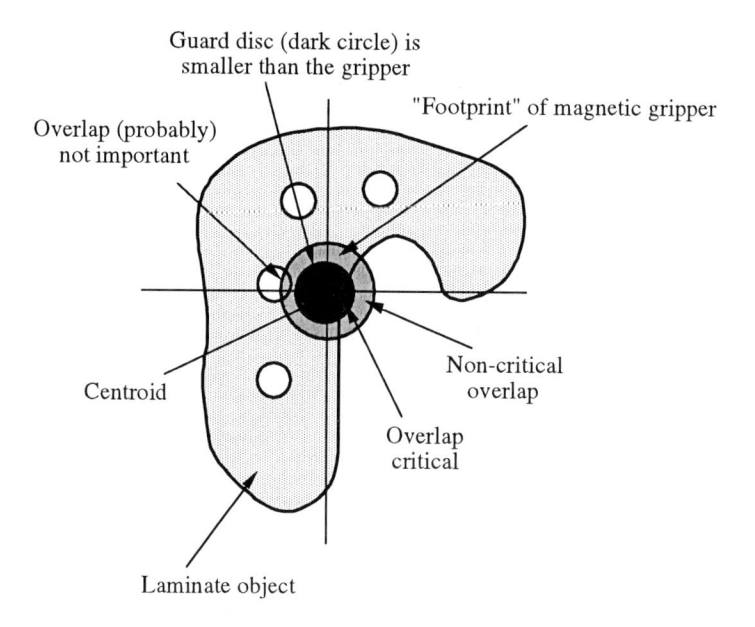

Figure 5.19 Using a magnetic gripper to hold a flat-topped object. The circular stippled area represents the gripper "footprint". Unlike the suction gripper, the magnetic gripper can tolerate a certain amount of overlap, without weakening the grasp unduly. Provided the dark stippled area is entirely covered, the grasp will be strong. However, the overlap can cause other problems: any object falling within the area labelled *"Non-critical overlap"* will be grasped and may be lifted or just moved a little as the gripper is raised.

5.9.4 Further Remarks

When discussing the magnetic gripper, we permitted the gripper to overlap the edge of the object silhouette. This is potentially dangerous but the solution is exactly the same as for the multi-finger gripper: superimpose the gripper "footprint" onto the image and test whether by so doing the gripper will overlap any other objects. There are several further factors to be considered. The first is the nature of the object itself. Clearly, if the object is not ferromagnetic, a magnetic gripper will not be able to lift it safely. On the other hand, if the object has a very rough, pitted, fibrous or "stepped" top surface, a suction gripper will not be able to provide a secure grasp.

If the centroid does not lie within the object silhouette then putting a magnetic or suction gripper there is pointless. For this reason, *locate_and_pick* would have to be modified to pick up horseshoes. It does not take much thought to realise that there can be no single "all purpose" solution to the task of picking up previously unseen objects. The use of Prolog+ allows the best of a finite set of gripping-point calculation methods to be sought and used for a given application. No doubt, the reader can suggest ways in which the various methods for calculating gripping points can be evaluated for their effectiveness. However, it should be understood that there is a limited amount that can be done with any vision system. Sooner or later, slip sensors, strain gauges and other mechanical sensors, will be needed as well, to ensure a firm but not destructive grip.

Clearly, the success of all vision-based methods for robot guidance depends upon the validity of the implicit assumption that the mass distribution of the object is nearly uniform. If it is not, there may be a high torque about the gripper when it is located at the object centroid. The result may be an insecure grip.

A multi-finger gripper suffers from these and other problems. One of the most important of these is that rigid, non-compliant fingers may not hold a rigid object securely. Compliant or rubber-tipped fingers can provide a more secure grasp. Calculating whether or not such a grip will be safe enough for practical use is very difficult, in the general case. However, in certain restricted cases it would be possible to define appropriate tests using Prolog+ programs. For example, the grass-fire transform, applied to the silhouette of the object, would provide some information about how far a compliant gripper would distort during lifting. This would in turn enable the friction forces to be estimated approximately.

5.10 Summary

In this chapter, we have discussed the use of Prolog in the unfamiliar role of controlling external devices, such as lamps, pneumatic devices, video cameras and an (X,Y,θ)-table. This task has been made easier by the use of a specialised hardware device, the MMB module. This is a versatile, low cost interfacing unit which provides facilities for operating a range of electrical, electro-mechanical, illumination and pneumatic devices. The physical layout, control and application of a Flexible Inspection Cell (FIC) have been described. The calibration procedure for the FIC and its use in a simple robotic task have been explained.

In the brief space available in this chapter, we have not been able to discuss in detail other important matters, such as the calibration of other types of robot (e.g. gantry, SCARA, multi-joint articulated arm, serpentine) used in conjunction with a visual guidance system. Nor have we been able explain in detail how to set up and align the FIC with its several cameras.

No practical machine vision system ever operates in complete isolation from the outside world. At the very least, the vision system must operate a warning bell or lamp. However, a more common requirement is for an inspection system which operates some simple *accept / reject* mechanism, such as a solenoid, air-blast, or pick-and-place arm. In another frequently encountered situation, the vision system is required to guide some robotic manipulator, such as a multi-axis articulated arm, or (X,Y,θ)-table. The latter may form part of a manufacturing device, such as a drilling-, milling-, or parts-assembly machine. An increasing number of applications are being found where the vision system generates data for a complex decision-making system which controls the whole manufacturing plant. Management information systems with vision system sensors providing inputs are also being installed. The point of this discussion is that *every* industrial vision system must be interfaced to some other machine(s) in the factory. While the latter may operate in some non-electrical medium (e.g. optical, mechanical, pneumatic, hydraulic, thermal, nuclear, acoustic, x-ray, microwave), they all require electrical control signals from the vision system. The interfacing requirements for industrial vision systems are very varied and, in this chapter, we have discussed some of these.

Of great importance for designers of machine vision systems is the ability to construct prototypes quickly and easily. The Flexible Inspection Cell and general purpose interface unit (MMB module) were developed specifically for this purpose, as was the HyperCard control software for the FIC. These, or facilities like them, are needed for a variety of reasons:

(a) To develop and prove new concepts.
(b) To act as a platform for the demonstration of new ideas to potential customers.
(c) To act as a vehicle for education and training.
(d) To provide an immediate "off-the-shelf" solution for a certain restricted class of vision applications, where the requirements for processing speed, precision of manipulation and image resolution are within defined limits.
(e) To gain familiarity with new system components (e.g. new lighting units, cameras, image processing sub-systems, etc.) in a realistic but controlled environment.

IMAGES

Image 5.1 Operation of *locate_and_pick*. (a) Original scene showing 6 objects. (b) The largest object ("club") has been identified. Its axis of minimum second moment has been calculated and the centroid has been located. The table has been rotated and shifted to normalise both orientation and position. The suction gripper of the pick-and-place arm is then lowered at the point indicated by the crossed lines. (c) Second largest object (sickle-shaped wrench) has been located. (Notice the small circular feature to the left of the white disc. This is one of four screws holding the table top in place.) (d) The third object to be picked up is the pair of scissors. (e) The fourth object to be picked up is the white disc. In this case, the orientation measurement is meaningless. (f) The fifth object to be picked up is a V-shaped plastic component. The rectangular component remaining is an electrical connector and cannot be grasped with this type of gripper.

Image 5.2 Attempting to pick up an automobile connecting rod (con-rod). (a) Silhouette of the con-rod after thresholding and noise removal. (b) Centroid and axis of minimum second moment. (c) Orientation determined by joining the centroid of the silhouette to the centroid off the largest "bay". It is purely fortuitous that this method of determining orientation very nearly coincides with that obtained using the axis of minimum second moment. (d) Circular suction gripper, represented here by a circle located at the centroid of the silhouette. (e) After normalising both position and orientation, a 2-finger gripper can be used to lift the con-rod. The two white bars have been drawn here to represent the gripper footprints on the object plane.

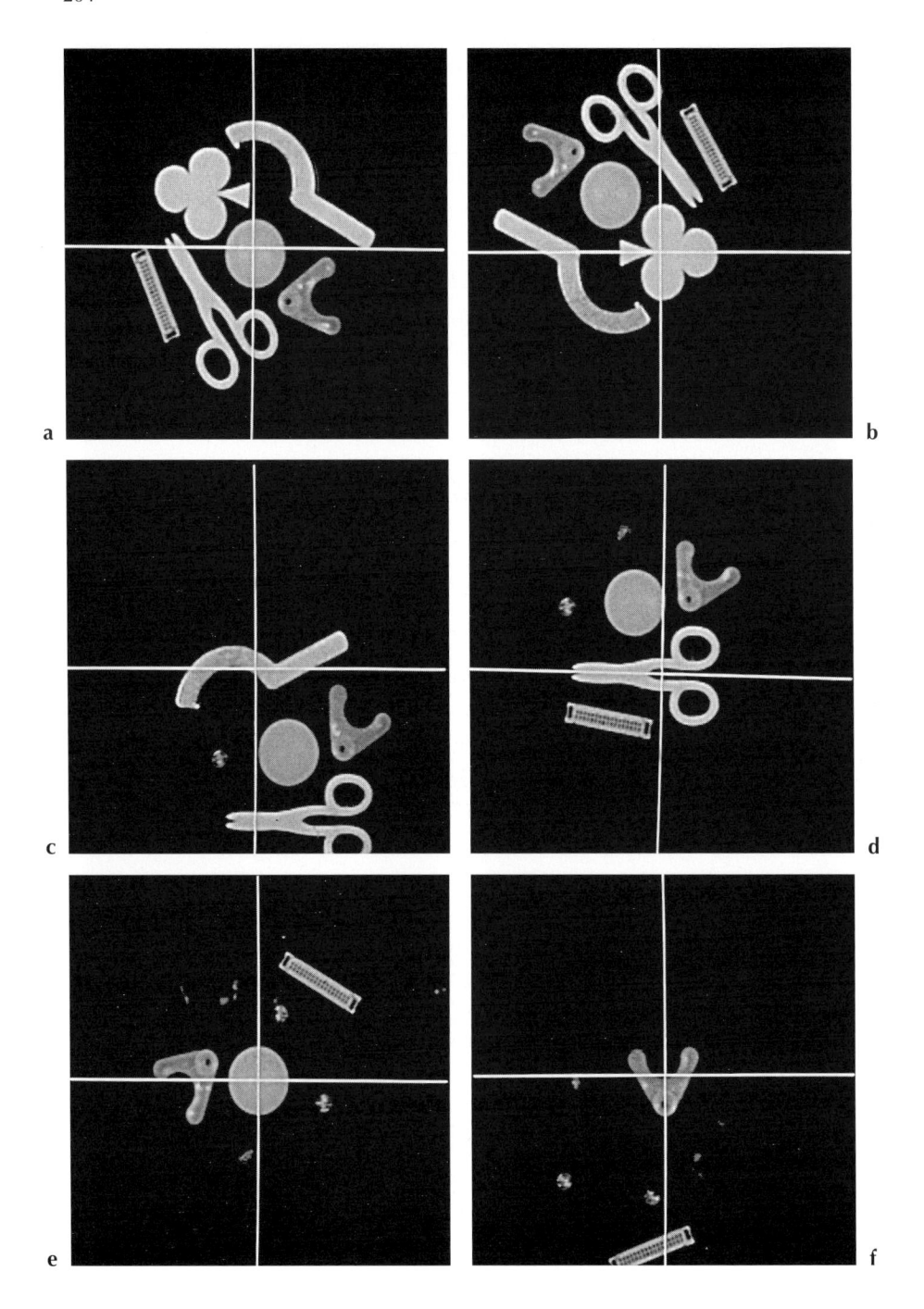

Image 5.1 a–f

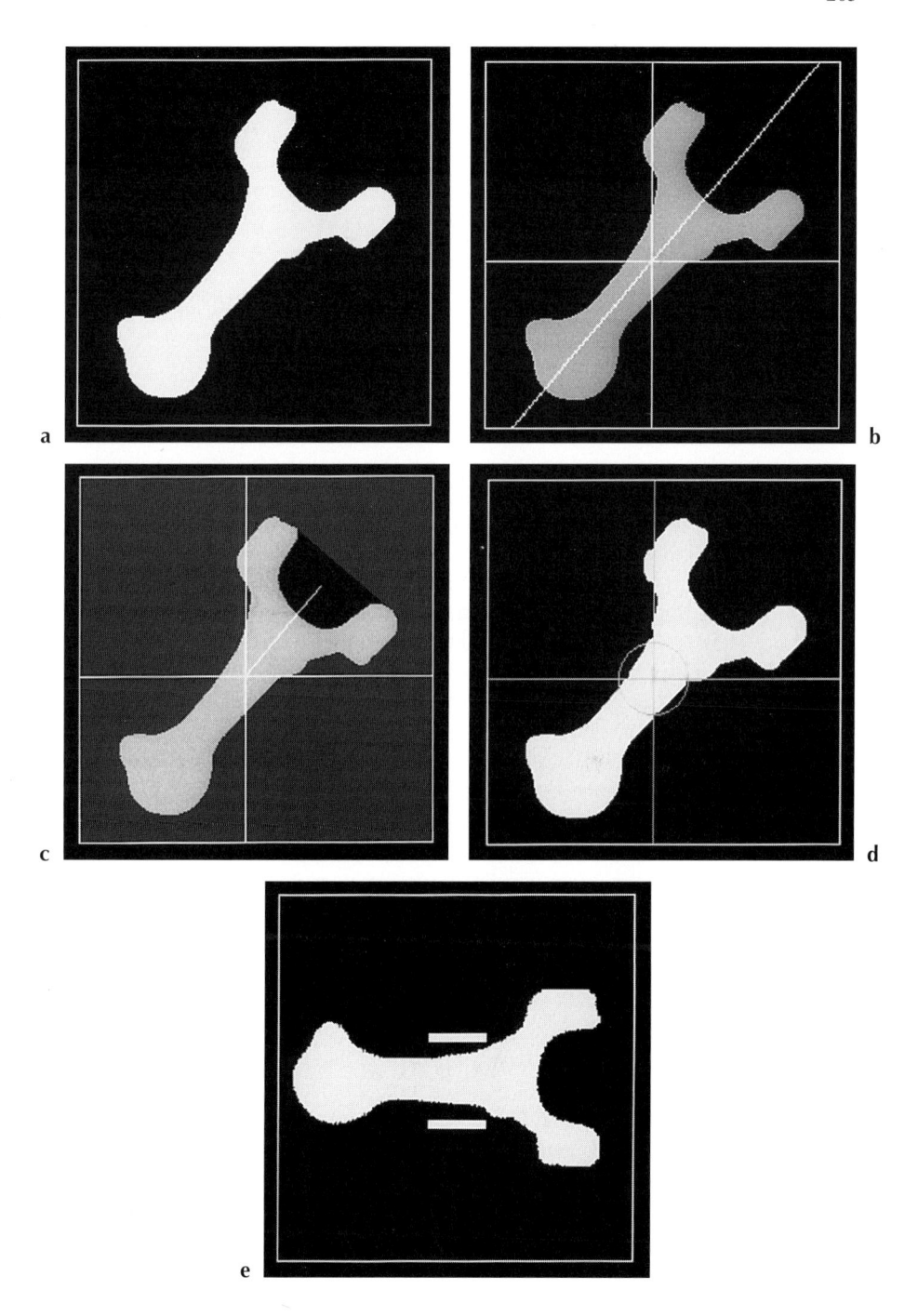

Image 5.2 a–e

6

Colour Image Recognition

6.1 Introduction

Colour vision is undoubtedly of very great value to an organism; the multitude of colours in nature provides ample evidence of this. Brightly coloured flowers signal the availability of food to bees. A great number of trees and bushes attract birds by displaying bright red fruit, so that the seeds will be carried off to some distant location. Butterflies indicate their identity by vivid wing markings. Wasps and many snakes warn of their venomous sting / bite with brightly coloured bodies. On the other hand, chameleons hide their presence by changing colour, while some animals (e.g. certain monkeys) signal sexual readiness, by changing the colour of parts of the body. Colour vision is so fundamental to animals and to us in our every day lives that we tend to take it for granted. We indicate both personality and mood with the clothes that we wear. Many different types of entertainment are enhanced by the use of colour. We attempt to make our food attractive by the careful balancing and selective use of colour. We reserve special coloured signals to warn of danger on the roads, or at sea. Colour is used very extensively to attract customer attention in shop, magazine, television and hoarding advertisements. Not only is colour vision of great value to human beings, it is a source of pleasure and a great many of our artefacts reflect this fact. Numerous industrial products are brightly coloured, simply to be attractive and eye-catching. The ability that we have to see colour makes our knowledge of the world much richer and more enjoyable than it would be if we had simple monochrome vision. The range of manufactured products reflects this fact. It is natural, therefore, that we should now perceive a need to examine the colour of industrial objects. To date, machine vision has been applied most frequently to monochromatic objects, or those in which colour differences are also visible as changes in intensity, when they are viewed using a monochrome camera.

Despite the obvious importance of colour in manufactured goods, machine vision systems that can process coloured images have been used only infrequently in the past. Colour image processing has received very little attention, as far as inspection systems are concerned. Three major reasons have been put forward for this.

(a) *Many industrial artefacts are monochromatic, or nearly so.* For example, uncoated metal surfaces are usually grey. Plastic mouldings, though often coloured, are usually monochromatic. Painting converts a multi-coloured surface into one that is monochromatic. There is a large class of applications in which variations of hue have no importance whatsoever, since they do not, in any way, reflect the fitness of a component to perform its function.

(b) *Some colour inspection problems are inherently very difficult, requiring very precise measurement of subtle colour differences.* Commercial colour image sensors (i.e. cathode ray tube and CCD cameras) are simply too imprecise to allow really accurate measurements to be made. Applications that demand very precise colour discrimination are to be found in the automobile, printing, clothing and food industries.

(c) *Colour image processing equipment is necessarily more complex and more expensive than the equivalent monochrome devices.* Colour cameras are more complicated, lighting has to be better controlled, and a colour image requires three times the storage capacity, compared to a monochrome image of the same spatial resolution.

The main point to note here is that many people have been discouraged in the past from investigating colour image processing systems for industrial inspection. We hope to demonstrate that these arguments are no longer valid. The four main lessons of this chapter are as follows:

(i) Colour recognition can be very useful and for this reason, should be considered for use in a far greater number of industrial applications than hitherto.

(ii) The cost of using colour image processing need not be prohibitively high.

(iii) There are numerous inspection applications that could benefit from the use of "coarse" colour discrimination. Subtlety of colour discrimination is not always needed. In business terms, there is a good living to be made from designing coarse colour discrimination devices

(iv) Novel signal processing techniques have recently been devised that make the recognition of colours simpler and easier to use than hitherto.

6.2 Applications of Coarse Colour Discrimination

In this section, we shall simply list a number of industrial inspection tasks that could benefit from the use of rather coarse colour discrimination. Consider first the task of verifying that there is a crayon of each desired colour in a cellophane pack. Subtlety of colour discrimination is clearly not needed and might well prove to be a disadvantage, because it would highlight colour variations that are of no interest to us. This is typical of the other tasks that we shall list below.

A manufacturer of confectionery wishes to verify that each carton contains no more than a certain number of sweets (candies) wrapped in red paper / foil. Clearly, it is important to monitor the number of expensive sweets and to maintain a good balance of colours (i.e. varieties of sweets). A manufacturer of electrical components wishes to inspect "bubble packs" containing coloured connectors. (These are frequently red, blue and yellow / green.) A common method of packaging that is used for small items like these uses a brightly printed card. A transparent "bubble" forms a case for the components, which can frequently move about quite freely. A large number of stationery goods, household and do-it-yourself items are sold in this type of packaging.

Sensitive paints and adhesive labels are used to indicate whether pharmaceutical / medical products have been opened or exposed to excessively high temperatures. Tamper-proof seals on certain food containers change colour when they are opened. A certain manufacturer of domestic goods has a requirement for a system that can count objects showing the company logo, which is a multi-coloured disc, against a high-contrast black, grey or white background.

Product cartons are often colour coded. For example, products with blue labels might be sold by a certain company in Europe, while red-labelled ones, with a slightly different specification, are exported to USA. Identifying the colour of the labels is of great importance to the pharmaceutical industry, where companies supply goods conforming to legally enforceable standards. In many other industries, it is important, for economic reasons, to identify and count cartons with *"Special offer"* overprinted labels.

A visit to a supermarket, or do-it-yourself store, reveals a host of potential applications for a colour inspection system. Verifying that containers of corrosive chemicals have their tops properly fitted is clearly of importance, in avoiding damage to clothing and personal injury. One plastic container that is sold in the UK and Ireland holds domestic cleaning fluid and is bright blue with a red top. Verifying that such a container has a complete and well fitting top is to the advantage of the manufacturer, retailer and customer. The identification of labels on bottles, jars, cans and plastic containers is another task that would be made easier with colour discrimination. Locating self-adhesive labels, in order to make sure that they are located on the side of a container is another task that is made easier through the use of colour image processing.

Inspecting food products is of very great interest to a large number of companies. For example, a company that makes pizzas is interested in monitoring the distribution and quantity of tomato, capsicum peppers, olives, mushrooms and cheese on the bread base. A certain popular cake is called a Bakewell tart. It consists of a small pastry cup, filled with almond paste, is covered in white icing (frosting) and has a cherry placed at its centre. Verifying that each tart has a cherry at, or near, its centre is a typical food inspection application for a colour recognition system. One of the authors (BGB) has been collaborating for several years with a company that makes large quantities of cake in the form of "unending" strips, with a view to inspecting the decoration patterns [CHA-95]. Colour machine vision systems have been applied to situations such as the harvesting of fruit and vegetables [LEV-88] and their subsequent grading for sale, or processing. Certain foods are often accepted / rejected merely on the grounds of their colour. The preference for foods of certain colours has important consequences to the food processing industry. It is of interest that some countries have different preferences for the colour of food. For example, British people like green apples, Americans like red ones and Italians prefer deep red. Even the colours of food packaging have to be carefully considered. For example, blue on white is a very popular packaging for dairy produce, because the colours suggest the coolness and hygiene of the dairy. The degree to which a product is cooked is often indicated by its colour. The food baking and roasting processes have to be precisely controlled, to ensure that the products emerge with the right colour. A colour sensitive vision system has been developed to monitor how well beef steaks have been cooked [KEL-86]. Another colour sensitive system has been devised which monitors the fat-to-lean ratio of raw meat. [HOL-92] Some applications exist where the *lack* of colour in a product needs to be checked. For example, the presence of minute flecks of bran (red-brown) in white flour can have a detrimental effect upon its baking properties.

6.3 Why is a Banana Yellow?

"In the absence of an observer, there is no such thing as colour." [CHA-80]

Let us begin by considering the physics involved in the question. The visible part of the electro-magnetic spectrum comprises wavelengths between 0.4 and 0.7 µm. (Figure 6.1) A light signal containing components with a non-uniform mixture of wavelengths will appear to be coloured to a human observer. If the energy spectrum is constant for all wavelengths in this range, the observer will perceive white light. The non-uniform spectrum of light reaching the eye of a human observer from a banana is perceived as being yellow. A similar but slightly different spectrum of light emanates from a canary, or lemon. The seemingly trivial question at the head of this section has a deeper meaning than we might at first think. Many people, when faced with such a question, begin, to think about Newton's demonstration of the multiple-wave composition of white light and the fact that bananas reflect light within a certain range of wavelengths, while absorbing others. Curiously, most people, even non-scientists, attempt to answer this question with reference to the *physics* of light absorption. However, light rays are not actually coloured. So-called "yellow light", reflected from a banana, simply appears to us to be coloured yellow, because the human brain is stimulated in a certain way that is different for the way that it is when the eye receives "orange light".

In fact, a banana is yellow because a lot of people collectively decided to call it yellow. As children, we are taught by our parents to associate the sensation of looking at lemons, bananas and canaries (and many other yellow things) with the word *"yellow"*. We do not know anything at that age about the physics of differential light absorption. Nor do we need to so as adults, to communicate the concept of yellow to a person who is learning English for the first time, or to children. A ripe banana "is yellow" because a lot of other people persistently call it yellow and we have simply learned to do the same. There is no other reason than that. In particular, the physics of light has got nothing whatsoever to do with the *naming* of a colour. Of course, physics can explain very well how our eyes receive the particular mixture of wavelengths that we have learned to associate with a certain name.

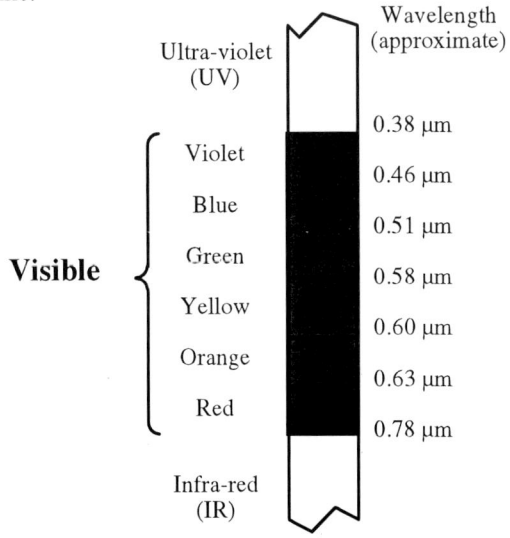

Figure 6.1 The visible part of the electro-magnetic spectrum.

From recent research, it appears that our eyes have three different colour sensors, although this was disputed by some researchers until quite recently. *Trichromacity* is the idea that any observed colour can be created by mixing three different *"primary colours"* (whether we are mixing paint or coloured light) and was commonplace by the mid-18th century, having been mentioned by an anonymous author in 1708. Colour printing, with three colour plates, was introduced by J. C. LeBlon, early in the 18th century, who apparently understood the additive and subtractive methods of mixing colours. By measuring the colour absorption of the human retina at a microscopic scale, it has been possible to determine that there are three different types of pigment in the colour sensing nerve cells in the retina (called *cone cells*), with absorption peaks at 0.43 µm, 0.53 µm and 0.56 µm. It is tempting to identify these peaks in the absorption curves with the Blue, Green and Red sensors, respectively. J. Mollon [MOL-90] explains that *"the use of these mnemonic names has been one of the most pernicious obstacles to the proper understanding of colour vision"*. He also argues that there are two quite distinct colour-recognition mechanisms. However, in the context of a discourse on machine colour recognition, such nice points, about the psychology and physiology of visual perception, need not concern us unduly.

Trichromacity is a *limitation* of our own colour vision and arises, we believe, because there are three different types of colour sensor in the retina. However, it is trichromacity that has facilitated colour printing, colour photography, colour cinema film and colour television. Briefly stated, the central idea of trichromacity is that physically different light sources will look the same to us, provided that they stimulate the three receptor cells in the same ratios. (Figure 6.2) Thus, a continuous spectrum, derived by passing white light through coloured glass, can produce the same sensation as a set of three monochromatic light sources, provided that the intensities are chosen appropriately.

There are also some physiological and psychological disturbances to human colour perception that a well designed vision system would avoid. Some of these disturbances are outlined below, while others are listed in Table 6.1.

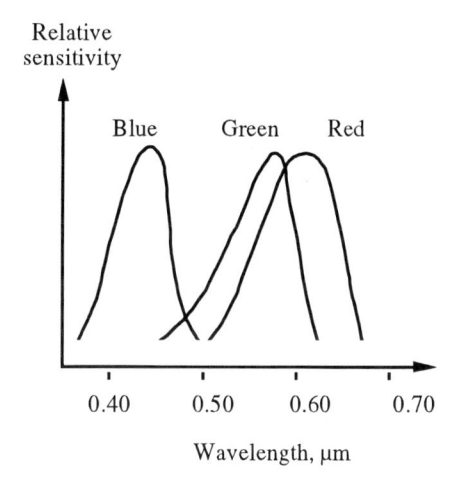

Figure 6.2 Spectral sensitivity curves of the cones in the human eye.

Disease / Chemical	Change in colour perception
Alcoholism	Blue defect
Brain tumour, trauma	Red-green or blue-yellow defects
Malnutrition	All colours
Multiple sclerosis	Red-yellow defects
Caffeine	Enhances red sensitivity & reduces blue sensitivity
Tobacco	Red-green defect

Table 6.1 Changes in the human perception of colours resulting from various diseases and chemical substances.

Dichroism occurs when two colours are present as peaks in the spectral energy distribution curve. One or the other peak dominates, according to the viewing conditions. The proximity of vividly painted objects nearby can affect the perception of the colour of the paler surface. *Metamerism* is the term used to describe the situation when two surfaces are viewed in the same illumination and produce different reflectance spectra but create the impression of being the same colour to a human observer. (See Figure 6.3) *Chromatic adaptation* occurs when the brain makes allowances for changes in illumination and subsequently changes in the colour perceived. *Dark adaptation* effectively switches off the human colour sensors and occurs because different retinal sensors (rods) are used from those used in day-light (cones). Nocturnal scenes appear to be monochrome, because the rods are insensitive to colour variations. It must be remembered that the dark-adapted eye is effectively colour blind, although bright points of light still appear to be coloured.

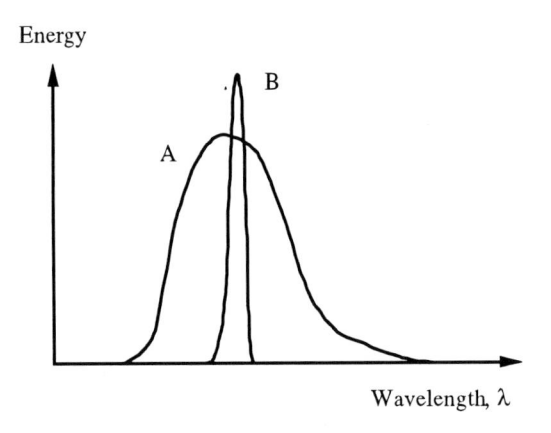

Figure 6.3 Metamerism. Curves A and B represent the spectra from two different surfaces but which appear to be the same colour to a human observer.

Human colour vision clearly varies from person to person, and can vary considerably at different times in a given person's life. A significant proportion (about 7%) of the population is colour blind, with men being more prone to this disability than women. Colour blindness is not always recognised by the sufferer, although it can be detected relatively easily, by simple tests.

A machine vision system would not be subject to such variations in its ability to perceive colours. However, we do require that a human being be available initially, to define the colours that the machine will subsequently recognise. It is important, therefore, that we realise that numerous external factors can alter the human perception of colour. Since the "definition" is so subjective, any two people are likely to disagree when naming intermediate colours, lying between the colour primaries. With this thought in mind, we begin to appreciate how important it is for a colour recognition system is able to learn from a human teacher. This is just the approach that we shall take in the latter part of this chapter.

A theory relating to the perception and reproduction of colour, in photography, cinema, television and printing, has been developed over many years and seemingly forms an integral part of the education for all people who work in these areas at a technical level. However, we shall not discuss this theory in detail, since to do so would merely introduce a major digression from our main theme. The "classical" theory of colour is largely superfluous to our rather specific needs, since we are not concerned at all with the *fidelity* of colour reproduction; we need only a small amount of theory, to put into context the conceptual and practical aspects of a relatively new but very powerful technique for colour recognition.

We are leading the reader towards a technique for recognising colour, that is both conceptually straightforward and conforms to the pragmatic, declarative approach to programming implicit in Prolog+.

6.4 Machines for Colour Discrimination

Apart from the general advantages that a machine vision system has over a human inspector, there are certain factors specific to colour inspection that render a machine more attractive. In any manufacturing industry, there is constant pressure to achieve ever higher production rates and mistakes are inevitably made when people work under pressure. This is particularly true, if a colour judgement has to be made. If people are asked to discriminate between similar colours, such as red and red-orange, they frequently make mistakes. On the other hand, a simple opto-electronic system should be able to make more consistent decisions even when working at high speed. We shall therefore introduce the topic of optical filters before moving on to discuss colour cameras.

6.4.1 Optical Filters

Colour discrimination has been widely applied in machine vision systems for a number of years, using optical filters. These devices can make a major difference to the image contrast derived from a dichromatic scene (i.e. contains just *two* colours, apart from neutral). High performance optical filters are probably still the best and most cost-effective method of improving contrast in scenes of this type and, for this reason, they will be discussed first. Then, we shall be in a better position to understand how information about colour can be obtained from a camera.

Figure 6.4(a) shows the spectral transmission characteristics of members of a typical family of optical filters. These are called *band-pass filters*, since they transmit wavelengths between certain limits, while all other wave-lengths, within the visible wave band, are heavily attenuated. Other band-filters are available that have much narrower and much broader pass bands. (Figure 6.4(b) and 6.4(c).) Long-pass and short-pass filters are also sold. (Figure 6.4(d) and 6.4(e).) So called *interference filters* have a multi-layer sandwich construction and have very sharp cut-off characteristics. Unfortunately, they are sensitive to moisture and heat, both of which cause them to delaminate, so great care must be taken to protect them, particularly in a hostile factory environment. A curious phenomenon occurs, if an interference filter is held up to a source of white light. As the filter is tilted, relative to the line of sight, it appears to change colour. This characteristic can be used to good effect, to fine-tune optical colour filtering systems. Interference filters provide excellent colour discrimination capability at a low cost. However, they provide fixed colour discrimination and cannot be reprogrammed, as we would like. Multi-colour discrimination using interference filters is cumbersome, requiring a carefully calibrated optical bench. In effect, a colour camera provides such a facility. (Figures 6.5 and 6.6.)

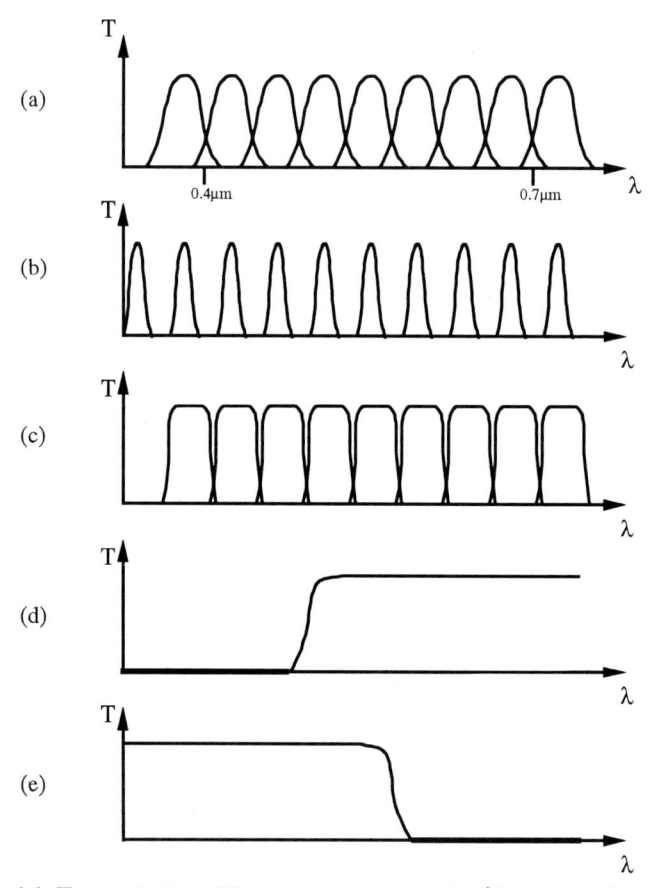

Figure 6.4 Transmission (T) versus wavelength (λ), for various sets of optical filters. (a) Middle-spread band-pass filters, (b) Narrow band-pass filters, (c) Broad-pass band filters, (d) Long-pass filter, (e) Short-pass filter.

6.4.2 Colour Cameras

The traditional *cathode-ray tube* (vidicon family) colour camera has the construction illustrated in Figure 6.5. The optical filters used in a colour camera typically have characteristics similar to those shown in Figure 6.2. Compensation must be made for the non-constant gain of the light sensing tubes, across the visible-light spectrum. In many 3-tube colour cameras, the output consists of three parallel lines, called R, G and B, each of which carries a standard monochrome video signal. Another standard, called *composite video*, exists in which the colour and intensity information is coded by modulating a high-frequency carrier signal. A composite video signal can readily be converted to RGB-format and vice versa. The details of the various video standards need not concern us here and can be found elsewhere. [HUT-71]

A *solid-state* colour camera can be constructed in a similar way to the CRT colour camera illustrated in Figure 6.5, by replacing the CRT sensors with CCD image sensors. Alternatively we can make good use of the precise geometry of the photo-sensor array; by placing minute patches of optical dye in front of the sensor to produce a cheaper colour CCD camera. (Figure 6.6) Some specialised cameras exist for use in machine vision systems applications (e.g. cameras with high spatial resolution, very high sensitivity, extended spectral response, etc.). Although they can produce some very impressive results in certain situations, none of these specialised cameras need concern us here, since we assume that a perfectly conventional colour camera is used by our vision system.

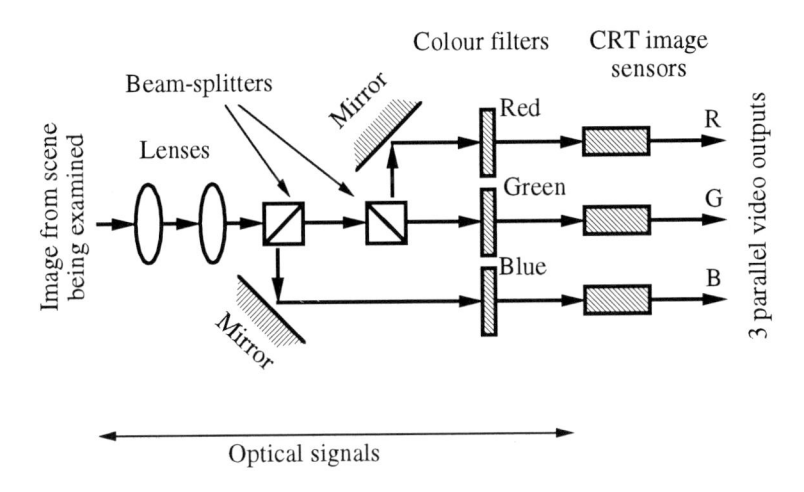

Figure 6.5 Construction of a 3-tube colour camera.

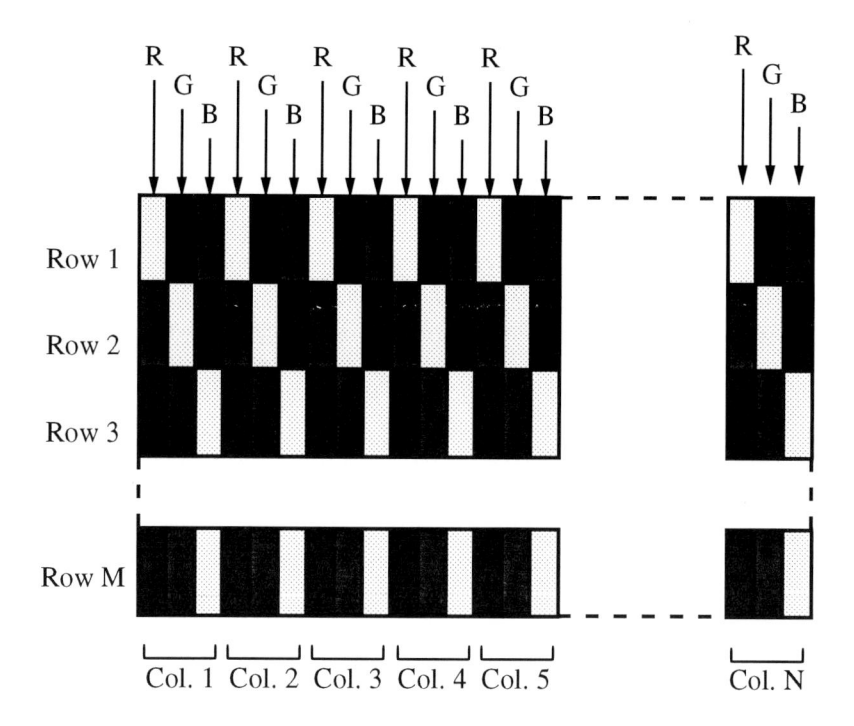

Figure 6.6 Colour mask structure for a single chip solid-state colour camera. Each shaded rectangle represents a photodetector with a coloured mask placed in front of it. R - passes red, G - passes green, B - passes blue.

6.4.3 Light Sources for Colour Vision

Images derived from a well lit object are almost invariably much more easily processed than those obtained from a scene illuminated using highly variable ambient light. Hence, it is important to obtain the optimum illumination, appropriate to the given vision task. There is no point in our taking great care to design a sophisticated image processor, if we do not exercise the same diligence when planning the illumination-optical sub-system. A small amount of effort when choosing the lighting might well make the image processing system much simpler (see Appendix A). We can express this point in the following rule:

Never compensate for a sloppy approach to the design of the optical / illumination sub-system, by increasing the sophistication of the image processing.

This maxim is of particular importance when applying colour image processing to industrial inspection, because the stability and purity of the lighting can make an even greater difference than it does for a monochrome system. *The design of the illumination sub-system is always of critical importance.* Here are some additional rules that we should always bear in mind, when choosing a light source for a colour inspection system.

(a) The brightness of the light source should be sufficient for a clear image to be obtained. It should be borne in mind that a colour camera is likely to be considerably *less* sensitive than a monochrome camera.
(b) The light source should not generate excessive amounts of *infra-red* radiation, since this may damage the object being viewed and / or the camera. High levels of infra-red radiation can also spoil the image contrast, particularly when we are using solid-state cameras. These are particularly sensitive to near infra-red radiation. (Wavelengths in the range 0.7 - 1.0 μm.) If in doubt, place an IR-blocking filter, between the object being viewed and the camera. Such a filter is inexpensive and highly effective.
(c) The light output from the lamp should not contain large amounts of *ultraviolet* radiation, since this can also reduce the image contrast. If in doubt, use a UV-blocking filter.
(d) Keep all optical surfaces clean. This includes the reflectors around the light source and the optical windows used on ruggedised enclosures, used to protect the camera and light source.
(e) Use a light source that is stable. The intensity of the light output from a lamp usually varies considerably during its life-time. Worse than this, the colour of many types of lamp changes, as it ages and as the power supply voltage varies. For this reason, it is essential that we use a regulated power supply, with closed-loop, feed-back control, sensing the light falling on a test target. The latter may be a standard surface (e.g. white paper) that is put in place once in a while (e.g. once every shift), to recalibrate the optical system. The vision system design engineer should always ask the manufacturer for data on lamp ageing. This should be available for any high-quality light source.
(f) Devise some means of calibrating the colour measuring system. Use the colour image processing system, if possible, to calibrate itself and its light source, by viewing some standard target, which must, of course be stable.
(g) Do not assume that a lamp that looks as though it is white actually is; our eyes have a remarkable ability to compensate for changes in the colour of the lighting, so human estimates of lamp colour are very unreliable. Remember, that during chromatic adaptation, the brain unconsciously compensates for minor changes in the colour of the illumination.

It is obvious that certain light sources yield unnatural colours when illuminating certain surfaces. This can occur, for example, if the spectrum of the source is discontinuous, or if the spectrum of the lamp is very different from that of a black-body radiator. Since ultra-violet radiation causes certain materials to fluoresce, this can cause unnatural looking colours to appear on some surfaces.
The NPL Crawford method [PER-91] for calibrating lamps compares the spectral power distribution of a test source with a reference source, by dividing the spectrum into 6 bands, and integrating the luminance over each band. These

values are then expressed as percentages of the total luminance. The excess deviance over a tolerance for each band is totalled and subtracted from a specified number, to give the value for colour rendering. A similar but more widely accepted method for characterising the colour rendering of lamps is the *Ra8 index,* defined by the CIE [CIE-31, PER-91]. The higher the value of the Ra8 index, the better the colour rendering. A value for Ra8 of approximately 90 or above is usually recommended for accurate colour appraisal. A source with a colour temperature of about 6500 K and a colour rendering index Ra8 of 90 or above is required for high quality colour vision. However, the British Standards Institute has produced a specification for illumination for the assessment of colour (BS 950). In addition to these requirements, the British Standards Institute specifies that the chromaticity and spectral distribution of the source lie within certain limits. This specification is intended for very accurate colour appraisal and therefore the tolerances given by the specification may, in some cases be relaxed slightly. Nevertheless, it serves as a good indication of the requirements of a source. Since the specification may not be met directly by certain lamps, or combinations of lamps, colour filters may be used to adjust the colour temperature of a source. The specific illumination set up chosen for a vision system is dependant very much on the situation.

6.4.4 Colour Standards

Many companies develop colour atlases for their own specific needs during manufacturing. For example, bakeries, manufacturers of cars, clothing and paint all require precise colour control and have well developed colour reference systems. For more general use, a colour atlas such as the *Munsell Book of Colour* [MUN] can be useful. The main reasons for using a colour atlas during experimentation are convenience, and the stability and objectivity of measurement. However, there are also a number of disadvantages, associated with the use of a colour atlas. The following points should be borne in mind:

(a) Colour atlases are limited in the number of samples they provide.
(b) Printed colour atlases are subject to fading and soiling. (Careful handling is needed to reduce errors to acceptable levels.)
(c) Colour atlases are limited by the range of dyes and pigments available.
(d) They are unable to represent colours of high saturation.
(e) The lighting must be accurately controlled and stable. Normal room lights are not.

An accurate and controlled source of coloured light is a *monochromator.* This is a device, like a spectrometer, in which white light is shone into a prism, or an interference grating. This separates the light into its spectral components. A thin slit is used to select the a desired narrow band of wavelengths. (See Figure 6.7.)

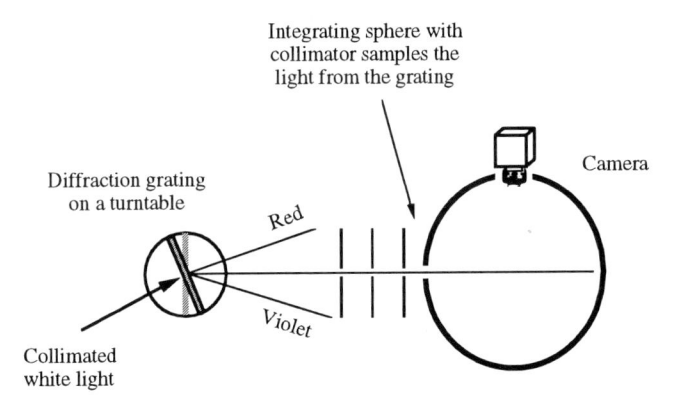

Figure 6.7 A monochromator provides a way of generating coloured light in a repeatable way. A human operator has no difficulty in rotating a diffraction grating (or prism), so that they can find the limits of named colours. The colours that can be produced by this equipment are called *spectral colours*. Notice that there are some colours that cannot be generated in this way, most notably purples and gold.

6.5 Ways of Thinking about Colour

As explained earlier, it is believed by most researchers that human colour perception employs three types of cones in the retina, each being tuned to a different part of the visual spectrum. Colour television cameras also contain three optical filters, which have been specially designed to approximate the spectral response of the cones in the human eye. In this way, the colour response of the camera is made to correspond approximately to that of the eye. It should be realised that almost all colour cameras used in machine vision systems have been designed to have this characteristic, even if their output signals are not displayed as images on a monitor. It seemed natural to investigate whether standard colour television equipment could be adapted to recognise colours in the same way that people do. The transmitted signals derived from a colour camera are formed by the encoding the RGB picture information into two parts: *luminance*, which carries the brightness information, and *chrominance*, which carries the colour information. The latter conveys both *hue* and *saturation* information. The method of coding colour information is different in the NTSC, PAL and SECAM broadcast television systems, although the details need not concern us here.

What is of greater relevance is that, in each of these broadcast standards, all of the colour information is carried in the chrominance signal. Without this, only information about intensity is carried. We are all familiar with the special effects shown on television programmes, especially advertisements, in which a monochrome (i.e. black and white) picture, representing some dull, unattractive scene, is suddenly transformed into a brilliantly coloured image. This is achieved in the editing room, by modifying the chrominance signal. The implication is clear: the information needed for colour perception is contained in the chrominance signal, not in the luminance.

A variety of other representations of colour have been devised, some of which are summarised below. In the following discussion R, G and B represent the outputs from a colour video camera.

6.5.1 Opponent Process Representation of Colour

The *Opponent Process Representation* emphasises the differences between red and green, between yellow and blue, and between black and white. A very simple *Opponent Process* model can be based upon the following transformation of the RGB signals.

$$\begin{bmatrix} (Red\_Green) \\ (Blue\_Yellow) \\ (White\_Black) \end{bmatrix} = \begin{bmatrix} 1 & -2 & 1 \\ -1 & -1 & 2 \\ 1 & 1 & 1 \end{bmatrix} \begin{bmatrix} R \\ G \\ B \end{bmatrix} \quad \ldots(6.1)$$

More complicated transformation models have been proposed in an attempt to account for certain psychophysical effects. The Opponent Process Representation of colour information is a good way to emphasise the difference between red and green regions, blue and yellow regions, and black and white regions in a scene. For example, it has been found to be able to discriminate between the green pepper and the red tomato on a pizza, and to isolate the black olives. [PER-91]

6.5.2 YIQ Colour Representation

Consider the following transformation of the RGB signals:

$$\begin{bmatrix} Y \\ I \\ Q \end{bmatrix} = \begin{bmatrix} 0.3 & -0.59 & 0.11 \\ 0.6 & -0.28 & -0.32 \\ 0.21 & -0.52 & 0.31 \end{bmatrix} \begin{bmatrix} R \\ G \\ B \end{bmatrix} \quad \ldots(6.2)$$

This equation is a simple representation of the NTSC encoding scheme, used in American broadcast television. If the so-called *chrominance angle,* Q, and *chrominance amplitude,* I, are plotted as polar co-ordinates, parts of the IQ plane can be identified with colours as classified by human beings. (See Table 6.2.)

Colour	Quadrant Number
Purple	1
Red, Orange, Yellow	2
Yellow/Green, Green	3
Blue, Blue/green	4

Table 6.2 Associating various colours with regions of the IQ plane.

6.5.3 HSI, Hue Saturation and Intensity

A technique for colour image processing that is enjoying increasing popularity, is based upon the concepts of *hue, saturation* and *intensity*. The HSI representation of colour is close to that method of colour description that is used by humans. When a person is asked to describe a colour, it is likely that they would first describe the *kind* of colour (hue), followed by the *strength* of colour (saturation) and the *brightness* (intensity).

Hue (H) defines the intrinsic nature of the colour. Different hues result from different wavelengths of light stimulating the cones of the retina. In other words, hue is related to the *name* of a colour that a human being might assign. We shall see much more of this in the following pages, where we will define the HSI parameters in mathematical terms. A saturated colour ($S \approx 1$) is deep, vivid and intense, due to the fact that it does not contain colours from other parts of the spectrum. Weak or pastel colours ($S \approx 0$) have little saturation.

6.5.4 RGB Colour Space: Colour Triangle

Consider Figure 6.8, which shows a diagram representing RGB space. The cube, defined by the inequality $0 \leq R,G,B \leq W$, where W is a constant for all three signal channels, shows the allowed range of variation of the point (R,G,B). The *colour triangle*, also called the *Maxwell triangle* is defined as the intersection of that plane which passes through the points (W,0,0), (0,W,0) and (0,0,W), with the colour cube. Now, the orientation of that line joining the point (R,G,B) to the origin can be measured by two angles. An alternative, and much more convenient method is to define the orientation of this line by specifying where the vector (R,G,B) intersects the colour triangle. (The (R,G,B) vector is extended if necessary.) Then, by specifying two parameters (i.e. defining the position of a point in the colour triangle), the orientation of the (R,G,B) vector can be fixed. It has been found experimentally that all points lying along a given straight line are associated with the same sensation of colour in a human being, except that very close to the origin (i.e. very dark scenes) there is a loss of perception of colour. Hence, all (R,G,B) vectors which project to the same point on the colour triangle are associated with the same colour name, as it is assigned by a given person (at a given time, under defined lighting conditions). Moreover, points in the colour triangle that are very close together are very likely to be associate with the same colour label. These are very important points to note and indicate why the colour triangle is so important.

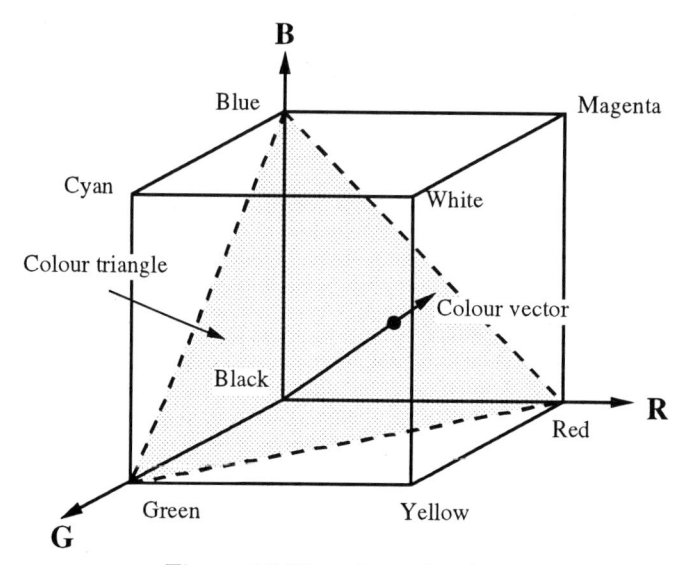

Figure 6.8 The colour triangle.

The following points should be noted:

(a) The quantity (R + G + B) is a useful estimate of the intensity as perceived by a human being when viewing the same scene as the camera. This is a reflection of a principle known to psychologists as *Grassman's Law*. [HUT-71] The length of the vector, given by $\sqrt{R^2 + G^2 + B^2}$, is not nearly so useful.

(b) The colour triangle allows us to use a graphical representation of the distribution of colours, showing them in terms of the spatial distribution of blobs in an image. This is a particularly valuable aid to our understanding the nature and inter-relationship between colours, since we are already very familiar with such concepts from our earlier work on monochrome image processing. It also permits us to use image processing software, to perform such operations as generalising colours, merging colours, performing fine discrimination between similar colours, simplifying the boundaries defining the recognition limits of specific colours, etc.

(c) Since the colours perceived by a human being can be related to the orientation of the (R,G,B) colour vector, a narrow cone, with its apex at the origin, can be drawn to define the limits of a given colour, e.g. "yellow". The intersection of the cone with the colour triangle generates a blob-like figure.

(d) Not all colours can be represented properly in the colour triangle. Trichromaticity is simply a useful working idea, but it does not guarantee that perfect colour reproduction, or recognition, is possible using just three primary colours. Hutson [HUT-71] says

"No three primaries exist for which all of the spectrum and all of the non-spectral colours can be produced. As a result, the colour triangle cannot distinguish between saturated and non-saturated cyans, purples and magentas. The situation is not as catastrophic as the chromaticity diagram appears to indicate. The colours of everyday scenes are generally rather unsaturated and have chromaticities lying near the centre of the chromaticity diagram."

Some colours that are distinguishable by eye are not easily differentiated by a colour camera. For example, "gold" and "yellow" are mapped to the same region of the colour triangle. Vivid purple is mapped to magenta, while vivid cyan is mapped to a paler (i.e. less saturated) tone. In relative terms, these are minor difficulties, and the colour triangle remains one of the most useful concepts for colour vision.

(e) Fully saturated mixtures of two primary colours are found on the outer edges of the triangle, whereas the centre of the triangle, where all three primary components are balanced, represents white. We shall refer to the centre of the colour triangle as the *white point*, since it represents *neutral* (i.e. non-coloured) tones. Other, unsaturated colours are represented as points elsewhere within the triangle. The hue is represented as the angular position of a point relative to the centre of the colour triangle, while the degree of saturation is measured by the distance from the centre.

(f) The following equations relate the HSI parameters to the RGB representation and are derived in [GON-92]. For convenience, it is assumed that the RGB components have been normalised (W=1). (Also see Figure 6.9.)

$$H = \cos^{-1}\left\{\frac{2R - G - B}{2\sqrt{(R-G)^2 + (R-B)(G-B)}}\right\} \qquad \ldots(6.3)$$

$$S = 1 - \frac{3 \cdot \min(R, G, B)}{R + G + B} \qquad \ldots(6.4)$$

$$I = \left[\frac{R + G + B}{3}\right] \qquad \ldots(6.5)$$

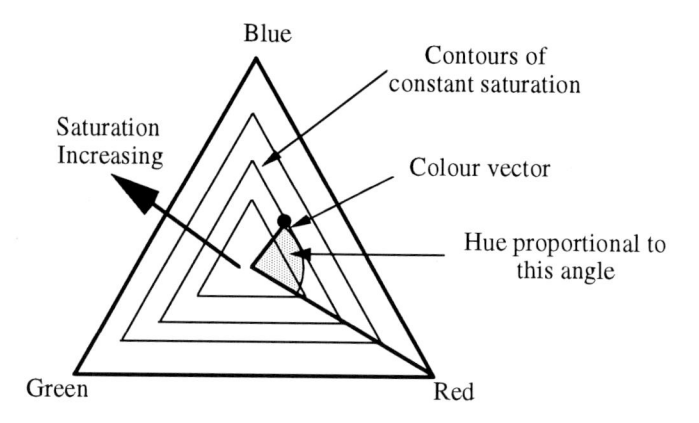

Figure 6.9 Hue and saturation plotted in the colour triangle.

6.5.5 1-Dimensional Histograms of RGB Colour Separations

It is possible to think of the RGB colour representation as being equivalent to separate monochrome images, known as the *RGB Colour Separations*. These can be processed individually, in pairs, or combined together in some way, using conventional image processing operators, such as those described in Chapter 2. One useful method of describing a grey scale image is, of course, the intensity histogram and we can apply the same technique to the R, G or B images. Various techniques have been devised for measuring the shape of histograms and these can quantify skewness, standard deviation, location of the peak. etc. All of these can be applied to any one of the three colour separations, resulting in a set of numbers describing the colour image.

Applications such as monitoring the cooking of beef steaks and pizza crust have been studied in this way. [KEL-86] Although the histogram analysis technique has distinct potential, it gives no indication about the spatial distribution of colours in an image. Another important consideration is the fact the this technique is, in no obvious way linked, to the human perception of colour, making it difficult to interpret the data.

6.5.6 2-Dimensional Scattergrams

The method of analysis about to be described provides an alternative, but generally less effective method of analysis than those based on the colour triangle. Since both methods rely upon the generation and use of 2-dimensional scattergrams, there is some possibility of confusion. It must be understood, however, that they operate in completely different ways and that we therefore need to be careful about our terminology. Later in this chapter, we shall use the term *colour scattergram*. We emphasise that colour scattergrams and 2-dimensional scattergrams are completely different, and must not be confused.

The *2-dimensional scattergrams*, about to be defined, are often able to yield useful information about the nature and distribution of colours in a scene. This is achieved without the specialised hardware needed by certain other methods. Figure 6.10 explains how the 2-dimensional scattergram is computed. In order to describe, in formal terms, how 2-dimensional scattergrams may be generated, let us consider two monochrome images $A = \{A(i,j)\}$ and $B = \{B(i,j)\}$. A third image C can be generated as follows:

Rule 1: Make C black (level 0) initially.
Rule 2: Plot a point of intensity $(Z + 1)$ at the position $[A(i,j), B(i,j)]$ in C, if this point is at level Z beforehand. (Notice that Z may equal 0 initially.)
Rule 3: Repeat *Rule 2* for all (i, j) in the image.

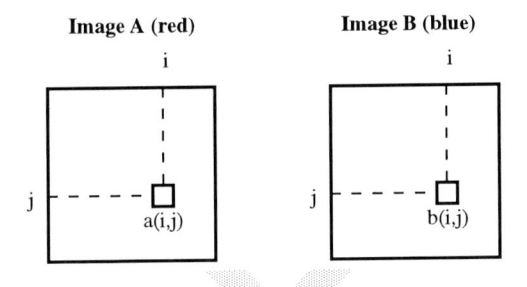

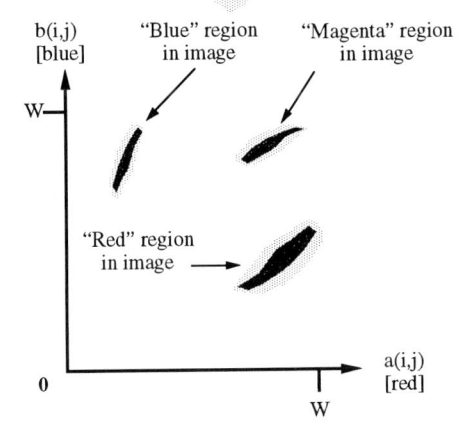

Figure 6.10 Generating the 2-dimensional scattergram.

Suppose that image *A* was generated by the *R* channel and that image *B* was derived from the *G* channel. The resulting (R,G)-scattergram can tell us a great deal about how the red and green signals are related to one another and therefore provide a useful aid to understanding the types colours in the image. It is often possible to identify bright spots in a scattergram. (As we saw in Chapter 2, 2-dimensional scattergrams are also very useful for texture analysis.)

6.5.7 Colour Scattergrams

Let us assume that a certain point (i,j) in the scene being viewed yields a colour vector $(R_{i,j}, G_{i,j}, B_{i,j})$. Furthermore, this vector, or its projection, will be assumed to intersect the colour triangle at that point defined by the polar co-ordinates $(H_{i,j}, S_{i,j})$ where $H_{i,j}, S_{i,j}$ are hue and saturation values calculated using equations (6.3) and (6.4). The point to note is that, each address (i,j) in the scene being viewed defines a point in the colour triangle. We are now in a position to be able to compute the *colour scattergram*:

Rule 1: Clear the current image.
Rule 2: Select a point (i,j) in the input image being viewed.
Rule 3: Compute values for the hue and saturation, using equations (6.3) and (6.4), respectively.
Rule 4: Transform $(H_{i,j}, S_{i,j})$ (polar co-ordinates) into Cartesian co-ordinates, (i,j).
Rule 5: Add 1 to the intensity stored at point (i,j) in the current image. (Hard limiting occurs if we try to increase the intensity beyond 255.)
Rule 6: Repeat Rule 5 for all (i,j) in the input scene.

It is clear that dense clusters in the colour scattergram are associated with large regions of nearly constant colour. (The colour plates, and the half tone images show several examples of colour scattergrams.) A large diffuse cluster usually signifies the fact there is a wide variation of colours in the scene being viewed, often with colours "melting" into one another. On the other hand, step-wise colour changes are typically associated with small clusters that are distinct from one another. The colour scattergram is a very useful method of characterising colour images and hence has a central role in programming the colour filters described in the following section. The reader should be aware of the distinctions between the 1-dimensional, 2-dimensional and colour scattergrams. Failure to appreciate the distinctions will be a severe hindrance to further understanding of colour analysis. The differences may be summarised as being variations in how integration is performed in the colour cube:

1-dimensional scattergrams: integrate one of the 2-dimensional scattergrams along one of its axes.
2-dimensional scattergrams: integrate along one of the axes in colour space.

Colour scattergrams: integrate along a series of lines, all radiating from the origin in RGB space.

6.6 Programmable Colour Filter (PCF)

The Programmable Colour Filter provides an electronic method of filtering colour images. It is fully under software control and as we shall see, is particularly well suited to the style of programming embodied in Prolog+. We shall place the PCF into the theoretical context that we have just discussed. First, however, we shall describe how the PCF may be implemented in electronic hardware. (See Figure 6.11.)

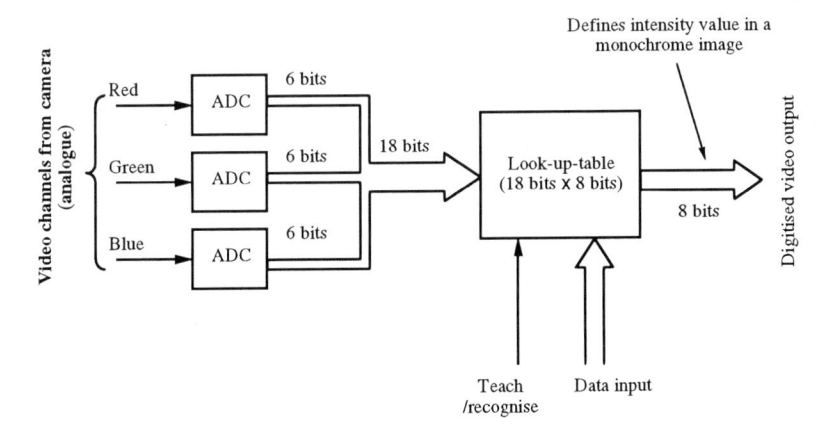

Figure 6.11 Programmable colour filter, block diagram. The output of a colour camera consists of three parallel analogue video signals, called the RGB channels. In effect, these signals define three monochrome images, each of which generates 6 bits / pixel when digitised. Altogether, the digitised RGB signals define a total of 18 bits and these form the address lines entering a random access memory, RAM. The latter implements a simple look-up table (LUT). The colour filter is programmed by changing the contents of the look-up table. An image processor might typically use a larger RAM to store several look-up tables, which can be selected at will, with additional input lines (not shown).

The PCF uses a standard RGB video input from a colour camera and digitises each channel with a resolution of n bits. Typically, n = 6. Thus, a total of 3n (18) bits of data is available about each pixel and together these 2^{3n} (262144) bits form the address for a random access memory, RAM. This RAM is assumed to have 8 parallel output lines and to have been loaded with suitable values, thereby forming a Look Up Table (LUT). By means that we shall discuss later, the contents of this LUT can be modified, enabling it to recognise any desired combinations of the incoming RGB signals. Thus, the filter can be programmed to recognise one, or more, colours. The LUT has a total of capacity of 2^{3n} bytes (256Kbytes) of data. Since the output of the LUT consists of 8 parallel lines, it can define the intensities in a monochrome video image.

Notice that there is no attempt to store a colour image. The digitised RGB video signal is processed *in real time* by the LUT, the output of which can be:

(a) redisplayed as a monochrome image, or
(b) passed through a set of three further Look Up Tables, providing a pseudo-colour display, or
(c) digitised, stored and then processed, exactly as a conventional monochrome signal from a camera would be.

The third of these options is particularly interesting, because it provides us with a very fast, powerful and convenient extension to a monochrome image processing system. The system we are about to describe permits all three of these options. (See Figure 6.12.)

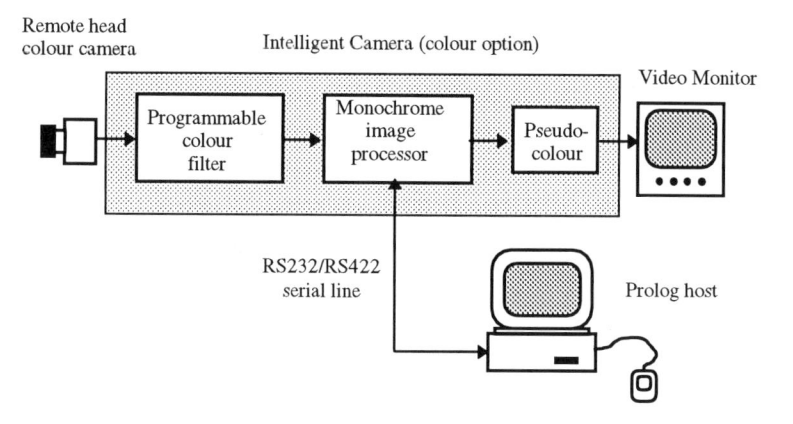

Figure 6.12 The block diagram of the colour image processing system used by the authors in the experiments reported in this chapter. The colour version of the Intelligent Camera [INT] consists of a programmable colour filter, forming a front-end processor, which supplies signals to a monochrome digital image processing system, with a pseudo-colour display unit. Apart from its ability to control these two units, the image processor is in other respects a standard Intelligent Camera.

6.6.1 Implementation of the PCF

One possible implementation of the PCF involves the use of the Intelligent Camera [INT, PLU-91]. The colour filter is normally programmed interactively, using a dialogue based upon a personal computer. First, a monochrome image is digitised and displayed. The user then draws a mask around a region of interest, within the input image. Colours in this region are then analysed. The (monochrome) intensity histogram is then displayed and the user is invited to define intensity limits, based upon his interpretation of the histogram. Pixels having intensities lying between these limits will later be taken into account when choosing the contents of the LUT. Pixels generating intensities outside these limits will simply be ignored. All pixels within the masked region and whose (monochrome) intensities lie between these limits are then analysed and a *colour scattergram* is generated. This typically consists of a set of bright points scattered against a dark background. The colour scattergram is displayed in the current image. The user then defines one set of colours to be recognised, by interactively drawing a closed contour, usually around the main cluster in the colour scattergram. Minor clusters and outlier points are normally ignored. (See Image 6.1.) Finally, the region within the contour drawn by the user is used, by the colour filter software, to define the LUT contents. A wide tolerance for the colours recognised is obtained, if the user draws a contour larger than the dominant cluster in the colour scattergram. A small enclosed region means that a smaller set of colours will be recognised. The procedure for programming the colour filter is, in our experience, far from easy to use and requires a great deal of skill on the part of the user to obtain good results. In its recognition mode, the output of the colour filter is in the format of a digitised video signal, representing a multi-level *grey-scale* image. The output "intensities" normally (but not always) consists of a set of discrete levels, each one representing a different recognised colour. Since the PCF is able to operate *in real time* on a digitised RGB video signal, it does not add to the processing time of *any* image processing operations.

6.6.2 Programming the PCF

Consider Figure 6.13. The position of a point in the colour triangle can be specified by the parameters[1] U and V, which can be calculated from R, G and B using the formulae:

U = (R - G) / [$\sqrt{2}$. (R+G +B)]

and

V = (2.B - R - G) / [$\sqrt{6}$. (R+G+B)]

To see how these equations can be derived, view the colour triangle normally (i.e. along the line QPO, the diagonal of the colour cube). When the vector (R,0,0) is projected onto the colour triangle, the resultant is a vector V_r of length

---

[1] These parameters are not to be confused with those used in the CIE Uniform Chromacity Scale (UCS)-system.

R√(2/3) parallel with the R' axis. In a similar way, when the vector (0,G,0) is projected onto the colour triangle, the result is a vector V_g of length G√(2/3) parallel to the G' axis. Finally, the vector (0,0,B) projected into the colour triangle forms a vector V_b of length B√(2/3) parallel to the B' axis. A given colour observation (R,G,B) can therefore be represented by the vector sum $(V_r+V_g+V_b)$. Finally, U and V can be calculated, simply by resolving V_r, V_g and V_b along these axes.

Let us now consider the mapping function Γ(.) given by:

$$\Gamma((R,G,B)) = ((R - G)/(\sqrt{2}.(R+G+B)), (2.B - R - G)/(\sqrt{6}.(R+G+B)))$$

Γ(X) projects a general point X = (R,G,B) within the colour cube onto the colour triangle. Clearly, there are many values of the colour vector (R,G,B) which give identical values for Γ((R,G,B)). The set of points within the colour cube which give a constant value for Γ((R,G,B)) all lie along a straight line passing through the origin in RGB space (O in Figure 6.13). Let us denote this set by Φ(U,V), where ∀X: X ∈ Φ(U,V) → Γ(X) = {U,V}. The colour scattergram is simply an image in which the "intensity" at a point {U,V} is given by the number of members in the set Φ(U,V).

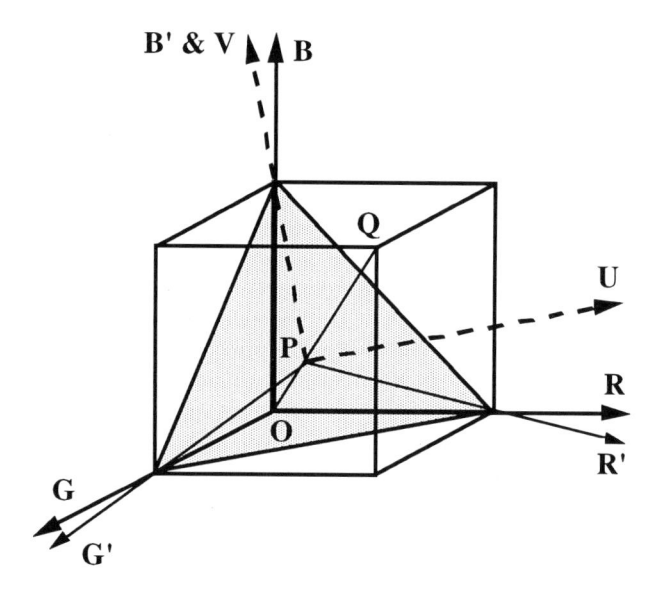

Figure 6.13. Showing the relationship between the RGB- and UV-coordinate axes. The vectors R', G' and B' all lie in the UV-plane, which also contains the colour triangle.

The details of the process of programming the PCF are as follows (see Figures 6.14 and 6.15):

(i) *Project all RGB vectors onto the colour triangle*, which of course contains the *colour scattergram*. (Use the Prolog+ predicate *plot_scattergram*.)

(ii) *Process the colour scattergram*, to form a synthetic image, S. (Image S is not a "picture of" anything. It is merely a convenient representation of the distribution of colours within the input).

(iii) *Project each point, Y, in image S back into the colour cube*. This process is called *Back-projection*. Every vector X, within the colour cube, that shares the same values of hue and saturation as Y, is assigned a number equal to the intensity at Y. The values stored within the look-up table are obtained by back-projecting each point within the colour triangle through the colour cube. Any points not assigned a value by this rule are given the default value 0 (black).

Back-projection is embodied within the Prolog+ predicate, *create_filter*. This predicate takes as its "input" the set of points lying within an equilateral triangle, T, within the current image. (Points outside T are simply ignored.) Triangle T corresponds in position to the colour triangle as it is mapped into the current image, by applying *plot_scattergram*. Understanding the details of how *create_filter* works is not essential for using the PCF. It is more important to realise that *create_filter* simply generates a set of values and stores them in the LUT. It should be noted that *create_filter* is able to program the PCF, using *any* image that may be shown to it. This may, be a colour scattergram, derived using *plot_scattergram*, from a complicated coloured scene. The scattergram may be "raw" or processed, for example, either by smoothing or removing outliers. Alternatively, certain patterns can be created using the image processor as a graphics generator and then applying *create_filter*. (Plate 1) As we shall see later, some particularly interesting and useful effects can be produced in this way. Access to PCF commands are available in Prolog+, through a set of items in a pull-down menu. (See Table 6.3.)

The standard colour filtering techniques listed in Table 6.3 could satisfy the needs of a significant proportion of applications. In many cases, an understanding of the theoretical basis of *create_filter* is unimportant. Certainly, the Prolog+ programmer who simply wants to use the PCF to recognise familiar "named" colours, such as "yellow", "orange" or "red" has no need to understand how *create_filter* works, since standard Prolog+ programs already exist. In some instances, however, more specific colour recognition is required. For example, it may be necessary to train a PCF to recognise application specific colours, such as "banana yellow" or "leaf green". The higher level operators, now embodied in Prolog+ programs have greatly simplified the task of training the PCF. Even so, the task is somewhat easier and certainly less mysterious, if the user understands the theoretical issues involved.

232

B — Z is on the outer edge of the colour cube

The value stored at Y is equal to the total number of points lying along OZ. Notice that Y lies in the colour triangle

(a)

(b)

(c)

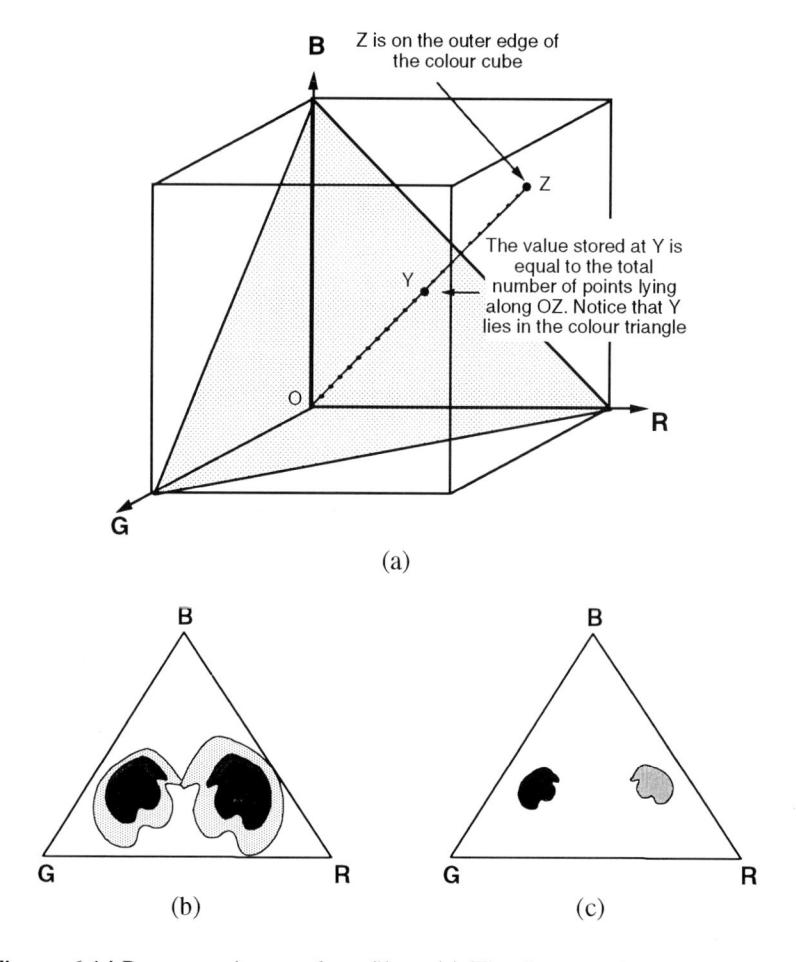

Figure 6.14 Programming a colour filter. (a) The first step is to compute the colour scattergram. The points lying along OZ are counted. This defines the value at Y in the colour scattergram. (b) The colour scattergram is presented to the user in the form of a grey-scale image in which intensity indicates how many pixels were found of each colour. (c) The second step is to process the colour scattergram. The steps represented diagrammatically here are thresholding and blob shading, using *label_blobs*. (d) The colour triangle is shown here within the colour cube. The process of programming the colour filter is to "back-project" each point, Y, through the colour triangle, so that each point lying along the line OY is given the same value as Y. The blobs shown in (c) each contain many points of the same intensity. The effect of "projecting" a blob through the colour cube is to set all points lying within a cone to the same intensity. (e) Two intensity limits (L1 and L2) are specified by the user. (L2 may well be set to 255, in which case, it has no practical effect.) When the LUT contents are being computed, points within the colour triangle are not "back-projected" into either of the two corners, shown shaded here.

(d)

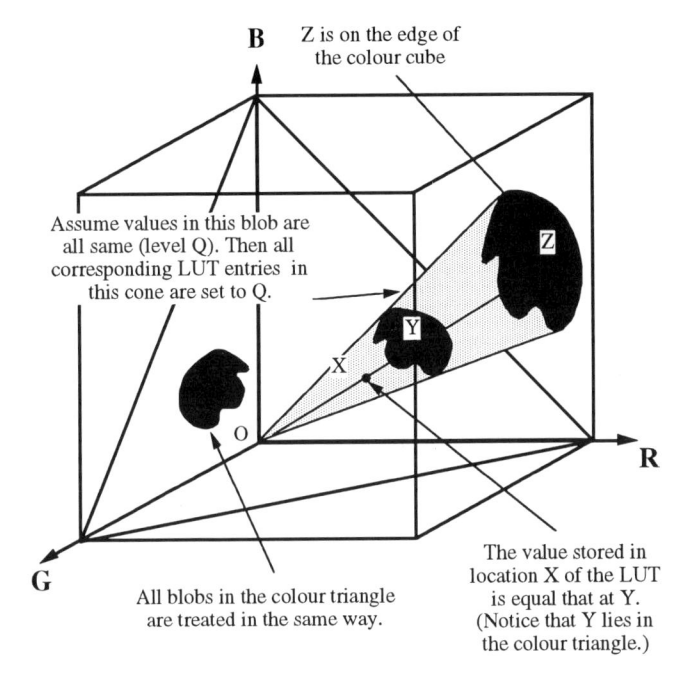

(e)

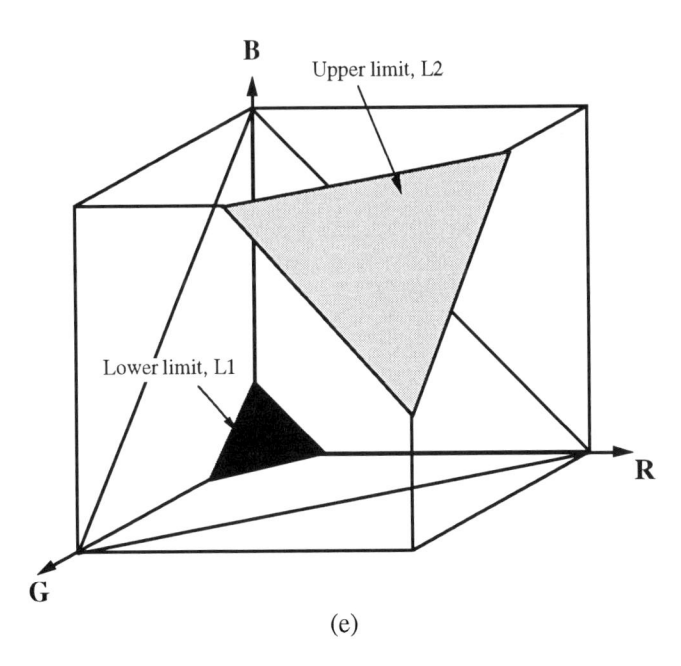

Figure 6.14 (Cont'd).

Menu item	Function and Prolog+ predicate name
Display one image - for photography	Display the current image only.
Reset colour processing system	Switch to normal (monochrome) mode. [*pcf_normal*, *pseudo_colour(off)*]
Live video image on colour monitor	Used to set up colour camera & monitor.
Pseudo-colour OFF	Switch pseudo-colour OFF. [*pseudo_colour(off)*]
Pseudo-colour ON	Switch pseudo-colour ON. [*pseudo_colour(on)*]
All 3 colour channels	Add R,G,B channels; monochrome image processing.
Red channel	Select RED colour channel for monochrome image input.
Green channel	Select GREEN colour channel.
Blue channel	Select BLUE colour channel.
Digitise 3 colour channels	Half resolution R, G, B colour separations placed in 3 quadrants of one image. [*colour_separation*]
Check colour camera light levels	Used during camera set up.
Two dimensional colour scattergram	2-D scattergram. User selects pair of colours.
Colour scattergram	Plot the colour scattergram in the current image. [*plot_scattergram*]
Draw colour triangle - reference only	Line drawing of colour triangle (*draw_triangle*).
Clear colour filter	[*initialise_pcf_lut*]
Colour filter ON	Load file & activate a named colour recognition filter.
Colour filter OFF	Deactivate colour filter.
Save colour filter	Give PCF a name and save it.
Learn colour within region	Learn colour(s) associated with a defined region of the current covered by blob scene. A binary image is used as a mask, to select the area to be used when training the PCF. [*learn_with_masking*]
Create PCF from current image	Design colour filter from a binary scattergram in the current image. [*create_filter*]
Design PCF by learning	Interactive design of a colour filter. Used for high precision colour recognition.
Colour generalisation	Generalise colours. [*generalise_colour*]
"hue" PCF	Filter recognises 256 colours [*hue*].
"hue" PCF plus pseudo-colour	Program "hue" PCF with pseudo-colour ON.
"Primary" colours	Filter recognises 7 colours: red, green, blue, yellow, cyan, magenta and neutral.
Saturation PCF	PCF which measures saturation by measuring the distance from the white point in the colour triangle.
Measure colour similarity	PCF which measures similarity to a single unique colour, defined as a point in the colour triangle. [*colour_similarity1* or 2, selected by user]
Approximate colour scattergram	Represent the colour scattergram as a set of
Rebuild colour scattergram	circles for storage & reconstruction in Prolog+

Table 6.3 Pull-down menus for operating the colour filter interactively. The names of Prolog+ predicates mentioned in the text are given in square brackets.

6.6.3 Recognising a Single Colour

Consider a very simple scene consisting of a single spectral colour, for example, vivid yellow, with uniform illumination everywhere. What happens when the PCF is trained on such an image? The colour scattergram, created during the training process, contains a single compact cluster. In its recognition mode, the programmed PCF may well be applied to the video signal derived from a more complex scene, containing the same shade of yellow and other colours. Only the particular shade of yellow encountered during training will be recognised. (Pixels of this shade of yellow are shaded white in the output image.) All other colours will be ignored. (Shaded black) Very dark and very bright yellow regions will also be ignored. (Remember that both upper and lower intensity limits are applied during the back-projection process outlined in Figure 6.14. (Take special note of Figure 6.14(e).) In the standard Prolog+ predicate, *create_filter,* the limits are set to 32 and 255. This choice of parameter values has not been found to be restrictive in any way.) When the PCF designed to recognise a particular shade of yellow is applied to a polychromatic scene, only that same shade of yellow will be detected; all other colours, including slightly different shades of yellow, will map to black. (Plate 2.)

6.6.4 Noise Effects

When a PCF that has been trained on one scene and reapplied to the same scene, it is often found that the filter output is noisy; some pixels in what appears by eye to be a region of uniform colour are mapped to black. (Plate 2) On a live video picture, some pixels are seen to scintillate. There are several possible reasons for this:

(a) Pixels which generate points in the colour scattergram close to the edge of the main cluster, will sometimes be displayed as black. Camera noise adds an unavoidable jitter in the values generated at the RGB signal outputs.
(b) Recall that hard upper and lower intensity limits are defined during training of the PCF. Some very dark pixels, below the lower intensity limit will be shaded black, even though they are of the particular tone that the PCF is supposed to recognise. A similar situation holds around the upper intensity limit.
(c) Specular reflection on wrinkles on the surface of the object being viewed is a prime cause of noise-like effects. The very high intensities produced by glinting cause black spots to appear in the PCF output image.
(d) The colour scattergram often consists of a compact cluster, with a diffuse "corona". Outlier points are often specifically excluded, by generating a small blob which covers only the dense centre of the cluster, prior to applying *create_filter.*

(e) Colour edges produce outlier points in the colour scattergram. As a result, sharp colour edges may become jagged, in the PCF output image.

6.6.5 Recognising Multiple Colours

The ability to recognise a single colour would be rather limiting. To avoid this restriction, there are two additional features of the PCF that we have not yet discussed. It is possible to teach the filter progressively, by applying *create_filter* to several different images in succession. This allows training of a PCF that can recognise multiple colours, to progress in manageable steps. For example, the user may wish to teach the PCF the general concept of "yellow", by training it successively on examples of grapefruit, bananas, lemons, canaries, etc. Alternatively, several blobs, possibly with different intensity values, can be placed in the same image, prior to applying *create_filter*. (See Figure 6.14(d))

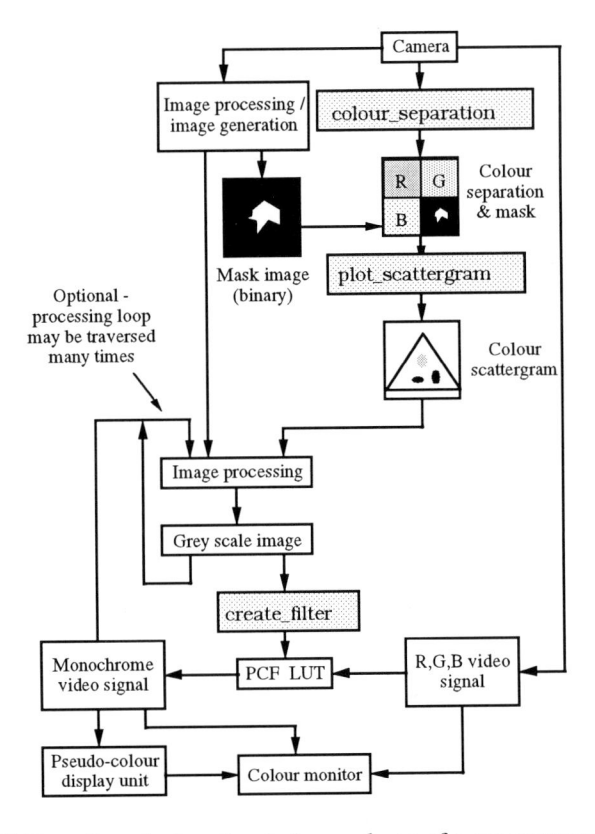

Figure 6.15 Data flow during the design and use of a programmable colour filter. The role of the mask is explained in Figure 6.20.

In order to distinguish different colours, it is possible to assign the PCF output to different levels. For example, the user might *arbitrarily* assign colours to the following intensity levels in the PCF output image:

blue	47	green	115
canary yellow	165	lemon yellow	175
red	215	etc.	

It must be emphasised that any such assignment is purely arbitrary. However, we shall see, in the next section, that some assignments are more useful than others, especially when they are used in conjunction with a pseudo-colour display.

6.6.6 Pseudo-Colour Display for the PCF

The role of the pseudo-colour display is illustrated in Figure 6.12. Now, consider Figure 6.16. The image that forms the input to the pseudo-colour unit is presented to three look-up tables, which define the RGB components in a colour image, displayed on a monitor. The contents of the look-up tables can be adjusted at will, to provide a pleasing / convenient mapping between intensity and colour. (See colour plates.)

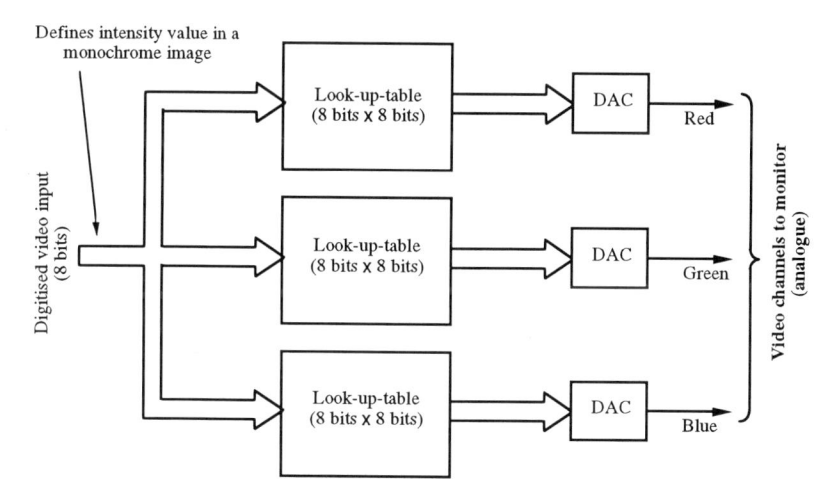

Figure 6.16 Organisation of the hardware for displaying images in pseudo-colour. Each box labelled "DAC" represents a digital-to-analogue converter.

The particular pseudo-colour mapping function shown in Plate 1 and Table 6.4, has one special merit. When the image sensor output is connected directly to the monitor (*ctm*), the user is able to determine when saturation occurs very easily. (The display monitor shows white pixels.)

Pseudo-colour can be specially helpful when working on colour recognition using a PCF. With care, it is often possible (and very useful), to match pseudo-colours to real colours. For example, all of the "red" points in the input image may be mapped by the PCF to a single value, such that, when it is applied to the pseudo-colour display, they all appear to be red. Hence, a display, which shows only a very few distinct pseudo-colours, can often approximate the original scene, which contains innumerable true colours. (Table 6.4) This is very convenient for the user, who can then relate the PCF output directly to the original scene being viewed. Several of the images in the colour plates are displayed in this way.

It may not be possible, or desirable, to use pseudo-colour in this way, if a subtle colour discrimination is required. For example, suppose that we wish to separate "canary yellow" from "lemon yellow". In this situation, there is no point in trying to make the pseudo-colours reflect the true colours, which are very similar. In this case, it is probably better to use quite different pseudo-colours, such as red and blue. (Plate 3)

The Prolog+ predicates for operating the pseudo-colour display system are *pseudo_colour(on)* and *pseudo_colour(off)*.

True colour	PCF output level	Pseudo-colour displayed
Very dark	0	Black
Blue	47	Blue
Cyan	95	Cyan
Green	115	Green
Yellow	160	Yellow
Red	215	Red
Magenta	250	Magenta
Neutral	255	White

Table 6.4 Showing the relationship between the true colours in a scene and the PCF output levels, which generate the corresponding pseudo-colours. The numbers given in the second column are the *approximate* values needed by the Intelligent Camera, to generate the pseudo-colours given in column three.

6.6.7 Recent Teaching of the PCF Dominates

An important feature of the PCF is the fact that the contents of the look-up table are defined in a serial manner, so that recently acquired data over-writes older information. To illustrate this point, imagine that we are training the PCF to recognise the colours of fruit. Suppose that we train first it to recognise lemons and that "lemon yellow" is represented by a set S_l of points in the colour scattergram. Next, we train the PCF to recognise grapefruit, for which the set of scattergram points is S_g. We shall assume (with some experimental evidence to support the hypothesis) that S_l and S_g have some common elements. While the recognition of grapefruit will be accurate, some parts of a lemon may be incorrectly attributed to grapefruit. If the PCF were trained to recognise several additional types of yellow fruit, we might well find that S_1 contains no members that are not also members of other sets. In this event, "lemon yellow" will never be identified. (Figure 6.17) Clearly, this raises two important questions:

(a) Is this likely to be a serious problem in practice?
(b) If so, how can the difficulty be overcome?

So far, the authors have not encountered any real difficulty, as a result of this phenomenon. However, it must be understood that this effect is problem specific; in some applications, it might be troublesome, while in many others, it simply does not occur at all. Clearly, it would be possible to write a Prolog+ program that is able to warn about the occurrence of scattergram overlap. So far, we have ignored the possibility of combining colour recognition with structural (e.g. shape) information. We will examine this issue later.

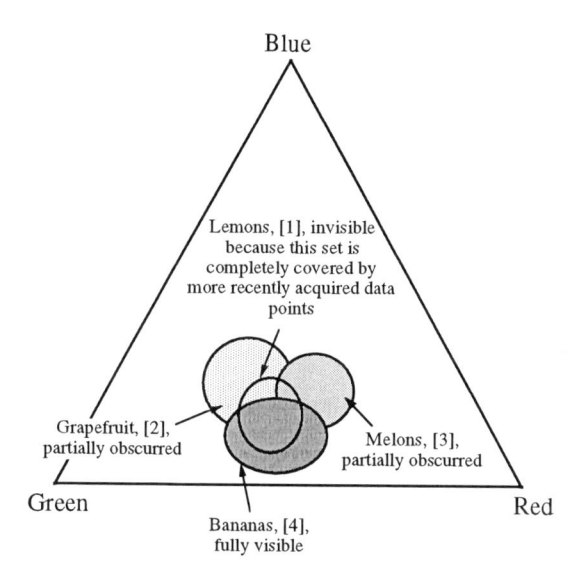

Figure 6.17 Recent learning dominates in a PCF. The colour recognition system is taught to distinguish several types of yellow fruit. These are presented in the following order: 1, lemons; 2, grapefruit; 3, melon's; 4, bananas. Of these, only bananas will be recognised correctly. Other information, such as object size and / or shape is needed to resolve this.

6.6.8 Prolog+ Software for Operating the PCF

The predicate *create_filter* is just one of many operators that have been written to control the PCF. A range of other facilities is provided in the form of pull-down menus. Even when programming the PCF using Prolog+, it is inevitable that there will need to be a high degree of interaction between it and the user. For example, when programming the filter to recognise "banana yellow", a human being must be available, as a teacher, to define what is / is not "banana yellow". Moreover, a human being can very quickly evaluate a PCF and can easily determine whether it achieves its design objective of recognising certain colours and distinguishing them from others. Since such processes as these are very difficult to automate, the authors devised a set of interactive tools, based on Prolog+, for programming and evaluating colour filters. (Figure 6.18)

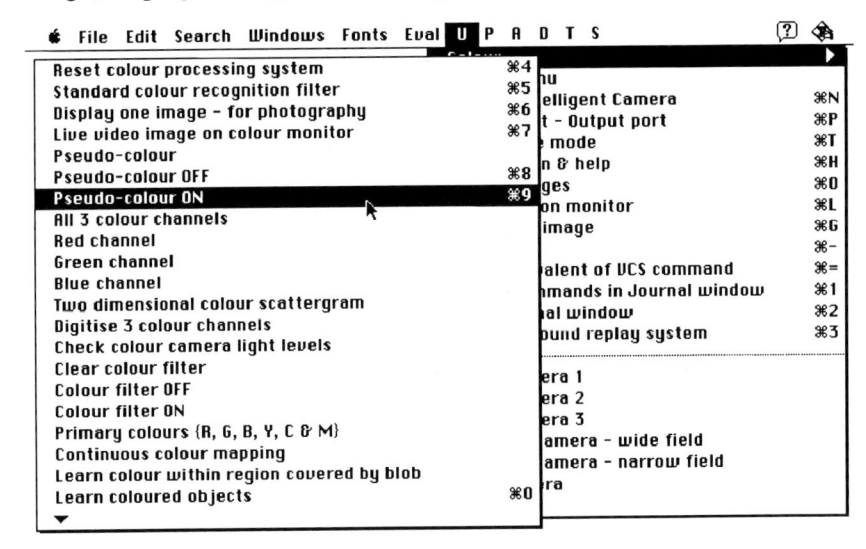

Figure 6.18 The *Colour* sub-menu, appears under the *Utilities* menu.

It should be understood that, once the PCF has been programmed, any subsequent calls to *grb* will result in an image being digitised via the PCF. For this reason, our programs, which appear later in this chapter, do not *appear* to contain any reference to colour recognition. Colour recognition via the PCF is implicit in the *grb* command. In the same way, a *real-time* display of live video, via the PCF is possible using the command *ctm*. We repeat the statement that the use of the PCF does not reduce the processing speed in any way whatsoever.

A suitable PCF could be programmed to recognise "yellow" objects, all other colours being ignored. To achieve this, the yellow regions in the scene being viewed would be mapped to white and other colours to black in the output image. Thus, by performing the operation *grb*, we obtain a *binary* image. (This process takes about 65 ms with our present computer configuration.) Hence, a simple program to detect a banana might look something like this:

```
banana :-
      grb,              % PCF already programmed to recognise yellow
      biggest,          % Ignore any small spots in PCF output image
      size(banana),     % Check blob size is within limits for banana
      shape(banana).    % Check that blob shape is OK for banana
```

No doubt, the reader can suggest various techniques for verifying that the shape and size of the biggest yellow object are both commensurate with that object being a banana. More programs like this will be presented later. It is possible to generate a pattern with *continuous* shading, as a prelude to applying *create_filter*. The Prolog+ sequence [*hic(128,92), enc, 3•sqr, create_filter*] draws an intensity cone centred at the white point in the middle of the colour triangle. The sub-sequence [*enc, 3•sqr*] simply rescales the image intensities. The resulting PCF is able to provide a crude measure of saturation. In addition to *create_filter*, there are several other dedicated predicates for controlling the PCF.

Plot Colour Scattergram

plot_scattergram plots the colour scattergram in the current image, from the RGB colour separations, plus a binary image, which acts as a mask. (The significance and use of the mask image will be discussed later.) Prior to evaluating *plot_scattergram*, it is necessary to load four half-resolution, sub-images into the four quadrants of the current image, as shown below. (See Images 6.11 and 6.12.)

Red	Green
Blue	Mask

The predicate *colour_separation* does just this.

```
% The Current Image Initially Contains The Binary Image Which Forms
% The Mask.
colour_separation :-
        shrink(50,50), % reduce image size to 50% along X & Y axes
        shift(bottom_right),
                       % place in bottom right quadrant of image C.
        video(red),    % select R video channel
        grb,           % digitise image
        shrink(50,50), % reduce image size to 50% along X & Y axes
        shift(top_left),
                       % place in top left quadrant of image C.
        video(green),  % select G video channel
        grb,           % digitise image
        shrink(50,50), % reduce image size to 50% along X & Y axes
        shift(top_right),
                       % place in top right quadrant of image C.
        video(blue),   % select B video channel
        grb,           % digitise image
        shrink(50,50), % reduce image size to 50% along X & Y axes

        shift(bottom_left),
                       % place in bottom left quadrant of image C.
        rea.           % read composite image back into image A
```

The colour scattergram can then be generated by calling *plot_scattergram*. Notice that *plot_scattergram* does not alter the contents of the PCF look-up table.

Draw Colour Triangle Outline

draw_triangle simply draws a geometric figure in the current image, thereby providing a means of calibrating the colour scattergram.

Clear LUT

The predicate *initialise_pcf* sets the contents of all elements in the PCF look-up table to zero.

Store Current LUT

store_lut stores the current LUT contents in battery-backed RAM, in a named file.

Reload Stored PCF

get_lut recovers a stored PCF from battery-backed RAM.

Reverting to Monochrome Operation

The predicate *pcf_normal* switches the PCF back to normal monochrome operation. The corresponding LUT effectively computes the function (R+G+B).

6.6.9 Programming the PCF using the Colour Scattergram

One naive procedure for programming the PCF is to plot the colour scattergram, using *plot_scattergram* immediately before calling *create_filter*. However, this is not a good idea, for the simple reason that the colour scattergram is a multi-level grey-scale image, in which intensity indicates frequency. It is far better to threshold the colour scattergram first. This will normally create a set of "blobs". (Image 6.1) Clearly, we would expect that, if the threshold value is well chosen, each cluster in the colour scattergram would generate just one major blob, plus perhaps a few small satellite blobs representing outliers. The result of applying *create_filter* to a multi-blob image is that all of the colours represented by those blobs will be mapped to white. Suppose that we shade the blobs first. A simple expedient is to do so using *label_blobs*, which has the effect of giving each of the blobs a different intensity value; big blobs are given high intensity values, while small ones become dark. Applying *create_filter* now will program the PCF to discriminate between colours. (Figure 6.19) The output levels generated by the PCF are quite arbitrary and can be chosen for the convenience of the user.

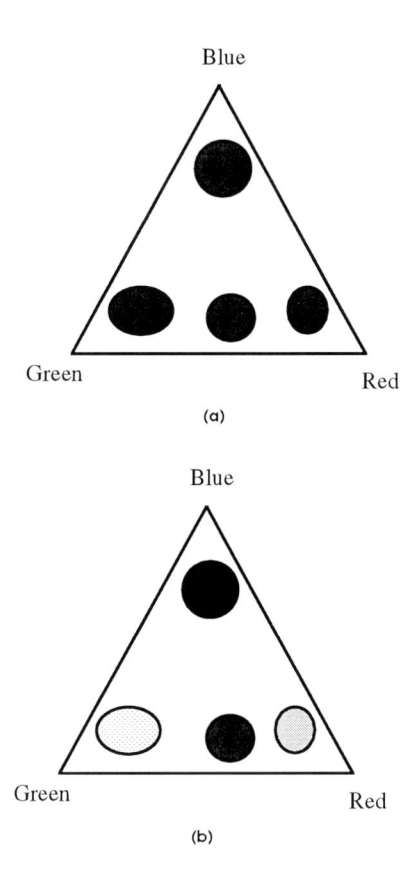

Figure 6.19 Using *label_blobs* to distinguish between different colours. (a) The PCF derived from this image, by applying *create_filter*, is unable to distinguish red, green blue and yellow. This occurs because all four blobs have the same intensity. (b) The PCF derived from this image is able to distinguish red, green blue and yellow. This image can be derived from (a) by applying *label_blobs*.

6.6.10 Programming the PCF by Image Processing

We may not want to generate the colour scattergram for the whole of the camera's field of view; we may prefer to concentrate instead upon certain regions that are of special interest and deliberately ignore others. The definition of *colour_separation* allows the use of a mask image specifically for this purpose. The mask simply controls *plot_scattergram*; pixels that are white in the mask image will contribute to the scattergram, while black pixels do not. (Figures 6.15 and 6.20) The following program trains the PCF to recognise whatever colours are "covered" by the white areas of the mask image.

```
% The mask is in the current image initially
learn_with_masking :-
    colour_separation,     % Generate 4 sub images of half
                           % resolution
    initialise_pcf,        % Set all LUT contents to zero
    plot_scattergram,      % Plot colour scattergram in current
                           % image (A)
    threshold,             % Create binary version of colour
                           % scattergram
    label_blobs,           % Shade blobs - optional
    create_filter.         % Set up LUT contents from current
                           % image
```

In the previous section, we mentioned that it is possible to generate blob-like figures by thresholding the colour scattergram. A cluster in the (grey-scale) scattergram may give rise to one large blob and several smaller "satellite" blobs. The latter can be a nuisance, because they generate "noise" effects. However these very small blobs are easy to eliminate, using *big_blobs*. Here is a program for learning the dominant colour in a scene, with automatic noise removal and masking.

```
pcf_with_noise_cleanup :-
    grb,                   % Digitise an image
    create_binary_image,   % Create the mask image in the current
                           % image
    colour_separation,     % Generate 4 sub images of half
                           % resolution
    initialise_PCF_lut ,   % Set all LUT contents to zero
    plot_scattergram ,     % Plot the colour scattergram
    blur                   % Low pass filter
    thr(16),               % Fixed value thresholding. Arbitrary
                           % choice
    big_blobs(20),         % Eliminate blobs with < 20 pixels,
    3•exw,                 % Expand white regions
    label_blobs,           % Shade blobs according to their sizes
    create_filter.         % Set up LUT contents from current
                           % image
```

6.6.11 "Hue" PCF

It is evident from the discussion earlier, that the HSI representation of colour has considerable merit. The principal reason is that hue can be related to the names that we give to colours. It is possible to use the PCF to measure hue, albeit with some modification of the term. The *hue* predicate, defined below programs the PCF so that its output gives a measure of hue. (See Plate 1(b).)

```
hue :-
    wgx,               % Intensity wedge
    cartesian_to_polar,
                       % "Bend" wedge into circular pattern
    hil(0,0,1),        % Black causes colour discontinuity when
                       % pseudo-colour is used. Avoid it.
    hil(255,255,254),
                       % White causes colour discontinuity when
                       % pseudo-colour is used. Avoid it.
```

```
psh(-64,-64),    % Shift image
shrink(50,50),   % Reduce image size to 50% along X & Y axes
psh(0,-36),      % Shift image
scroll_menu(['Choose white level parameter?'], ['0', '4',
'8', '12', '16', '20', '24', '28', '32'],['24'],X),
X =[Y],          % Decode answer from previous line
pname(Z,Y),      % Convert character string to number
draw_disc(128,92,Z,255),
                 % Draw white disc at (128,92), radius = Z
create_filter.
                 % Set up LUT contents from current image
```

Mask image, generated using normal image processing methods, or by user drawing a closed contour

RGB colour separations:

Red Green Blue

Mask B / A

Apply **plot_scattergram**

Blue

Contribution to the intensity here received from B but not A

Colour scattergram

Green Red

Process the colour scattergram and then apply **create_filter**

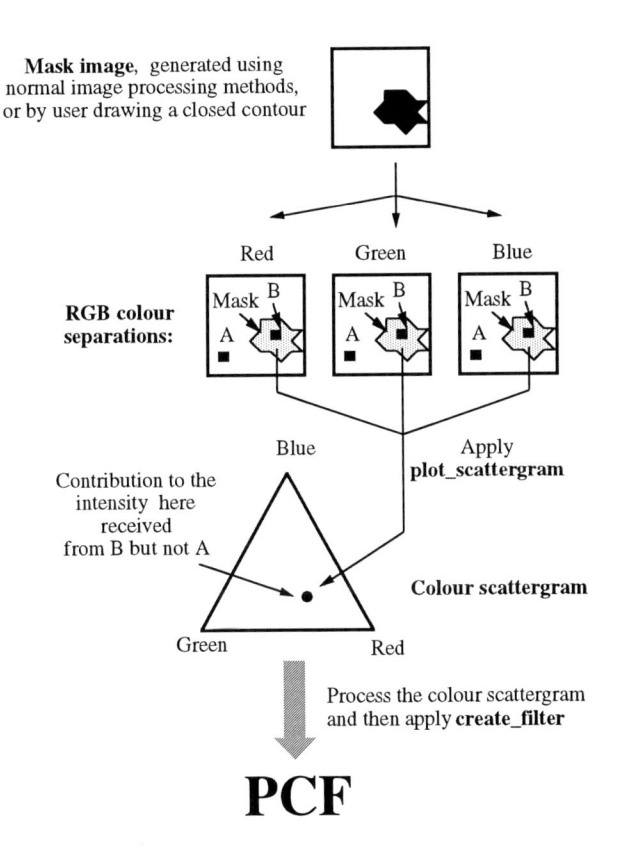

PCF

Figure 6.20 A mask may be used to limit the area of the input image which contributes to the colour scattergram. Hence, the PCF will learn to recognise only those colours within the region of interest covered by the mask.

The intensity in a circular wedge pattern, like the one generated by the first six lines of *hue*, is proportional to angular position, measured relative to a horizontal line through the centre of the colour triangle (co-ordinates (128,92)). Lines 3 and 4 eliminate values 0 and 255, which are represented by black and white respectively in pseudo-colour. If black and white were not suppressed in this way, one row of each shade appears at the 3 o'clock position in the colour triangle and would cause noise-like effects when the PCF is applied. A white disc is drawn at the centre of the colour triangle and is responsible for the PCF mapping neutral shades (very pale, non-saturated colours) into white. For obvious reasons, the colour filter generated by *hue* will be called the *"hue* PCF".

Let P denote the output of the *hue* PCF and H the value of the hue, as defined by Equation 6.3. These two quantities are related as follows:

$$P = \frac{255}{360}((H-30) \bmod 360)$$

The *hue* PCF output has a discontinuity in the *magenta* region of the colour triangle. (Plate 1(b)) This can occasionally cause some minor problems, although these can often be overcome in a straightforward way. (Plate 3)

In many situations, the *hue* PCF is able to provide an adequate basis for colour recognition, without the need to resort to learning. (See colour plates.) It should be noted that surfaces of the same hue but with different (high) degrees of saturation are indistinguishable to this filter. Black and dark grey are mapped to black, while brighter neutral shades are mapped to white by the *hue* PCF.

Pseudo colour-triangle is a term which will be used to refer to a synthesised pattern, superimposed on the colour triangle. (See Plate 1 and Image 6.3.) Applying *create_filter* to a pseudo colour-triangle generates a fixed-function PCF. A variety of interesting and useful functions, in addition to measuring hue, can be implemented thus.

6.6.12 Analysing Output of the Hue PCF

Figure 6.21(b) shows, in diagrammatic form, the histogram of an image generated by the *hue* PCF. (Also see Plate 1(e).) This is an example of what will be called a *colour histogram*. Well defined peaks in the colour histogram indicate that the image contains a number of distinct, well-defined colours, rather than continuously varying colours, which blend into each other. A series of intensity thresholds can then be chosen, so that the PCF output can be quantised into discrete levels, thereby enabling the various colours in the input scene to be separated. A multi-peak histogram, like that shown in Figure 6.21(b), is easy to analyse, either by eye or using a Prolog+ program, to select appropriate threshold parameters. Fixed-value thresholding, applied to the PCF output, is "safe", in the sense that the resulting image does not vary significantly with changing illumination. This is in sharp contrast to the use of fixed-value thresholding applied directly to the camera output.

Another stratagem is to apply the histogram equalisation operator (*heq*) to the *hue* PCF output. This often produces a very interesting effect, in which regions of nearly constant colour are all "stretched", so that their colour variations are all made more obvious. With some justification, the latter process could be termed *colour equalisation*, since it enhances subtle changes of hue within regions of nearly constant colour, while contracting the differences between such regions.

This can be a powerful tool for observing subtle colour changes in a scene. In a similar way, the "linear" contrast enhancement operator (*enc*) could be applied to the PCF output, in lieu of *heq*. It is certainly well worth investigating both of these possibilities, in practice. Similarly, we might apply various other contrast enhancement operators, such as *sqr, [neg, sqr, neg], log, exp*, etc. to the PCF output image. The *hue* PCF and, as we shall see later, certain other colour filters based upon fixed pseudo colour-triangles, are able to yield valuable insight, when they are used interactively. In view of this reliance on interaction, it is difficult to explain all of the possibilities that exist.

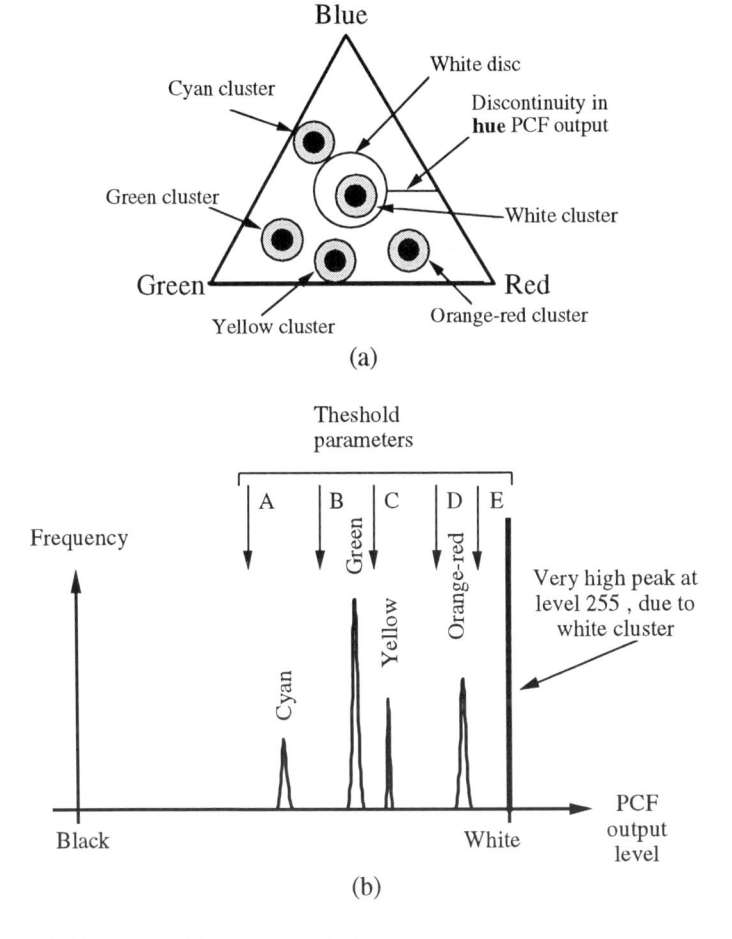

Figure 6.21 Colour histogram. (a) Colour scattergram for an hypothetical polychromatic scene. There is one cluster in the scattergram for each colour in the input scene, including white (neutral). (b) The colour histogram is the result of applying *hgi* to the output of the *hue* PCF. Placing intensity thresholds at A, B, C, D, E provides an excellent basis for separating these five colours.

6.6.13 "Segmented" PCF

The *hue* PCF provides a convenient and straightforward means of analysing colour variations. The *segmented* PCF is even simpler, providing for the recognition of only six broad categories of colour (i.e. red, green, blue, yellow, cyan and magenta), plus neutral. It does not provide the user with the same opportunities for experimentation as the *hue* PCF does. As a result, it is less demanding of the user but it is very coarse in its ability to discriminate colours. Nevertheless, it is well worth trying the *segmented* PCF in the study of any new application, since it is very easy to use.

The pseudo colour-triangle for the *segmented* PCF can be generated by drawing a series of polygons (using *vpl*), filling (*blb*) and shading them (*hil*). (Image 6.3) Finally, the central white disc is drawn using *draw_disc*. The Prolog+ program for drawing the pseudo colour-triangle is straightforward and does not warrant detailed attention.

6.6.14 Measuring Colour Similarity and Saturation

Given a suitable grey-scale image as input, the predicate *create_filter* will create a PCF that is able to recognise up to 256 different colours. As an example of this, suppose that we perform the following sequence of operations.

```
hic(128,92),      % Draw an intensity cone
heq,              % Histogram equalisation
create_filter.
```

The resulting PCF will produce an image in which the output level indicates the degree of saturation. The scale is not quite the same as that defined by Equation 6.4. Another more accurate approximation of the saturation is provided by the predicate *saturation*, defined thus:

```
saturation :-
      zer,            % Black image
      neg,            % Picture is all white now
      vpl(128,92,129,92,0),
                      % Draw single black spot at [128,92]
      gft,            % Grass-fire transform
      create_filter.  % Program the PCF
```

Next, consider the pseudo colour-triangle shown in Image 6.11(g). Intensity in this image measures the Euclidean distance from the centre point of the base of the colour triangle. This point corresponds to "archetypal" yellow. Hence, the intensity in this pseudo colour-triangle indicates the "yellow-ness" and the output of the corresponding PCF indicates how much yellow is present at each point in the scene being viewed. Clearly, the same concept can be applied to measure other colours, such as red, green, blue, cyan, etc. that can be easily located in the colour triangle.

It is, of course, possible to generalise the idea, so that the distance from any arbitrary point, (X,Y), in the colour triangle is represented by the intensity in the pseudo colour-triangle. Perhaps the simplest way of doing this is to draw an intensity cone (*hic*) with its centre at the point (X,Y).

```
colour_similarity1(X,Y) :-
    hic(X,Y),       % Draw intensity cone centred at (X,Y)
    sqr,            % Optional. Possibly use other transforms
    create_filter.  % Generate the PCF LUT from the current image
```

An alternative is to use the grass-fire transform (*gft*). Notice the similarity between the following program and *saturation*:

```
colour_similarity2(X,Y) :-
    zer,            % Create black image
    neg,            % Negate. Makes image all white
    vpl(X,Y,X,Y,0),
                    % Make (X,Y) black.
    gft,            % Grass fire transform
    create_filter.  % Generate the PCF LUT from the current image
```

When a new image is digitised, the PCF output measures the "similarity" between colours in the input scene and that single colour represented by the point (X,Y). Image 6.11(g) shows a pseudo colour-triangle for measuring the "purity" of the three primary colours R, G and B.

6.6.15 Detecting Local Colour Changes

It is possible to extend the ideas implicit in *colour_similarity2*, so that local colour changes in a complex coloured scene are made clearly visible. Here is the program for generating a pseudo-random, pseudo colour-triangle. (Image 6.4(a)) The parameter, N, defines the complexity of the pattern created in the pseudo colour-triangle. An essential feature of the image generated by *subtle_colour* is that the pseudo colour-triangle has a high intensity gradient almost everywhere.

```
subtle_colour(N) :-
    zer,            % Generate black image
    random(N),      % N points at random positions in current
                    % image
    thr(1),         % Remove intensity variations in random image
    3.exw,          % Ignore dense local clusters. Adjust the
                    % looping parameter to taste
    condense,       % Reduce blobs to their centroids
    neg,            % Negate
    gft,            % Grass fire transform
    enc,            % Enhance contrast in current image
    create_filter.  % Generate the PCF LUT
```

Let us consider how the *subtle_colour* PCF might be used. It is best applied to a scene which is nearly constant in time and with *local* changes of colour. Consider, for example, the task of printing sheets of paper, or flattened cardboard cartons. The pattern on the sheets may be quite complicated and involve a number of colours. It is important to understand that each sheet is inspected in exactly the same position and orientation.[2] Precise registration is important but it is not necessary to assume that the intensity of the lighting is constant in time, provided that it does not alter significantly in colour. When the *subtle_colour* PCF is applied to a printed sheet, the result is a monochrome image in which the intensity has a complicated pattern, indicating the colour, not brightness, variations. Now, suppose that a second sheet, identical to the first, but with a small local variation of colour is examined. The resulting image will be the same as before, except in the region of the colour change, where the intensities will be significantly different. By subtracting the two PCF output images, the differences in the original colour scenes can be highlighted. (See Image 6.4(d-f).)

The program *subtle_colour* PCF that may be used to detect colour changes simply by subtracting successive pairs of images. Here is a program to do this.

```
subtle_colour_changes :-
     grb,                    % Digitise an image. PCF already
                             % programmed
     cpy,                    % Copy image A to image B
     swi(a,c),               % Switch images A and C
     sub,                    % Subtract images A and B
     abs,                    % Absolute value of intensities
     thr(25),                % Threshold. Adjust level to taste
     big_blobs(10),          % Remove blobs with <10 pixels. Adjust
                             % to taste
     cwp(N),                 % Count white points
     N > 25.                 % Are differences significant
subtle_colour_changes :-
     subtle_colour_changes.
                             % Repeat until changes are found
```

At first sight, it would appear that repeated patterns, such as printed cloth, stamps, bank-notes and other roller-printed web products could be inspected using the *subtle_colour* PCF. In practice, however, these particular applications may present considerable difficulties, due to the very high precision needed in the registration of the two images. A more likely range of applications is likely to be found in manufacturing, for example, monitoring the packing of cakes, chocolates, pharmaceuticals, toiletries, etc. into boxes, and looking for splashes of coloured food materials on pies, packaging, etc.

6.6.16 Colour Generalisation

Consider a colour scattergram in which there are six distinct and compact clusters. (Image 6.11(j)) This form of scattergram is generated by polychromatic scenes in which there are several regions, each containing nearly constant and

---

[2] The technique for detecting colour changes suffers from the same restrictions as template matching, to which it is closely related.

perceptually distinct colours. (Plate 3) After applying thresholding and noise reduction to the colour scattergram, there are several small blobs, which can be shaded, using the operator *label_blobs*. A PCF, designed by applying *create_filter* to an image containing small blobs, will often be rather noisy. In particular, some of the points in the original image are mapped to black, incorrectly suggesting that they have not been seen beforehand. The reason is that the small blobs do not cover all points in the colour scattergram. If we were to make the blobs larger and then design a new PCF, the noise level would be reduced. Enlarging the blobs can be achieved by applying the *lnb* operator several times. This can be repeated as many times as we like, provided that the blobs do not merge. Here is a program which achieves this:

```
generalise_colour :-
     wri,             % Save scattergram image
     thr(1),          % Threshold at level 1. Very dark grey.
     count(blobs,A),
                      % Count blobs in colour triangle
     rea,             % Recover image saved earlier
     generalise_colour(A).
                      % Lower level predicate, defined below.

generalise_colour(A) :-
     rea,             % Recover image saved earlier
     lnb,             % Spread bright regions
     thr(1),          % Threshold at level 1. Very dark grey.
     count(blobs,N),
                      % Count blobs in colour triangle
     A is N,          % Check that no. of blobs is unchanged
     swi,             % Switch images
     wri,             % Read image saved earlier
     !,               % Inhibit backtracking
     generalise_colour(A).
                      % Repeat until blobs touch

generalise_colour(_) :-
     rea.             % Recover saved image
```

generalise_colour is applied after the colour scattergram has been generated and thresholded and before *create_filter* is applied. The program sequence is as follows.

```
     plot_scattergram,      % Generate the colour scattergram
     threshold,             % Threshold - creates small blobs
     label_blobs,           % Shade blobs in some arbitrary way
     generalise_colour,     % Apply colour generalisation
     create_filter          % Program the PCF
```

A rather better colour generalisation procedure has been devised around the grass-fire transform. (Section 2.3.) This procedure is superior to the version of *generalise_colour* given above, since it does not simply terminate when the first two blobs merge as they are being dilated. The process generates a map resembling the territorial waters surrounding a group of separate and independent island nations. (See Images 6.11(j) and 6.12(b).) The colour triangle is sub-divided on a nearest neighbour basis and it is possible to place a limit on the extent of "territorial waters" surrounding any given "island".

It should be understood that when a PCF is designed with colour generalisation to distinguish between two colours (e.g. "yellow" and "red"), strange effects may occur when it is applied to other colours (e.g. orange, green or blue). It is good practice to apply a PCF with colour generalisation only to those colours that it is was designed to discriminate. Colour generalisation is a very useful procedure but it does need to be applied intelligently and the results scrutinised carefully. Nevertheless, it is particularly useful in reducing "noise" effects, when designing PCFs to recognise the colours in polychromatic scenes. (See [BAT-95b] for more details.)

6.7 Colour Recognition in Prolog+ Programs

We are now in a position to use the PCF for colour recognition and to present Prolog+ programs indicating how this facility can be used in practice. For the remainder of this chapter, the subtleties of designing PCF's can safely be ignored. Indeed, our programs will not refer explicitly to any of the predicates listed above. We simply need to remember that *grb* and *ctm* both make *implicit* use of whatever PCF was last programmed. The programs listed below frequently contain instructions to switch the pseudo-colour display unit on. This is often of considerable help to the *user* when interpreting images generated with a PCF but it has no effect whatsoever on the processing.

6.7.1 Counting Coloured Objects

An obvious and important application requirement is that of counting objects having a certain range of colours. For example, we might want to count all of the "yellow" and "turquoise" objects in a scene, whilst ignoring "orange" and "magenta" items. As we shall see, this is often a relatively straightforward task and usually does not require the user to program a colour filter explicitly. The reason is that the *hue* filter, or some other standard PCF, will frequently provide the necessary discrimination, in conjunction with simple fixed-level thresholding. The user can usually decide what threshold levels to use by adopting a simple procedure, based upon the colour histogram. Peaks in the colour histogram can usually be identified with specific colours in the scene being viewed. For example, the simple polychromatic scene in Plate 3(a) generates 6 peaks, which a user of the Prolog+ system can identify with little difficulty.

The Prolog+ program presented below is quite general and is able to accommodate several bands of colour. For example, it can count objects which are "green", "yellow", *or* "red". Hence, it could, for example, count tomatoes, at any stage of ripeness. It was assumed that the *hue* PCF, or some similar filter has already been programmed. Recall that 256 different colours are recognised by the *hue* PCF; the filter outputs are numbers (i.e. intensities) in the range [0, 255].

Hence, we can use integers in this range to represent colours. Using this simple notation, a single integer represents a very narrow band of colours. (For convenience, we shall refer to this as a *"single colour"*.) We can extend this notation, so that a pair of integers, [P,Q] denotes a broader, continuous range of colours, with limits P and Q. In addition, a list of integer pairs will be used to denote a more general (i.e. discontinuous) set of colours. For reasons which will soon become apparent, we shall reserve colour 255 for a special purpose. Hence, we shall assume that the PCF does not generate any output pixels with intensity 255. If it does, these pixels will be assigned to level zero and hence may be "lost". (The program works quite happily with the *hue* PCF, except that neutral shades, cannot be counted.) Here is the program for counting coloured objects.

```
% Instantiates B to the number of objects having the colours
% specified by the list A.
count_coloured_objects(A,B) :-
      create_filter, % Program/load whatever PCF is to be used
      hil(255,255,0),% Remove any pixels at level 255
      grb,           % Digitise an image using this filter
      isolate(A),    % Isolate all regions specified by list A
      remove_noise,  % Optional noise removal operator
      count(blobs,B).% Count regions. Instantiate B

isolate([]) :- thr(255).
                      % Keep all regions with any of specified
                      % colours

isolate([[A,B]|C]) :-      % Consider colour band [A,B]
      hil(A,B,255),  % Map pixels in range [A,B] to level 255
      !,             % Added to improve efficiency of recursion
      isolate(C).    % Repeat for other colour bands, if necessary
```

[*grb, isolate(A)*] generates an image consisting of a set of white blobs, representing the areas whose colours are included in the "input" list, A. The reader might like to contemplate how the above program could be modified to count objects of *any* colour, ignoring those which are of a neutral shade. Another variant can be devised, in which colours are specified by name, rather than by number. No changes are needed to *count_coloured_objects*, modifications are needed only to *isolate*. Assuming that the *hue* PCF is being used, *isolate* may be redefined thus:

```
isolate([]) :- thr(255).     % Terminate recursion.

% This clause deals with a list of colours, such as
% [sulphur_yellow, tangerine, cyanide_blue, leaf_green].
isolate([A|B]) :-
      colour_limits(A,C,D),
                      % Consult db for limits corresponding to A
      hil(C,D,255),   % Map "yellow" pixels to white
      !,
      isolate(B).     % Repeat for all colours in tail of list

% This clause deals with single colours
isolate(A) :-
      colour_limits(A,B,C), % Consult database for limits (A)
      thr(B,C).             % Select colours in band [B,C]
```

```
% Sample of the database. This clause defines limits for "yellow"
% in the PCF output
colour_limits(yellow,135,185).
```

Of course, *isolate* has a far wider range of applications than has been explained so far. The following program counts pink and blue sheep, but only if they have green eyes.

```
count(sheep,A) :-
      create_filter(hue),    % Program the 'hue' PCF
      grb,                   % Digitise image.
      wri(temp),             % Save image for use later
      isolate([pink, blue]), % Keep pink and blue objects
      keep(sheep),           % Discard all non-sheep
      blb,                   % Fill any holes (where eyes are)
      wri(sheep),            % Save for use later
      rea(temp),             % Recover input image
      isolate(green),        % Keep green objects
      keep(eyes),            % Discard all non-eyes
      rea(sheep),            % Recover sheep image
      touches,               % Keep sheep if they have ≥ 1
                             % green eyes
      count(blobs,A).        % Count the sheep.
```

It is assumed that *keep(sheep)* and *keep(eyes)* are both based upon the size and/or shape of blob-like objects in a binary image. *touches* compares two images: if a blob in image A overlaps a blob in image B, the blob in A is retained. If a blob in A does not overlap at all with any white pixels in B, then the blob in A is discarded. [BAT-91] Defining *touches* is left as an exercise for the reader. *(Hint: Use label_blobs,* mask one image with the other and make use of recursion to identify which blobs have overlap.) Also see Section 3.6.2.

6.7.2 Recognising a Polychromatic Logo, Program 1

The familiar logo associated with Apple™ Computers Inc. contains six nearly monochromatic regions. The task that we shall consider is that of recognising such a pattern, independently of its scale. The program that we shall discuss first simply calculates the proportion of each of the six colours, relative to the total coloured area. The estimated proportions are then compared to values determined experimentally and written explicitly into the program. Later, we shall describe several improvements on this naive approach, adding self-adaptive learning and taking the positions of the coloured stripes into account.

Here is our first program for recognising the Apple™ Computer logo.

```
apple_logo:-
      create_filter(hue),    % Program the 'hue' PCF
      grb,                   % Digitise image
      wri,                   % Save image for use later
      thr(1,254),            % Find coloured regions; ignore B & W
      biggest,               % Find biggest blob. This ignores the
                             % green leaf
```

```
        cwp(N0),                % Measure area of main part of the logo
        rea,                    % Recover original image
        min,                    % Mask to remove black and white areas
        wri,                    % Save masked image for use later
        thr(13,57),             % Threshold to select magenta
        cwp(N1),                % Calculate its area
        swi,                    % Switch current and alternate images
        thr(57,128),            % Threshold to select blue
        cwp(N2),                % Calculate its area
        swi,                    % Switch current and alternate images
        thr(128,185),           % Threshold to select green
        cwp(N3),                % Calculate its area
        swi,                    % Switch current and alternate images
        thr(185,206),           % Threshold to select yellow
        cwp(N4),                % Calculate its area
        swi,                    % Switch current and alternate images
        thr(206,223),           % Threshold to select orange
        cwp(N5),                % Calculate its area
        swi,                    % Switch current and alternate images
        thr(223,250),           % Threshold to select red
        cwp(N6),                % Calculate its area
        swi,                    % Switch current and alternate images
        % Calculate proportions of the various colours
        M1 is 100*N1 // N0,     % M1 is percentage of magenta
        M2 is 100*N2 // N0,     % M2 is percentage of blue
        M3 is 100*N3 // N0,     % M3 is percentage of green
        M4 is 100*N4 // N0,     % M4 is percentage of yellow
        M5 is 100*N5 // N0,     % M5 is percentage of orange
        M6 is 100*N6 // N0,     % M6 is percentage of red
        % Write parameter list for the user to peruse
        writeseqnl(['Parameter list:',[M1,M2,M3,M4,M5,M6]]),
        % Calculate Euclidean distance between [M1,M2,M3,M4,M5,M6] &
        % stored vector
        euclidean_distance([M1,M2,M3,M4,M5,M6],[16, 13, 13, 19, 18,
        19],0,Z),
        writeseqnl(['Distance measure:',Z]), % Tell user how far
        Z < 100,                % Is distance small enough?
        writenl('Apple logo was detected'),  % Printed message
        say('Found Apple Logo').             % Spoken message
apple_logo:-
        writenl('Apple logo was NOT visible'),  %Printed message
        say('Apple Logo NOT found').            % Spoken message

euclidean_distance([],_,A,A) :- !. % First terminating clause

euclidean_distance(_,[],A,A) :- !. % Second terminating clause

euclidean_distance([A|B],[C|D],E,F) :-
        G is (A-C)*(A-C) + E,        % Sum of squares of differences
        !,     % Included for faster/more efficient recursion
        euclidean_distance(B,D,G,F).
               % Repeat until one/both of input lists is empty
```

The predicate *apple_logo* is unsophisticated, being intended to recognise the Apple™ Computer logo and no other pattern. The program simply counts the proportions of pixels lying within certain colour bands, defined by applying various thresholds to the output of the *hue* PCF. The threshold parameters were chosen by finding the valleys in the colour histogram, of the image generated by the *hue* PCF.

The predicate *euclidean_distance* calculates the square of the so-called *Euclidean distance*, which is defined as follows. Let X = {X_i, i = 1,...,N} and Y = {Y_i, i = 1,...,N} be two N-dimensional vectors. Then, the Euclidean distance between them is given by D(X,Y), where

$$D(X,Y) = \sqrt{\sum_{i=N}^{N}(X_i - Y_i)^2}$$

If X and Y have almost identical values, D(X,Y) is small. In the special case when these vectors are identical, D(X,Y) = 0. On the other hand, if X and Y are very different D(X,Y) is large. D(X,Y) therefore measures the *dissimilarity* between X and Y.

apple_logo succeeds if the measured vector [M1,M2,M3,M4,M5,M6] is very similar to the stored reference vector: [16, 13, 13, 19, 18, 19]. The significance of [M1,M2,M3,M4,M5,M6] is explained above. The reference vector represents the measurements obtained using the same program on a pattern that was known to be an (ideal) logo. The idea of comparing the Euclidean distance to a fixed threshold is explained in Figure 6.22(a). Later, we shall adopt the more advanced approach in which several stored reference vectors are used.

6.7.3 Recognising a Polychromatic Logo, Program 2

The following program adopts a slightly more sophisticated approach to that explained above. Here, the *vertical order* of the colour stripes is taken into account. The program accepts a pattern as being an Apple™ Computer logo, if the stripes are located in the following order (moving upwards): blue, magenta, red, orange, yellow, green. Only differences from the earlier version of the program are annotated.

```
apple_logo :-
      create_filter(hue),
      grb, wri, thr(1,254),
      biggest, cwp(N0),
      rea, min, wri,
      thr(13,57),
      cgr(_,Ymagenta),      % Y co-ordinate of magenta band
      cwp(N1), swi,
      thr(57,128),
      cgr(_,Yblue),         % Y co-ordinate of blue band
      cwp(N2), swi,
      thr(128,185),
      cgr(_,Ygreen),        % Y co-ordinate of green band
      cwp(N3), swi,
      thr(185,206),
      cgr(_,Yyellow),       % Y co-ordinate of yellow band
      cwp(N4), swi,
      thr(206,223),
      cgr(_,Yorange),       % Y co-ordinate of orange band
      cwp(N5), swi,
      thr(223,250),
      cgr(_,Yred),          % Y co-ordinate of red band
```

```
        cwp(N6), swi,
        M1 is 100*N1 // N0, M2 is 100*N2 // N0,
        M3 is 100*N3 // N0, M4 is 100*N4 // N0,
        M5 is 100*N5 // N0, M6 is 100*N6 // N0,
        writeseqnl(['Parameter list:',[M1,M2,M3,M4,M5,M6]]),
        Yblue < Ymagenta, writenl('1'),      % Blue is below magenta
        Ymagenta < Yred, writenl('2'),       % Magenta is below red
        Yred < Yorange, writenl('3'),        % Red is below orange
        Yorange < Yyellow, writenl('4'),     % Orange is below yellow
        Yyellow < Ygreen, writenl('5'),      % Yellow is below green
        euclidean_distance([M1,M2,M3,M4,M5,M6],[16, 13, 13, 19, 18,
        19],0,Z),
        writeseqnl(['Distance measure:',Z]),
        Z < 100,
        writenl('The Apple logo has been detected'),
        say('Found Apple Logo'),
        !.
```

This program is, of course, also specific to this one application; it is necessary to write a new program, if objects other than the Apple™ Computer logo are to be detected. The program, can be modified slightly to allow it to recognise scenes in which there is a continuous variation of colour, rather like that in a rainbow. For example, a program has been written that is able to recognise the 3M™ Company logo, which consists of a multi-coloured disc.

Let us consider two further points that arise here. The first is that the shapes of the coloured bands have been ignored. Given that a crude measure of the shape of a blob can be obtained by computing the ratio of its area to the square of its perimeter, it is possible to enhance the program given above. The reader might like to contemplate how this could be done. (*Hint*: Add two lines of Prolog+ code for each coloured stripe. One computes the shape measure, while the second compares its value to stored tolerance limits.) The second point to note is that the second version of *apple_logo* makes no use of the abstract relationship *above*, discussed in Chapter 3. A third and much clearer approach to recognising the Apple™ Computer logo is therefore represented by the program given in the following section.

6.7.4 Recognising a Polychromatic Logo, Program 3

The two earlier definitions of *apple_logo*, are both "linear" (procedural) in structure. Better programming style is to be seen in the following program which performs the same operations as the second definition.

```
apple_logo :-
        get_image(N),   % Equivalent to first lines of earlier defs.
        area(red, N, Ared),   % Normalised area of red pixels is Ared
        area(green, N, Agreen),
        area(yellow, N, Ayellow),
        area(orange,Aorange),
        area(magenta, N, Amagenta),
        area(blue,N, Ablue),
        check_areas(Amagenta,Ablue,Agreen,Ayellow,Aorange,Ared),
                        % Matching sizes
```

```
            above(green,yellow),    % Verifies that green is above yellow
            above(yellow,orange),
            above(orange,red),
            above(red,magenta),
            above(magenta,blue).

/* "get_image" is identical with the first few lines of the two
earlier definitions of "apple_logo". */
get_image(A)  :-
            create_filter(hue),    % Program the 'hue' PCF
            grb,                   % Recall that "hue" PCF has been programmed
            wri,                   % Save image for use later
            thr(1,254),            % Ignore black & neutral regions
            biggest,               % Keep only the biggest one
            cwp(A),                % Compute its area
            rea,                   % Read grey-scale image back again
            min,                   % Remove pixels corresponding to non-colours
            wri.                   % Needed by "isolate" (not defined here)

area(A,B,C)  :-
            isolate(A),
            cwp(D),                % Count number of pixels of colour A
            C is 100*D/B.          % Rescale by dividing by total area of logo

% Performs match between measured sizes and stored values
check_areas(A,B,C,D,E,F)  :-
            euclidean_distance([A,B,C,D,E,F],[16, 13, 13, 19, 18,
            19],0,Z),
            Z < 100.

% Definition of "above" for coloured stripes
above(A,B)  :-
            isolate(A),              % Isolate pixels of colour A
            locate(A,_,Xa),          % Centroid of colour band A. Not
                                     % defined here
            isolate(B),              % Isolate pixels of colour B
            locate(B,_,Xb),          % Centroid of colour band B
            Xa > Xb.
```

The observent reader will note that this definition of *above* is perfectly standard; we have merely annotated it in such a way that its relevance to colour recognition is evident.

6.7.5 Multiple Exemplar Approach to Recognition

The simple approach to the recognition of coloured objects, exemplified by the first two versions of *apple_logo*, requires that a new Prolog+ program be written for every pattern that is to be recognised. A more general program, called *crude_color_recognition*, makes better use of the declarative nature of Prolog+ and is listed below. This program calculates the Euclidean distances from the measurement vector (X) to a *set* of stored vectors, held in *stored_vector*. If any of these distance values is less than some pre-defined threshold (taken to be 100 here), *crude_color_recognition* will succeed. (The name of this predicate was chosen to emphasise the point that a more sophisticated colour recognition program will be presented later.)

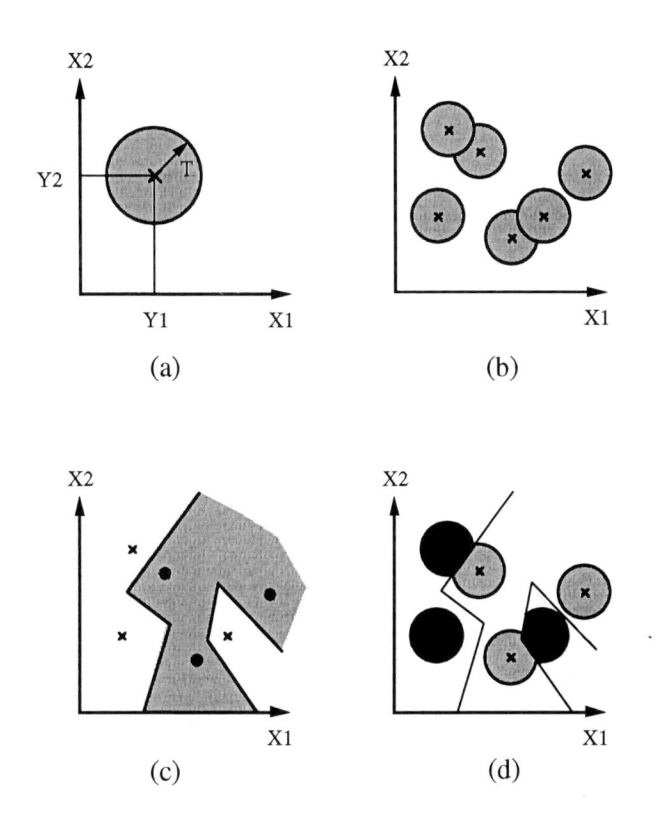

Figure 6.22 Decision surfaces drawn in two dimensions. (a) A single point [Y1,Y2] and distance threshold (T) defines a circular region. Any point [X1,X2] which falls inside the shaded area is associated with the decision *YES* (*apple_logo* succeeds), while all other points are associated with the decision *NO* (*apple_logo* fails). (b) Several circular sub-regions can be superimposed. If [X1, X2] falls inside any circle, the decision is *YES*. If [X1, X2] falls outside all circles, the decision is *NO*. Notice that we have to store 3 parameters for each circle. (c) The Nearest Neighbour decision rule. Representatives of more than one class are stored. Here, there are just two. A point is associated with a certain class, if [X1, X2] is closer to one of the stored representatives of that class than it is to all of the representatives of all other classes. (d) A modified version of the Nearest Neighbour decision rule. The decision is *"Don't know"*, if the distance to the nearest neighbour is greater than some pre-defined limit. (Also see Figure 7.2.3.)

```
crude_color_recognition :-
     create_filter(hue),    % Program the 'hue' PCF
     grb,                   % Digitise image
     process;               % Processing, optional, adjust to taste
     get_parameters(X),     % Calculate measurement vector
     stored_vector(Y),      % Consult database.
     euclidean_distance(X,Y,0,Z),
                            % Z = Euclidean distance between X & Y
     Z < 100,               % Is Z small enough? Adjust to taste.
```

```
          writenl('Object was recognised'),
                              % Object similar to stored pattern
          !.                  % Inhibit back-tracking
crude_color_recognition :-
      writenl('Object was NOT recognised'),
                              % Message to user
          fail.               % Force failure - object not recognised

% Stored vector, appropriate for recognising the 3M Company logo.
stored_vector([15, 6, 11, 32, 26]).
```

There are, of course, many possible ways to define *get_parameters*. This program was presented as if there were only one *stored_vector* fact in the database. Suppose there are more several / many. What effect does this have? Figure 6.22(b) demonstrates the potential improvement in power of recognition that this provides. The program simply makes use of back-tracking over the three lines set in italics to perform a search for a stored reference vector (Y) that is sufficiently similar to the measured vector (X) to satisfy the test Z <100, where Z = D(X,Y). If any Y is discovered that satisfies this test, the vector X is recognised as belonging to that class of objects represented by the set of stored reference vectors. This is the basis of a method of decision making known as a *Compound Classifier*. [BAT-74] So far, we have not indicated how the Y vectors can be computed. One possible way is to measure the parameters [M1,M2,M3,M4,M5] for each member of a carefully selected set of objects, forming what is known as a *training set*. Another method is to store Y vectors progressively in the database (i.e. asserting new *stored_vector* facts), subject to the constraint that a new fact is only added, if it is sufficiently different from all of the vectors already stored.

In the following section, we describe a program which develops these ideas and which permits several classes of object to be represented by vectors stored in the database.

6.7.6 Learning Proportions of Colours in a Scene

The following program calculates eight parameters measuring the proportions of the picture in eight colour bands, ignoring black and neutral. In this respect, it is similar in operation to *apple_logo*. However, these numbers are then used in a different way.

```
% Top level predicate for learning to recognise coloured objects
learning_coloured_objects :-
    yesno(['Do you want to initialise the colour recognition
    filter and database? If in doubt, select YES']),
    retractall(colour_vector(_,_)),
                          % Initialise the database
    pseudo_colour(on),    % Switch pseudo-colour ON
    create_filter(hue),   % Program the 'hue' PCF
    learn_coloured_objects.
                          % Learning colours.
```

```
learn_coloured_objects :-
     ctm,              % Allow user to set up the camera and
                       % lighting
     yesno(['Click on YES when you are ready to continue, or
     select NO to finish']),
     grb,              % Digitise image. We are using 'hue' PCF
     wri,              % Save the image for use later
     thr(1,254),       % Keep colours only - eliminate black &
                       % neutral
     big_blobs(50),    % Eliminate blobs with less than 50 pixels
     blb,              % Fill any holes
     skw,              % Eliminate edge artefacts
     cwp(N0),          % Count total number of white points
     rea,              % Read image saved earlier
     min,              % Apply binary image as a mask
     wri,              % Save masked image for use later
     thr(1,32),        % Keep colours coded by int. in range [1,32]
     cwp(N1),          % Count number of pixels (magenta colour band)
     swi,              % Switch images
     thr(33,64),       % Keep colours coded by int. in range [33,64]
     cwp(N2),          % Count number of pixels (blue colour band)
     swi,              % Switch images
     thr(65,96),       % Keep colours coded by int. in range [65,96]
     cwp(N3),
     swi,              % Switch images
     thr(97,128),      % Keep colours coded by int. in range [97,128]
     cwp(N4),          % Count number of pixels
     swi,              % Switch images
     thr(129,160),     % Keep colours coded by int. in range
                       % [129,160]
     cwp(N5),          % Count number of pixels
     swi,              % Switch images
     thr(161,192),     % Keep colours coded by int. in range
                       % [161,192]
     cwp(N6),          % Count number of pixels
     swi,              % Switch images
     thr(193,224),     % Keep colours coded by int. in range
                       % [193,224]
     cwp(N7),          % Count number of pixels
     swi,              % Switch images
     thr(225,254),     % Keep colours coded by int. in range
                       % [225,254]
     cwp(N8),          % Count number of pixels
     swi,              % Switch images
     M1 is 100*N1 // N0, M2 is 100*N2 // N0,
     M3 is 100*N3 // N0, M4 is 100*N4 // N0,
     M5 is 100*N5 // N0, M6 is 100*N6 // N0,
     M7 is 100*N7 // N0, M8 is 100*N8 // N0,
     learn_coloured_objects1([M1,M2,M3,M4,M5,M6,M7,M8],Z),
     !,
     learn_coloured_objects(Q).

learn_coloured_objects(_).

learn_coloured_objects1(X,Z) :-
     colour_vector(A,B),          % Consult database
     euclidean_distance(X,A,0,Z), % Euclidean distance X to A
     writeseqnl(['Distance from',B,'is',Z]),
     Z < 50,                      % Is A close enough to X?
     writeseqnl(['Object was recognised as',B]).
                                  % Yes! So tell user so
```

```
% Failed to recognise the object, so the user guides the program
% through learning
learn_coloured_objects1(X,Z) :-
    yesno(['No known object has been seen. Do you want to expand
    the database?']),
    grb,                        % User sees what he is talking about
    prompt_read(['What do you want to call this object?'],Z),
    assert(colour_vector(X,Z)).
                            % Add vector to database

learn_coloured_objects1(_,unknown_object).
```

6.7.7 Superior Program for Learning Colour Proportions

The decision-making mechanism used in the above definition of *learning_coloured_objects* is rather weak. The following program improves matters by using the Nearest Neighbour decision rule. [BAT-74] The theoretical basis is explained in Figure 6.22(c). (Also see Section 7.2.4.)

```
learning_coloured_objects :-
    yesno(['Do you want to initialise the colour recognition
    filter and database? If in doubt, select YES']),
    retractall(colour_vector(_,_)),
                            % Initialise the database
    pseudo_colour(on),      % Switch pseudo-colour ON
    create_filter(hue),     % Program the "hue" PCF
    learning_coloured_objects .
                            % Keep going

learning_coloured_objects :-
    ctm,                    % Live video
    yesno(['Do you want to perform (any more) learning?']),
    get_parameter_vector(X),
                            % Calculate list of image descriptors,
                            % X
    learn_coloured_objects1(X,_),
                            % Apply learning
    !,                      % Included for efficient recursion
    learning_coloured_objects .
                            % Repeat process

learning_coloured_objects.  % User indicated learning finished

% Recognition and learning
learn_coloured_objects1(A,B) :-
    nnc(A,B,C),     % Nearest neighbour classifier
    B < 100.        % Is nearest neighbour distance small enough?

learn_coloured_objects1(X,Z) :-
    yesno(['No known object has been seen. Do you want to expand
    the database?']),
    grb,            % Digitise image. Remind user about object
    prompt_read(['What do you want to call this object?'],Z),
    assert(colour_vector(X,Z)).
                    % Store details of object in DB

learn_coloured_objects1(_,unknown).
                    % User decided not to expand database
```

```
% Nearest neighbour classifier. (Also see page 309.)
nnc(_,_,_) :-
     remember(nnc,[1000000,nothing]),
                              % Initialise database
     fail.                    % Force this clause to fail

nnc(X,_,_) :-
     remember(nnc,[1000000,nothing]),
                              % Initialise database
     colour_vector(Y,Z),      % Consult database for descriptor
                              % vector
     euclidean_distance(X,Y,0,D),
                              % D = Euclidean distance from X to Y
     recall(nnc,[E,_]),       % E = smallest distance encountered so
                              % far
     E > D,                   % Is D smaller than E?
     remember(nnc,[D,Z]),     % It is, save new value & associated
                              % vector
     fail.                    % Force backtracking This clause always
                              % fails

% No more stored vectors to be considered. Return identity of
% nearest neighbour and distance
nnc(_,D,X) :-
     recall(nnc,[D,X]),       % Get NN identity (X) and distance D
     !.                       % Not resatisfied on backtracking
```

The authors have successfully used this learning program to distinguish between coloured printed packages. However, some difficulty was encountered, when trying to use the program to distinguish certain cartons of this general type, since they were found to contain large light brown regions (i.e. cake and pastry) and only small areas of other colours (fruit / filling). The program could, of course, be modified to look for known proportions of colours within certain limited areas of the image, and hence ignore problem regions like this.

6.7.8 Teaching the PCF by Showing

Although we have used the *hue* PCF in Prolog+ programs, we have not yet made use of the ability of the colour filter to learn. To understand why this is important, consider the task of recognising apples from their colours. It is clearly not sufficient to say that apples are always green. Nor are they always red. Clearly, the unripe fruit are green and some ripe apples are too. However, the ripe fruit can be red, brown (russets), yellow or yellow-green, depending upon the variety. It is impossible to define accurately, *in words*, what is meant by the term "apple coloured". Apart from a set of photographs of apples, there is no known object in existence anywhere that contains all of the possible colours that are encompassed by this term and no others. In a situation such as this, we have to rely upon (machine) learning. We therefore need a Prolog+ program that can learn, by progressively updating the contents of the PCF LUT. Such a machine should then be able to learn what the concept of "apple coloured" means.

The program generates the colour scattergram of the first scene shown to the camera. This is then stored and the second scene is analysed, in the same way.

The new and stored colour scattergrams are then merged (using *max*) and the composite scattergram is stored. Subsequent views are treated in the same way; each new colour scattergram is merged with the previously stored composite scattergram and the result is retained for the next learning cycle.

```
naive_colour_learning :-
      zer,              % Create all black image
      wri,              % Initialising intermediate results store
      naive_colour_learning1,
                        % Auxiliary predicate
      rea,              % Recover saved image
      blur,             % Smooth composite colour scattergram-optional
      thr(8),           % Adjust threshold parameter to taste
      create_filter.    % Program the PCF

naive_colour_learning1 :-
      grab_3_images,    % Digitise RGB colour separations
      colour_scattergram,
                        % Generated from RGB separations
      rea,              % Get composite scattergram image
      max,              % Merge composite and new scattergrams
      wri.              % Save enhanced composite colour scattergram
      yesno(['Do you want to perform more learning']),
      !,
      naive_colour_learning1.

naive_colour_learning1.
```

It should be noted that this extremely simple learning program does not have any provision for synchronising the image acquisition and learning with external events, nor for the user to confirm / cancel self-adaptation. Issues such as these are clearly very important in practice but their inclusion here would merely obscure the program structure. The algorithm implemented in *naive_colour_learning* can only learn a single colour. It cannot, for example, learn to distinguish between apples and bananas, whereas the following program can do so.

```
% Top level predicate for improved colour learning
learning_colour :-
      pseudo_colour(on),    % Easier to work with pseudo-colour on
      ctm,                  % Live video, facilitates setting up
                            % the camera
      ((yesno(['Set up the camera. Do you want to initialise the
      colour scattergram?']),
      zer,                  % Clear image
      keep);                % Save image
      true),                % Force success locally, even if
                            % "yesno" failed
      repeat,               % Beginning of loop
      learning_colour1,     % Auxiliary predicate where learning is
                            % done
      ctm,                  % Live video
      not(yesno(['Do you want to perform more learning?'])),
      ((yesno(['Do you want to use the colour generalisation
      procedure?']),
      generalise_colour_recognition);
                            % Optional colour generalisation
```

```
            true).                  % Force success locally, even if
                                    % "yesno" failed
learning_colour :-                  % Finished, so finish off tidily
            pseudo_colour(off).     % Switch pseudo-colour off

% Second level predicate
learning_colour1 :-
            learning_colour2,       % A third level predicate
            yesno(['A filter for a single colour has been created. Does
            it achieve a satisfactory discrimination?']),
            rea,                    % Recover image saved
            get_colour(X),          % Ask user for name / identity no. of
                                    % pseudo-colour
            hil(1,255,X),           % Shade blob to appropriate pseudo-
                                    % colour level
            fetch,                  % Get composite scattergram
            max,                    % Superimpose new scattergram on
                                    % composite
            create_filter,          % Program the filter
            cpy,                    % Make copy of composite scattergram
            ctm,                    % Show the user result of new composite
                                    % filter
            yesno(['The single-colour filter has been added (temporarily)
            to the existing multi-colour filter Is this OK?']),
            swi,                    % Recover composite scattergram
            keep,                   % Save new composite scattergram
            !.                      % Avoid back-tracking

learning_colour1 :- !.              % Force success & avoid back-tracking

% Third level predicate
learning_colour2 :-
            grab_3_images,          % Digitise RGB colour separations
            colour_scattergram,     % Generated from RGB separations
            blur,                   % Smoothing
            thr(32),                % Cut off scattergram tails
            wri,                    % Save the binary scattergram
            create_filter,          % Program the PCF
            ctm,                    % Live video
            !.                      % Avoid possibility of backtracking

% Find out what pseudo-colour to shade the blobs in the colour
% triangle
get_colour(X) :-
            findall(X,pseudo_colour_value(X,_),Q),
                                    % Find list of known colours
            scroll_menu(['Selecting pseudo-colour to be displayed. Choose
            just ONE item '], [other | Q],[other],Y),
            Y = [Z|_],              % Select head if there is more than
                                    % item selected
            not(Z = other),         % Abandon this clause if Z = other
            pseudo_colour_value(Z,X),
                                    % Convert from named colour to number
            !.                      % No back-tracking allowed.

/* Getting ready for clause 3. Clause 2 simply draws a wedge
(displayed in pseudo_colour) & a series of vertical black lines to
indicate which pseudo-colours are already in use. */
get_colour(_) :-
            wgx,                    % Intensity wedge
            pseudo_colour_value(_,Z),
                                    % Consult database
```

```
           Z > 0,                    % Ignore black - this would cause an
                                     % error
           vpl(Z,1,Z,256,0),         % Draw vertical black line
           fail.                     % Go through database. Then force
                                     % failure

% This clause always follows clause 2. User selects pseudo-colour
% with the cursor
get_colour(X) :-
           pseudo_colour(on),        % Switch pseudo-colour ON
           prompt_read(['Choose a pseudo-colour with the mouse. Avoid
           the vertical black lines. What do you want to call this
           colour?'],Y),
           cur(_,_,X),               % Cursor. User selects colour
           assert(pseudo_colour_value(Y,X)),
                                     % Expand the database
           pseudo_colour(off),       % Switch pseudo-colour OFF
           !.

% Database converting named colours to numbers
pseudo_colour_value(red,215).                   % Standard colour
pseudo_colour_value(green,115).                 % Standard colour
pseudo_colour_value(blue,47).                   % Standard colour
pseudo_colour_value(yellow,160).                % Standard colour
pseudo_colour_value(cyan,95).                   % Standard colour
pseudo_colour_value(magenta,250).               % Standard colour
pseudo_colour_value(black,0).                   % Standard colour
pseudo_colour_value(white,255).                 % Standard colour
pseudo_colour_value(violet,16).                 % Standard colour
pseudo_colour_value(orange,186).                % Standard colour
pseudo_colour_value(vanilla_ice_cream,150).     % Item added by user
pseudo_colour_value(cobolt,30).                 % Item added by user
```

6.7.9 Template Matching of Colour Images

There is a common requirement in industry to recognise scenes that are repeated in time. Consider for example, the task of examining brightly coloured printed cartons and containers, such as those used for food products, toiletries, stationery, automobile parts, etc. The cartons are moved along, either by indexing, or by continuous motion. In the latter case, it often happens that some timing signal can be generated to indicate the arrival of the new carton and thereby allow image digitisation to be synchronised to the production process. The essential point is that, in either situation, the objects being inspected are always viewed in very nearly the same orientation and position, lighting and magnification. *Template matching* has traditionally been used in this type of situation, when monochrome images are being processed.

The process of template matching is illustrated in Figure 6.23, and is clearly very closely related to N-tuple filtering (Section 2.2.6) and Morphology (Sections 2.4 and 2.5). Since a PCF maps colour into intensity, it is possible to apply template matching to the colour images, as well. The following program performs a crude template match, using a stored image, which can be either monochrome, or the output of a PCF.

```
template_match :-
    grb,            % Digitise an image
    fetch,          % Recover stored mask
    sub,            % Subtract images
    avr(X),         % Compute average intensity
    tolerance_band(P1,P2),
                    % Consult database for tolerance parameters
    X ≤ P2,         % Check upper limit
    X ≥ P1.         % Check lower limit
```

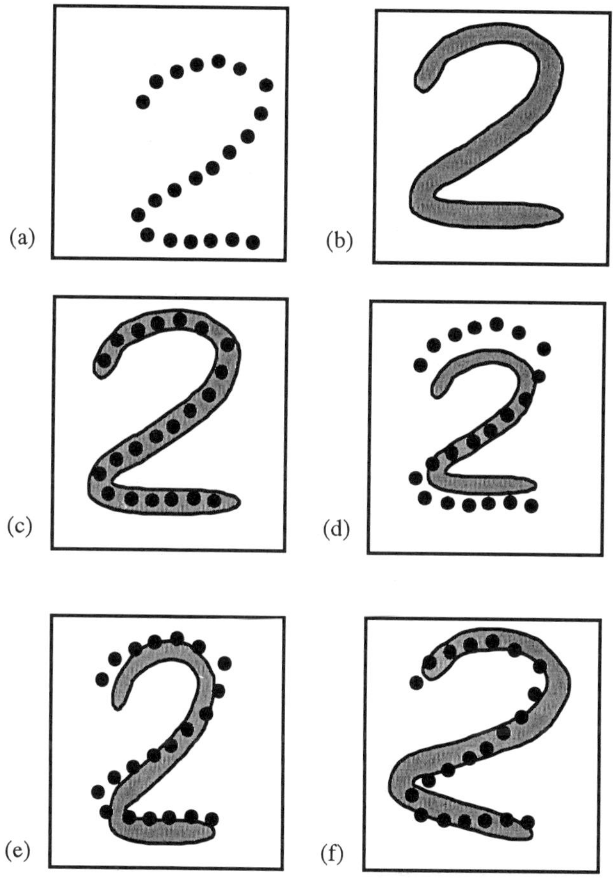

Figure 6.23 Template matching (a) The template. This might consist of several disjoint parts, or it might be a single connected shape, possibly containing "holes". (b) Pattern to be compared to the template. (c) By shifting the [X,Y] position, the template can be made to fit the pattern exactly. (d) When the pattern is made smaller or larger, the template will not fit exactly. (e) When the pattern has a different aspect ratio, it will not fit the template exactly. (f) When the pattern has been rotated, it will not fit the template exactly.

If preferred, the maximum difference of intensity can be used as the criterion for establishing a match:

```
template_match :-
     grb,
     fetch,
     sub,
     gli(_,X),           % Compute maximum intensity difference
     tolerance_band(P1,P2),
     X ≤ P2,
     X ≥ P1.
```

A third variant is to use the Qth percentile of the intensity difference:

```
template_match(Q) :-
     grb,
     fetch,
     sub,
     pct(Q,_),           % Threshold at Q'th percentile of int.
                         % difference
     min,                % Select darkest Q% of the picture
     gli(_,X),           % Find Q'th percentile intensity
     tolerance_band(P1,P2),
     X ≤ P2,
     X ≥ P1.
```

These three programs are all slightly different variations of the basic template matching scheme. However, the following program is fundamentally different, since it allows the image to be shifted and rotated before the matching is attempted.

```
template_match(X,Y,Z) :-
     grb,           % Digitise an image
     psh(X,Y),      % Shift image by [X,Y]
     tur(Z),        % Rotate image by Z degrees
     fetch,         % Recover stored mask
     sub,           % Subtract images
     avr(X),        % Compute average intensity
     tolerance_band(P1,P2),
                    % Consult database for tolerance parameters
     X ≤ P2,        % Check upper limit
     X ≥ P1.        % Check lower limit
```

Various methods can be used to calculate the shift and rotation parameters. In many instance, of course, the centroid and principal axis (i.e. the axis of minimum second moment) could be used to achieve this. Alternatively, certain key features could be located first. To illustrate how colour can help to achieve this, consider the task of calculating the position and orientation of a *red* picture playing card. (See Image 6.6.)

```
normalise_card :-
     grb,           % Digitise image. "hue" PCF programmed
     wri,           % Save image for use later
     thr(255),      % Select white (i.e. neutral) parts of the
                    % image
```

```
        biggest,        % Ignore any smaller bits
        blb,            % Fill any holes
        cgr(P,Q),       % Centroid
        rea,            % Recover image stored earlier
        isolate(red),   % Equivalent to "thr(235,254)"
        biggest,        % Isolate red suit symbol at corner of card
        cwp(U),         % Count white points
        swi,            % Switch images
        V is 0.75*U,    % 75% of area of red suit symbol
        big_blobs(V),   % Keeps only 2 suit symbols at corners of
                        % card
        condense,       % Reduce them to single points (centroids)
        cwp(N),         % Count white points
        !,              % Avoid backtracking if next goal fails
        N is 2,         % Check that there are exactly two points
        get_points([[X1,Y1],[X2,Y2] |_]),
                        % Get co-ordinates of the centroids
        angle(X1,Y1,X2,Y2,R),
                        % Find angle of line joining [X1,Y1] & [X2,Y2]
        rea,            % Read image saved earlier again
        P1 is 128 - P,  % Calculate shift along X axis
        Q1 is 128 - Q,  % Calculate shift along Y axis
        psh(P1,Q1),     % Shift image by [P1,Q1]
        R1 is -R,       % Inverse of orientation
        tur(R1,128,128),
                        % Rotate by -R degrees
        wri,            % Save image for use again later
        thr(1),         % Keep everything but black
        biggest,        % Make sure there is only one blob
        blb,            % Fill any holes in it
        dim(A,B,C,D),   % Find max/min X and Y values
        A1 is -A +1,    % Calculate X shift parameter
        B1 is -C + 1,   % Calculate Y shift parameter
        C1 is 100*(1- (B - A)/256),
                        % Rescaling parameter for X axis
        D1 is 100*(1- (D - C)/256),
                        % Rescaling parameter for Y axis
        rea,            % Recover image saved earlier
        min,            % Recover image from disc
        psh(A1,B1),     % Shift it by [A1,B1]
        rescale_axes(C1,D1).
                        % Rescale [X, Y] axes by [C1, D1]
```

6.7.10 Using Colour for Object Orientation

The program listed below was designed to recognise the VISA™ logo, used on credit cards. This consists of a broad blue stripe above the word "VISA", which is printed in blue, with an orange stripe below it. There may well be other information in these and other colours on a credit card. In its present form, the predicate *visa_card* uses the ubiquitous *hue* PCF.

```
% A naive program for recognising the VISA logo
visa_card :-
        grb,            % Digitise an image. "hue' PCF is being used
        isolate(blue),  % Isolate blue regions, discard all others
        wri,            % Save image showing blue regions
        swi,            % Revert to PCF output
        isolate(orange),
                        % Isolate orange regions, discard all others
```

```
    hin,              % Halve intensities
    rea,              % Recover image showing blue regions
    max,              % Superimpose images
    wri,              % Save image for use later
    thr(120,130),     % Find orange regions again
    biggest,          % Process biggest orange blob only
    cwp(N),           % Count points in orange stripe
    lmi(X,Y,Z),       % Find its orientation
    X1 is 128 - X,    % Calculate X shift parameter
    Y1 is 128 - Y,    % Calculate Y shift parameter
    Z1 is -Z,         % Calculate rotation parameter
    rea,              % Read image saved earlier
    psh(X1,Y1),       % Shift it …
    tur(Z1),          % …and rotate it
    wri,              % Save image for use later
    N1 is 0.9*N,      % Lower limit: orange stripe size - 10%
    N2 is 1.1*N,      % Upper limit: orange stripe size + 10%
    thr(1),           % Select all non-black points
    big_blobs(N1),    % Keep blobs with ≥ N1 pixels
    big_blobs(N2),    % Keep blobs with ≤ N2 pixels
    xor,              % Blue stripe same nominal area as orange
                      % stripe
    chu,              % Convex hull around blue and orange stripes
    blb,              % Solid figure enclosing blue & orange stripes
    rea,              % Recover image saved earlier
    min,              % Retains blue & orange stripes & word VISA
    rescale,          % Rescale so that the logo fills the image
    get_parameters(L1),
                      % Calculate parameter list. Example given
                      % below
    consult_db(L2),
                      % Consult database for reference vector
    euclidean_distance(L1, L2,0,E),
                      % E is Euclidean distance between L1 & L2
    writeseqnl(['Euclidean distance =',E]),
                      % Message for user
    ((E < 5000,       % Small difference between L1 & L2
    writenl('A VISA card was found'));
                      % Announce logo found
    writenl('No VISA card was found')).
                      % Announce logo was NOT found

% Compute the average intensity in each 64*64 square in the image.
get_parameters(_) :-
    remember(par_list,[]),
                      % Initialise list of feature values
    member(X,[1,65,129,161,193]),     % Select a value for X
    member(Y,[1,65,129,161,193]),     % Select a value for Y
    swc(X,Y,32,32),
                      % Place 32*32 processing window at [X,Y]
    recall(par_list,L),          % Get intermediate results list
    avr(Z),           % Calculate average intensity
    remember(par_list,[Z|L]),
                      % Save enlarged intermediate results list
    fail.             % Step through image

get_parameters(L) :-
    recall(par_list,L),          % Get result list
    swc(1, 1, 256, 256).
                      % Reset processing window, 256*256, at [1,1]
```

In this form, *visa_card* does not demonstrate good Prolog+ programming style, since it is a simple linear list of operations to be performed. The following version is probably easier to understand.

```
visa_card :-
    grb,                % Digitise image, PCF creates 3-level image
    wri,                % Save image for use later
    thr(128,128),       % Keep orange pixels; ignore blue for now
    biggest,            % Keep orange stripe in logo only.
    cwp(N),             % Area of orange stripe in logo is N
    normalisation_parameters(X,Y,Z),
                        % Find centroid [X,Y] & orientation, Z
    rea,                % Recover "raw" image, saved earlier
    size_selector(N),
                        % Keeps blobs with areas = N ± 0.1*N pixels
    solid_convex_hull,
                        % Draw solid figure enclosing logo stripes
    rea,                % Recover image saved earlier
    min,                % Apply binary image as mask to keep only 2
                        % stripes and word "VISA" in logo
    translate(X,Y,Z),
                        % Normalise position and orientation of logo
    rescale,            % Rescale so that the logo fills the image
    recognise(visa_card).
                        % Possible to use template matching here
```

The first point to note here is that we have assumed that the PCF has been specially pre-programmed to recognise only orange (mapped to level 128) and blue (level 255). This simple change makes the remainder of the program rather easier to understand. Further simplifications are achieved through the use of three perfectly standard predicates: *normalisation_parameters, translate* and *rescale*. Only recognise is specific to this application. Its function is similar to that embodied in *template_match*.

6.7.11 Approximating an Image by a Set of Overlapping Discs

Suppose that the colour scattergram of a certain scene has been calculated and that we need to find some suitable representation of it, so that, at some time in the future, we can reconstruct it. (It will be assumed that we do not have sufficient storage space to retain the scattergram in the form of an image.) It is necessary therefore to reduce the scattergram to some parametric form. One possible method of doing this is to use a set of overlapping discs. (Image 6.7) The parametric representation of the scattergram is then in the form of a list of lists, having the following structure: [[X1,Y1,Z1], [X2,Y2,Z2], [X3,Y3,Z3], ..., [Xn,Yn,Zn]], where [X_i, Y_i] denotes the centre of a white circular disc of radius Z_i, i = 1,...,n.

The program *approximate_colour_scattergram* calculates these parameters and operates according to the procedure explained below.

(i) A scattergram in binary form is first created, by thesholding the colour scattergram at some suitable level. (The user might need to adjust the threshold parameter experimentally, to obtain the best results. This is usually quite straightforward.)

(ii) Initialise the parameter list. The initial parameter list could simply be the empty list, [ ]. Alternatively, we may wish to extend an existing parameter list, for some reason.

(iii) The grass-fire transform of the binary scattergram image is obtained, using the command *gft*. (See Section 2.3, Figure 2.10)

(iv) The brightest point in the image is then found. Suppose that its intensity is Z and its position is [X,Y].

(v) Superimpose a black disc, centred at [X,Y] and with radius Z, onto the image.

(vi) Append [X,Y,Z] to the parameter list. Notice that [X,Y,Z] defines the disc completely.

(vii) Perform steps (iv) to (vi), until the brightest pixel in the image has an intensity less than some pre-defined limit. (This has been arbitrarily set to 3 in the program listed below.)

```
/* Approximate white regions in a binary image by a set of discs,
which may but need not overlap. Big circles are put into place
initially, followed by progressively smaller ones. The grass-fire
transform is used to find out where large circles can be placed. */
approximate_colour_scattergram :-
      yesno(['Do you want to retain any previously stored details
      about colour histogram approximations?']),
      cover_image(A),
                     % Cover white region with black discs
      asserta(disc_parameters(A)).
                     % Save position & size parameters in database

approximate_colour_scattergram :-
      retractall(disc_parameters(_)),
                     % Clear the database
      cover_image(A),
                     % Cover white region with black discs
      asserta(disc_parameters(A)).
                     % Save position & size parameters in database

/* Perform the approximation. Large discs will be fitted first. A
is the list of disc position and size parameters. */
cover_image(A) :-
      gft,           % Grass fire transform.
      wri,           % Save image for use later
      reduce([],A).  % Approx. white regions with overlapping discs

/* This predicate is the one that does the hard work. It
progressively "nibbles away" the white regions, by superimposing
black discs onto it. */
reduce(A,B) :-
      rea,           % Get image stored earlier
      gli(_,Z),      % Find maximum intensity
      Z > 3,         % Ignore very small discs. Adjust to taste
      thr(Z),        % Threshold at maximum intensity
      top_left(X,Y), % Get address of top-left most white pixel
      swi,           % Revert to grey-scale image
```

```
        draw_disc(X,Y,Z,0),
                        % Draw black disc at [X,Y] with radius Z
        wri,            % Save image
        reduce([[X,Y,Z]|A],B).
                        % Disc parameters added to list. Continue

reduce(A,A).            % End recursion; no more big discs can be
                        % added

% Sample of the database where disc parameters are stored.
disc_parameters([[100,100,25], [110,125,30], [95,126,16]]).
```

Given a parameter list in the same format, *rebuild_colour_scattergram* allows us to reconstruct the scattergram. The program allows each blob (formed by a set of overlapping circles) in the reconstructed image to be assigned a different intensity and then programs the PCF.

```
% Rebuild the colour scattergram by drawing a set of overlapping
% discs
rebuild_colour_scattergram :-
    disc_parameters(A),
                        % Get disc parameter list from database
    zer,                % Create black image to initialise the process
    draw_discs(A),      % Draw discs with parameters defined by A
    yesno(['Do you want to program the colour filter?']),
    label_blobs,        % Shade blobs
    3·lnb,              % Optional: may make PCF more robust
    create_filter.      % Program the PCF from the current image

rebuild_colour_scattergram.
                        % Force this predicate to succeed

/* Drawing a set of white discs, which may but need not overlap.
The position and size parameters are defined by the "input" list
[A|B]. */
draw_discs([]).         % Terminate recursion, no more discs to draw

draw_discs([A|B]) :-
    A = [X,Y,Z],        % Decompose A into three components
    draw_disc(X,Y,Z,255),
                        % Draw disc at [X,Y], radius Z, intensity 255
    !,                  % Included for the sake of efficiency of
                        % recursion
    draw_discs(B).      % Repeat, draw all discs defined in database
```

6.7.12 Interpreting Resistor and Capacitor Colour Codes

In developing a Prolog+ program capable of interpreting resistor and capacitor colour codes, (Figure 6.24) there are several sub-problems that must be solved.

(a) Obtaining a good image. Resistors are small and shiny. Solder joints can cause serious glinting problems. This problem can be solved by paying careful attention to the optics and lighting.
(b) Recognising resistors. In the general case, this is may present considerable difficulties, since resistors and capacitors are highly variable in appearance. The PCB also forms a highly variable and complex background. However, a

great deal of help can be obtained by using the fact, that in most cases, the layout of the PCB is predictable. In these instances, this sub-problem reduces to a trivial level.

(c) Deciding the component polarity. (i.e. which way round the colour code is to be read.) There only two alternatives for resistors and only one option for capacitors like that sketched in Figure 6.24. The spacing of the colour bands can be of assistance in this decision.

(d) Identifying the colour code bands. (This and step (e) might be merged.) The resistor body may be coloured and must be ignored. In some cases, when the colour of the resistor body is known beforehand, this task becomes trivially easy.

(e) Recognising colours in the code. This is the task for which we might well use a colour filter, although there may be some difficulties. This arises because some of the code colours are ill suited for automatic recognition using a PCF. Brown, grey, silver and gold are especially difficult. It would be possible to select a much better set of colours, as far as the colour recognition system is concerned, but of course, the whole electronics industry would be reluctant to adopt a new standard simply for our benefit !

(f) Interpreting the code; calculating the resistance/capacitance value, decoding the tolerance and working voltage.

The general unconstrained task of reading resistor/capacitor colour codes clearly requires a considerable amount of program intelligence. Since it is not our intention here to try to describe a complete solution for this challenging problem, let is suffice to say that the combination of colour recognition and intelligent image interpretation that is embodied in Prolog+ is *exactly* what is needed. We conclude by presenting a simple little program that performs the last mentioned task, (f), of interpreting the resistor colour code. *resistor* is the top level predicate for calculating numeric values, for resistors with only three colour bands. (i.e. tolerance = ± 20 %). The goal *resistor(brown, black, green, Z)* will instantiate Z to the resistance, expressed in Kilo-ohms (KΩ). If the computed resistance is not a preferred value, *resistor* will fail.

```
resistor(A,B,C,D) :-
      value1(A,A1),        % Interpret Band 1 colour as a number
      value1(B,B1),        % Interpret Band 2 colour as a number
      value3(C,C1),        % Interpret Band 3 colour as a number
      !,                   % Force failure; not preferred value
      Z is 10*A1 + B1,     % Combine Bands 1 and 2
                           % List of preferred values follows
      on(Z,[10,11,12,13,15,16,18,20,22,24,27,30,33,36,39,43,47,51,
      56,62,68,75,82,91]),
      D is Z*C1.           % Compute final value

% Interpreting Bands 1 and 2 as numbers
value1(black,0). value1(brown,1). value1(red,2). value1(orange,3).
value1(yellow,4). value1(green,5). value1(blue,6).
value1(violet,7). value1(grey,8). value1(white,9).
```

```
% Interpreting Band 3 as a number
value3(silver, 0.00001). value3(gold, 0.0001). value3(black,0.001).
value3(brown,0.01). value3(red,0.1). value3(orange,1).
value3(yellow,10). value3(green,100). value3(blue,1000).
```

The reason for including this program here is to emphasise the general point that the interpretation of colour images may well require a high level of intelligent activity on the part of the program. In other words, the mere recognition of colours, or any other features in an image, is insufficient to meet the needs of many inspection and other machine vision tasks.

Resistors

Tolerance band, red, gold or silver, not present on ±20% resistors

Band 1	First digit	Band 2	Second digit	Band 3	Multiplier (Ohms / pF)
Black	---	Black	0	Silver	0.01
Brown	1	Brown	1	Gold	0.1
Red	2	Red	2	Black	1
Orange	3	Orange	3	Brown	10
Yellow	4	Yellow	4	Red	100
Green	5	Green	5	Orange	1000
Blue	6	Blue	6	Yellow	10,000
Violet	7	Violet	7	Green	100,000
Grey	8	Grey	8	Blue	1,000,000
White	9	White	9		

Tolerance band, white or black

Working voltage, red or yellow

Capacitors

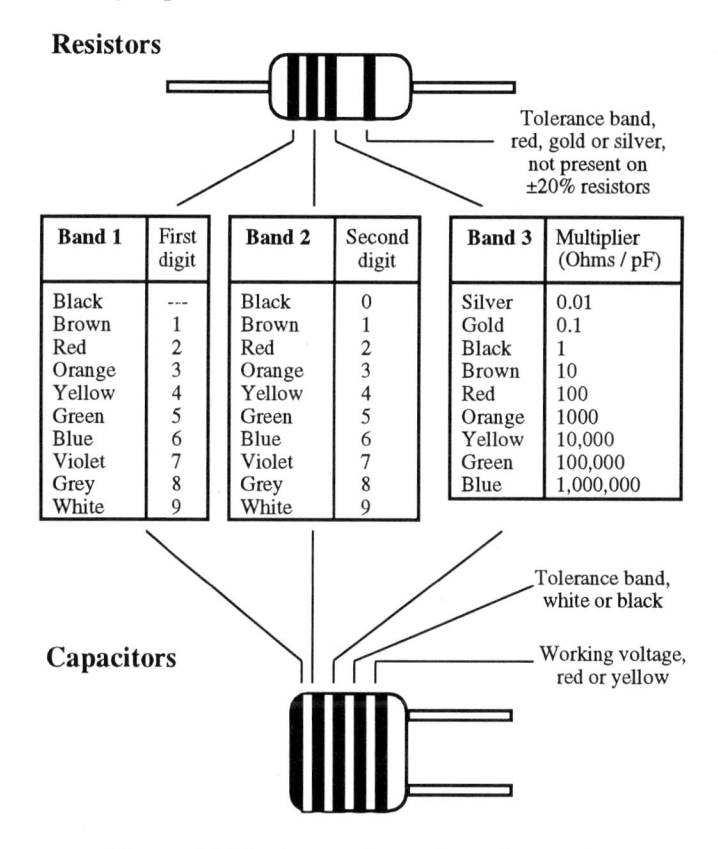

Figure 6.24 Resistor and capacitor colour codes.

6.8 Discussion and Conclusions

The results of a series of varied experiments involving colour recognition are presented in the coloured plates. Notes describing these applications are given in the legends. When working on colour recognition, it is important that we use a stable light source. (Image 6.9) There can be a distinct shift of the colour perceived by a colour recognition system, if the light source is changed from one type of lamp to another.

Evidence of the importance that we place on colour is to be found in the fact that a large proportion of manufactured goods are coloured. Despite this, relatively little work to date in machine vision has concentrated on the development of *industrial* inspection systems capable of detecting colour. There exists a new and exciting technique for recognising colours, based upon the programmable colour filter. The PCF is simple to implement and fast in operation and it can perform *any* task that is possible using the RGB, opponent process, YIQ and HSI representations of colour. It is capable of recognising almost all the "named" colours that are familiar to us in everyday life. Moreover, it can learn new colours, such as "banana yellow", "strawberry red", "cucumber green", etc. A number of programs have been presented that make use of colour recognition in a variety of ways, including self-learning. However, it is the use of colour in declarative programming that is most exciting.

Let us perform a simple thought experiment. Suppose that we want you find a certain object. Let us call it XXXX. (There are no prizes for guessing what class of objects we have in mind) We might describe XXXX in the following way:

1. An XXXX is yellow-green, yellow, or yellow with brown spots.
2. An XXXX is between 60 and 300 mm long.
3. An XXXX is between 15 and 60 mm in width.
4. An XXXX is curved in a simple arc.

Your task now is to find an XXXX. Look around the room where you are sitting. Can you see an XXXX? You might find an XXXX in your lunch box ! Of course, there are many objects in the universe that conform to rules 2, 3 and 4. A small cucumber does, for example. However, there are far fewer objects in existence that conform to all four recognition rules. By adding information about the colour of an XXXX, we are able to be very much more specific. It is very unlikely that, unless you have a banana for lunch today, or you are reading this book in the kitchen, that you will find an XXXX. We must not confuse ourselves by believing that XXXX and banana are synonymous. Our rules are perfect for recognising XXXXs but are prone to producing false positive responses for bananas. By using colour, we have simply made the number of false positive responses rather smaller.

In order to emphasise the potential value of colour recognition in declarative programming, consider Figure 6.25. Suppose that a "general purpose" machine vision system is to be built, to monitor the manufacture and packaging of household and industrial chemicals, such as cleaning fluid, polish, detergent, etc. A new product line is about to be introduced and will be distributed in bright blue plastic containers, with red tops. (See Figure 6.25) Colours on products like these are carefully chosen, both to project the corporate image and to provide a warning that the fluid in the bottles is corrosive. The task before us is to reprogram the supposedly "general purpose" vision system, so that it will recognise the new bottles and distinguish them from other types and from malformed bottles. In many cases like this, we would like to avoid reprogramming the vision system using low-level computer languages, such as C, Pascal, or even Prolog+. Nor do we want to have the task of programming a PCF on the factory floor, since this is a fairly complicated procedure, requiring skilled labour. We simply want to be able to use low-skill labour, communicating with the machine in a way that is both natural and straightforward. In Chapter 4 we discussed the rôle of natural language for programming machine vision systems. The point to be made here is simply that it is legitimate to include terms relating to everyday colours in the vocabulary of the language, since the means exists for recognising a wide range of tones. A description of the bottle portrayed in Figure 6.25 in constrained English might be something like this:

1. *A is a red rectangle*
2. *A is at the top of the image..*
3. *A has area Z4 ±10% and height Y4 ± 5%.*
4. *B is a blue rectangle*
5. *B has area Z3 ±10% and height Y3± 5%.*
6. *B is below A.*
7. *C is mixed_colour*
8. *C has area Z2 and height Y2.*
9. *C is below D. etc.*

For many purposes, this provides an adequate basis for recognition. The use of colour makes the definition much more specific than it would be with only monochrome image processing at our disposal.

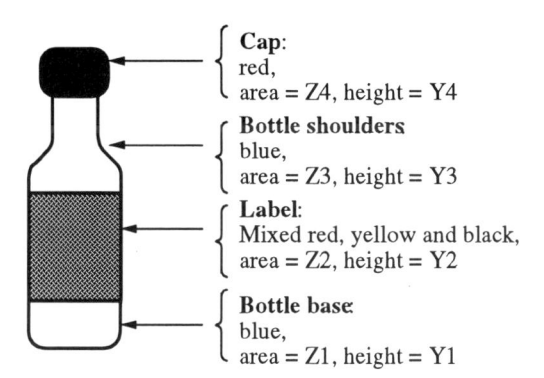

Cap:
red,
area = Z4, height = Y4

Bottle shoulders
blue,
area = Z3, height = Y3

Label:
Mixed red, yellow and black,
area = Z2, height = Y2

Bottle base
blue,
area = Z1, height = Y1

Figure 6.25 Using colour in declarative programming. The object represented diagrammatically here is a plastic bottle containing household bleach.

IMAGES

Image 6.1 Colour analysis of a multi-colour scene using various techniques. (a) Monochrome image. The original artwork consisted of 6 well-defined and well-separated colours. (b) Image derived by digitising the R channel. (c) G channel. (d) B channel. (e) Colour scattergram, thresholded at level 128. (f) Thresholded at level 4. (g) Thresholded at level 2. (h) Thresholded at level 1. (i) [*lnb, thr(4), big_blobs(20)*] applied to the colour scattergram. (Compare to (e).) (j) Colour generalisation applied to image (i). (k) 2-dimensional colour scattergram. Vertical axis, R. Horizontal axis, G. (l) 2-dimensional colour scattergram. Vertical axis, G. Horizontal axis, B.

●

Image 6.2 Colour scattergrams of various natural objects. In each case, thresholding and a noise-reduction filter has been applied, to remove outlier points. (a) Green foliage (lime tree). (b) Courgette. (c) Cucumber. (d) Banana (e) Corn cob. (f) Red apple.

●

Image 6.3 Using computer-generated pseudo colour-triangles to *pre-program* the PCF. (Also see Image 6.11.) (a) The circular wedge which forms the basis of the *hue* filter. The pseudo-colour display of this image (with a white disc superimposed) is shown in Plate 1(b). (b) A PCF generated from this image distinguishes between neutral (mapped to black) and coloured regions (mapped to white). (c) Segmentation of the colour triangle. The wedge shaped sectors are all of equal area. This filter performs a very crude discrimination of six *primary* colours (red, magenta, blue, cyan, green, yellow) and neutral. Compare this to Plate 1(b). (d) Superimposing a colour scattergram derived from a scene containing red, yellow and green regions onto the pseudo colour-triangle explains why the PCF derived from (c) does always not distinguish between the *primary* colours very accurately. The blob at the left of centre corresponds to green and overlaps the boundary of two sectors in (c). (e) A colour triangle shaded so that the intensity indicates *distance* from its centre. The PCF derived from this image, by applying *create_filter*, provides a measurement of saturation and hence forms the basis of the saturation PCF. (Also see Plate 1(c).) (f) The saturation detection filter based on (e) was used to analyse a scene containing four different yellow regions. Bright areas are close to saturation.

●

Image 6.4 Detecting subtle, local colour changes in a complex scene that is fixed in space. (a) Pseudo colour-triangle. This image was generated by scattering a number of points at random in a binary image. The image was negated and the grass-fire transform [*gft*] applied. (b) Monochrome image derived from a children's game. Notice the very faint dark triangle, just to the right of the word *kite* in the top-right corner. This is the *defect* referred to below. (c) The PCF derived from (a) by using *create_filter* and applied to the scene without the defect. (d) The same PCF applied to the scene with the *defect*. Notice how the *defect* stands out here. (e) Images (c) and (d) subtracted and the resulting image thresholded. (f) Image (e) after noise removal.

●

Image 6.5 Compound colour scattergrams, corresponding to more than one colour. (a) Scattergram corresponding to two different shades of yellow (black blobs) and red (white blob). Notice that the red scattergram is fragmented into one large blob and several very small ones. (b) Colour separation achieved by a filter that was trained on the colour scattergram shown in (a) . The black spots are due to noise and indicate that the blobs in (b) are too small. The input image consisted of a red rectangle at the bottom left, a yellow rectangle at the bottom right and another yellow rectangle, of a slightly different shade, at the top. (c) The colour triangle shown in (b) was processed by expanding the blobs (separately) using [*6•exw*]. (d) Colour separation achieved by the PCF obtained from the colour triangle in (e). Notice that the noise level is much reduced, compared to (b). (e) Colour scattergams for two different shades of yellow (merged into a single blob at the centre bottom), red (bottom right), green (left of centre) and blue (above left of centre). (f) Colour generalisation procedure applied to the colour triangle in (e). Notice the very small region at the bottom right. This arises because the noise removal procedure was imperfect.

●

Image 6.6 Using colour to determine the orientation of a picture playing card. (a) Monochrome image. (b) Colour separation. (c) Thresholded output of the *hue* PCF. (d) The orientation of the line that joins the centroids of the two largest blobs in (c) determines the orientation of the card.

●

Image 6.7 Representing a colour scattergram by a set of overlapping discs. (a) Colour scattergram derived from the logo of a well-known company. This logo consists of six well-defined and well-separated colours. The scattergram has been converted to binary form by thresholding and a simple noise removal procedure has been applied, to remove very small white regions. The outline of the colour scattergram has been omitted here and in (b), for convenience. (b) Approximating the colour scattergram by a set of 12 circles.

●

Image 6.8 Recognising the colours of wires on a UK standard mains plug. (a) Monochrome image. The wire connecting to the terminal at the bottom-left (*neutral* wire) is blue. That one connecting to the brass terminal on the right (*live*, only partially in view) is brown. The wire connecting to the top terminal (*earth* wire) has yellow and green stripes. The body of the plug is matt white. (b) PCF output. (c) Multi-level thresholding and noise removal applied to image (b) isolates the three coloured wires.

●

Image 6.9 Showing how the colour scattergram shows the effects of varying the colour of illumination. The scene being viewed was a piece of white paper, under the following lighting conditions: *1.* 8-foot fluorescent tube, type Philips 125W/36. The colour scattergram forms a single very compact cluster. (Upper white blob.) *2.* Fluorescent desk lamp, type Osram 11W/41. Again the colour scattergram forms a single compact cluster. (Lower white blob.) *3.* Volpi 150W fibre optic light source, bulb type Philips EFR A1/232. The colour scattergram is virtually identical to that generated for case (1). *4.* Filament desk lamp. 60W bulb. Once again, the colour scattergram forms a single compact cluster. (Black blob)

●

Image 6.10 Pattern with a continuously varying colour, resembling a rainbow. (The original artwork was the logo of a well-known company.) (a) Colour scattergram. (b) After thresholding and noise removal.

●

Image 6.11 Inspecting coloured packaging. The original image is shown in Plate 3(a). (a) Sum of the three colour channel outputs (R+G+B). (b) *Top-left*: Image derived by digitising the R channel. *Top-right*: G channel. *Bottom-left*: B channel. (c) 2-dimensional scattergram. Horizontal axis: G channel. Vertical axis: R channel. (d) Histogram of the R channel output. (e) Pseudo colour-triangle for measuring *purity* of the primary colours (R, G & B). (f) Histogram of the output of a PCF which measures *purity* of the primary colours. (g) Pseudo colour-triangle for measuring *yellow-ness*. (h) Histogram of the output of the PCF derived from (g). (i) Output of the filter described in (g). (j) Thresholded colour scattergram (black) and the watershed, which forms the basis of the colour generalisation procedure (white). In effect, the blobs in the colour scattergram are extended until they fill the *cells* defined by the *watershed*.

●

Image 6.12 Examining dress fabric. The original colour image is shown in Plate 4(a). (a) *Top-left*: Image derived by digitising the R channel. *Top-right*: G channel. *Bottom-left*: B channel. (b) Colour scattergram (black spots). The white lines indicate the *watershed* contours separating the black spots. The areas they enclose define the blobs generated during colour generalisation.

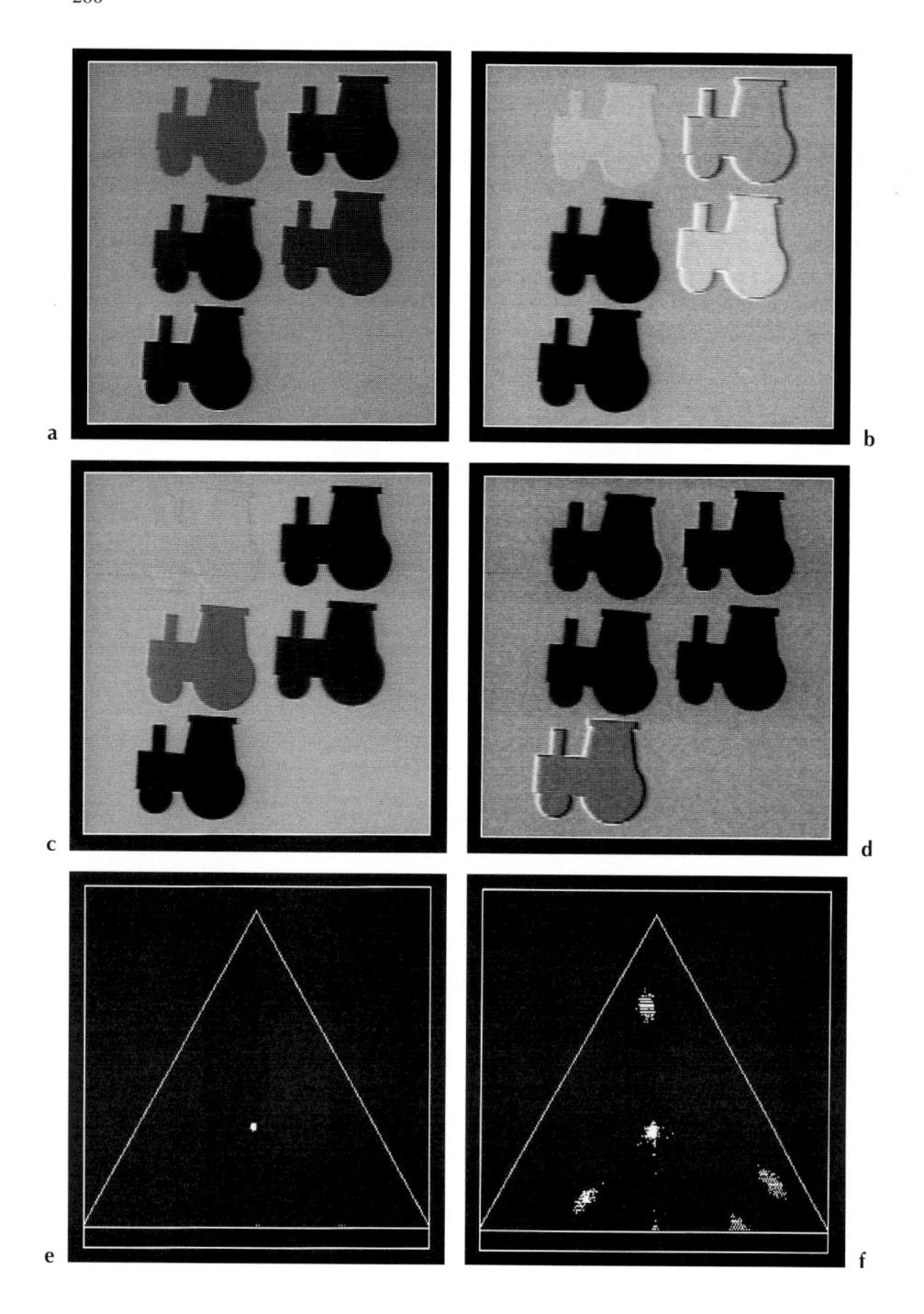

Image 6.1 a–f

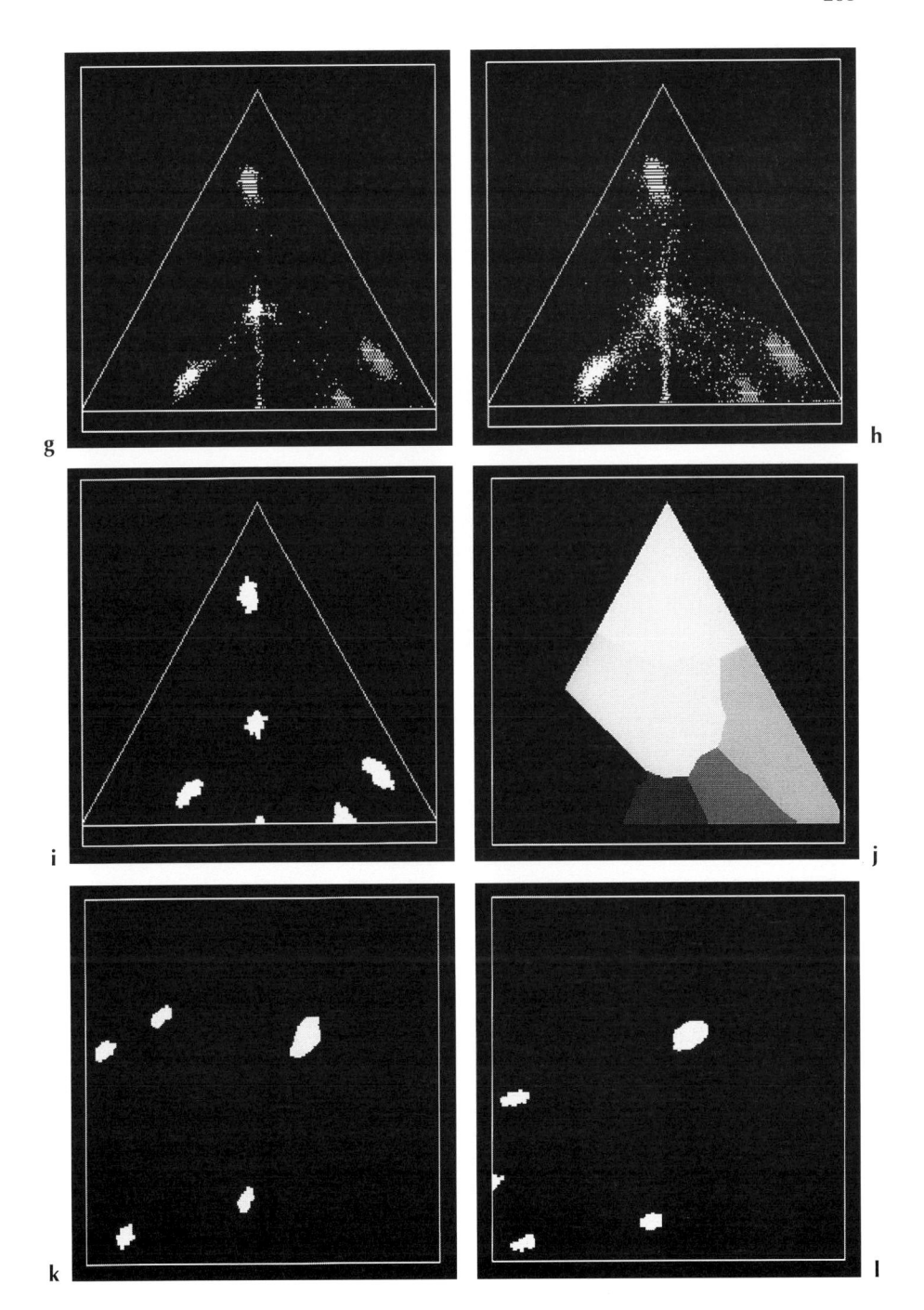

Image 6.1 g–l

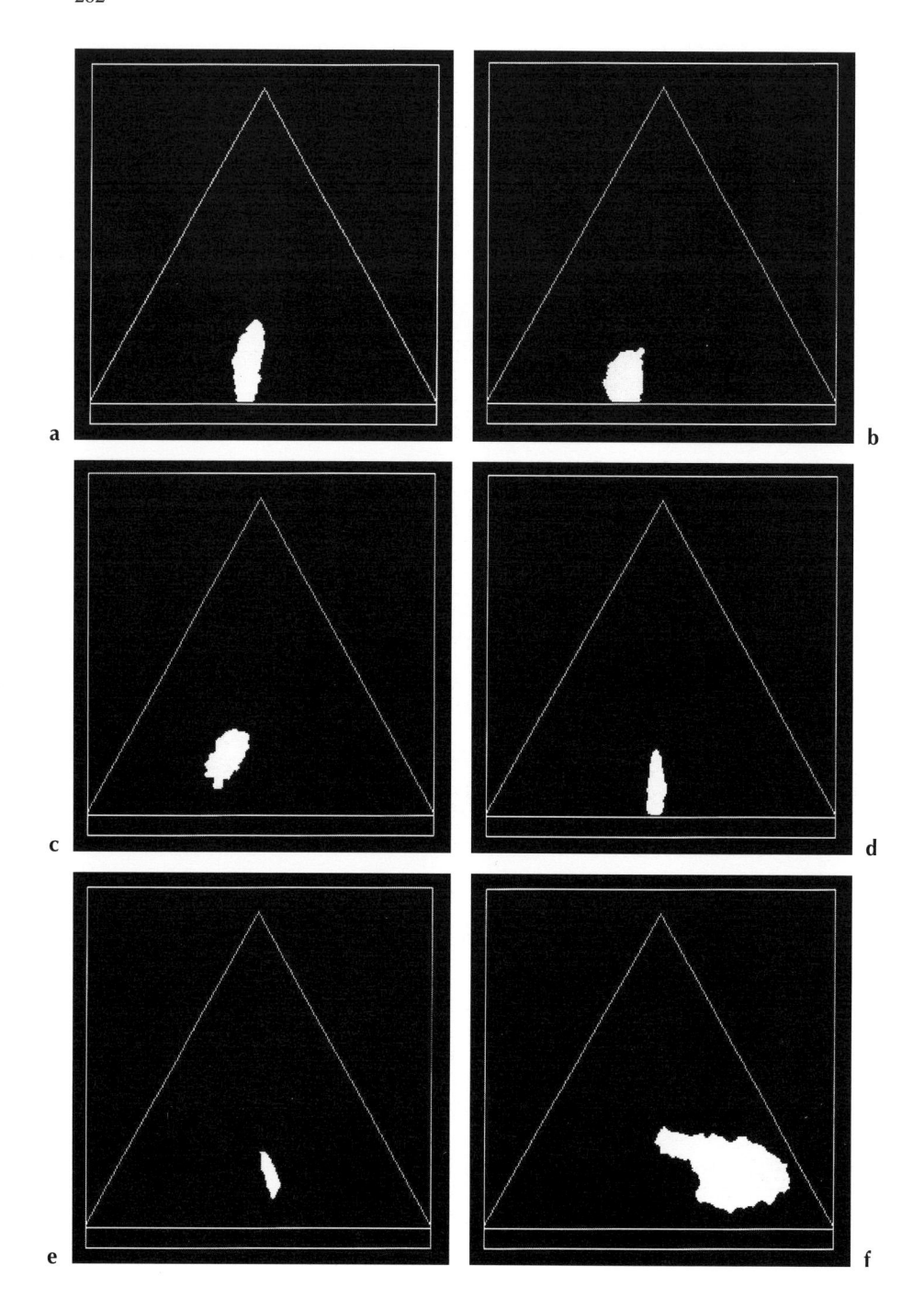

Image 6.2 a–f

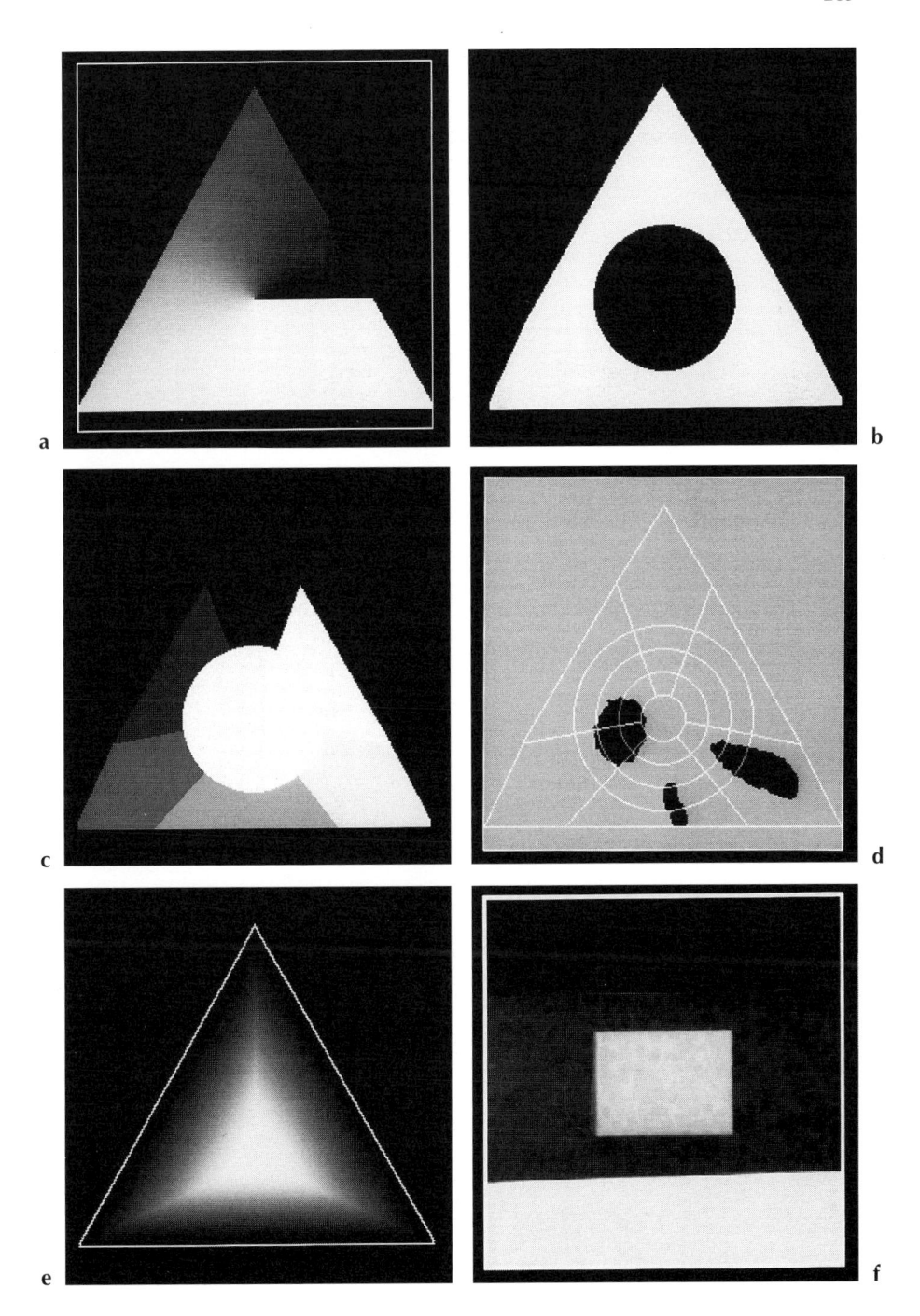

Image 6.3 a–f

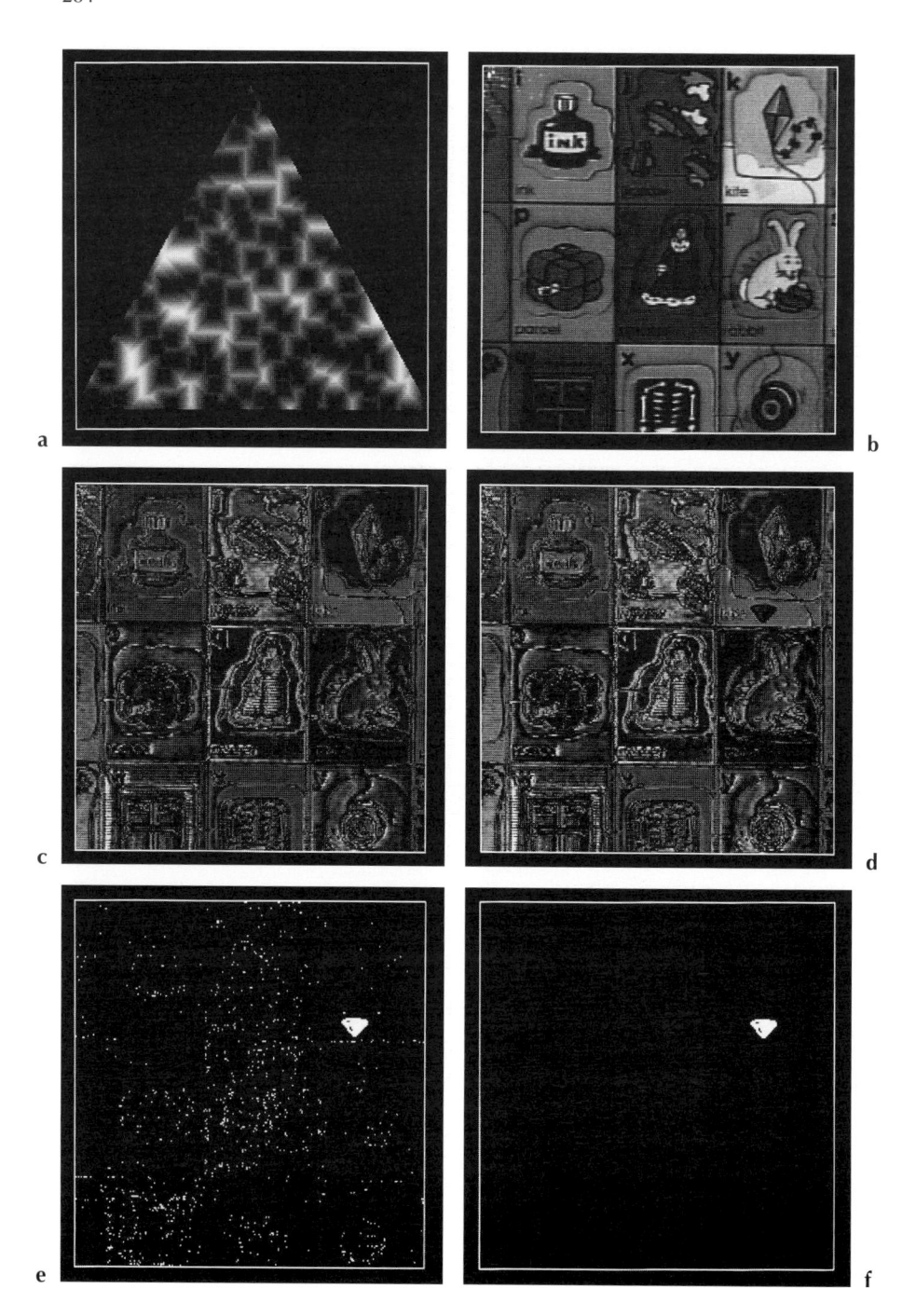

Image 6.4 a–f

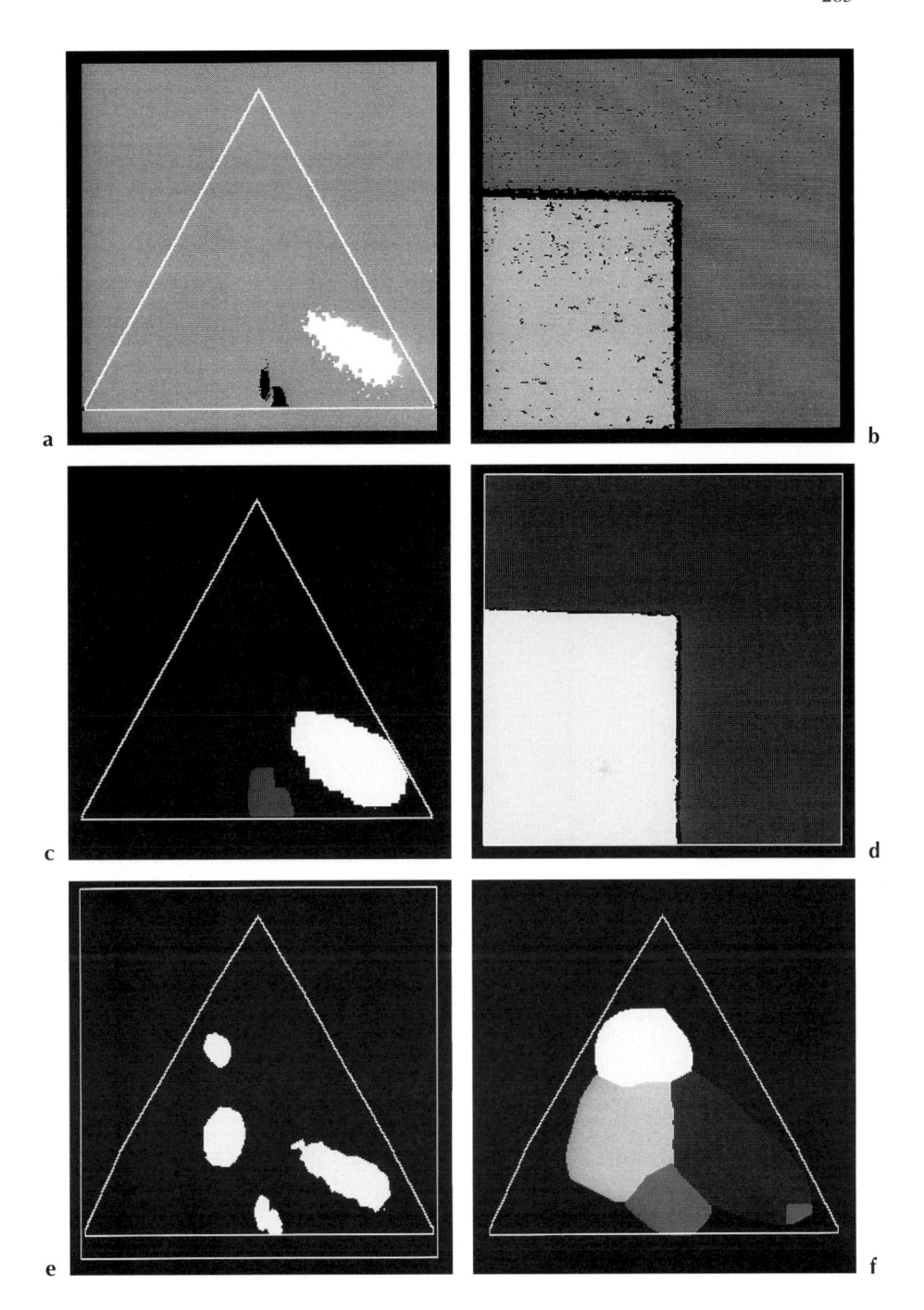

Image 6.5 a–f

Image 6.6 a–d

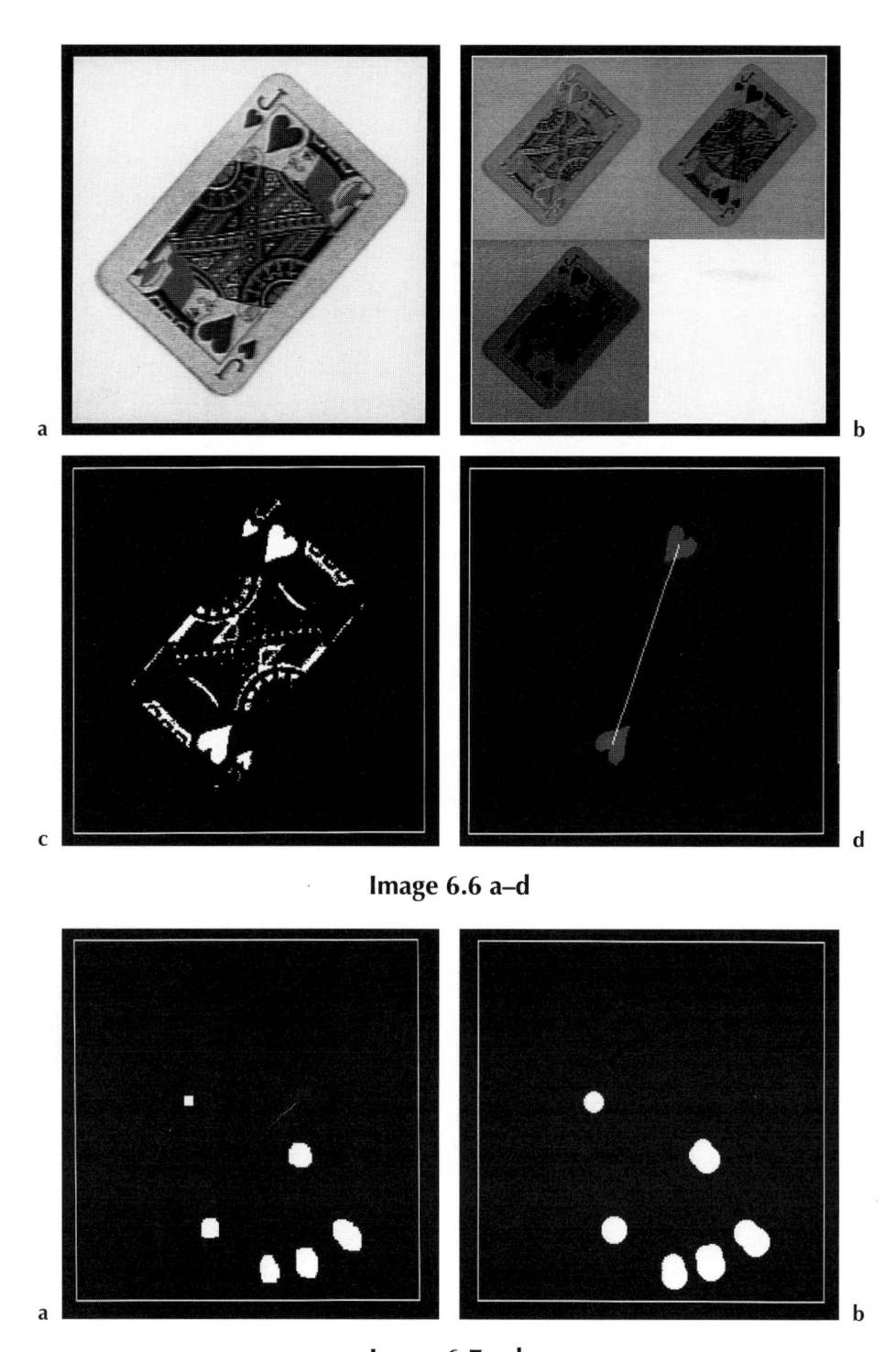

Image 6.7 a–b

287

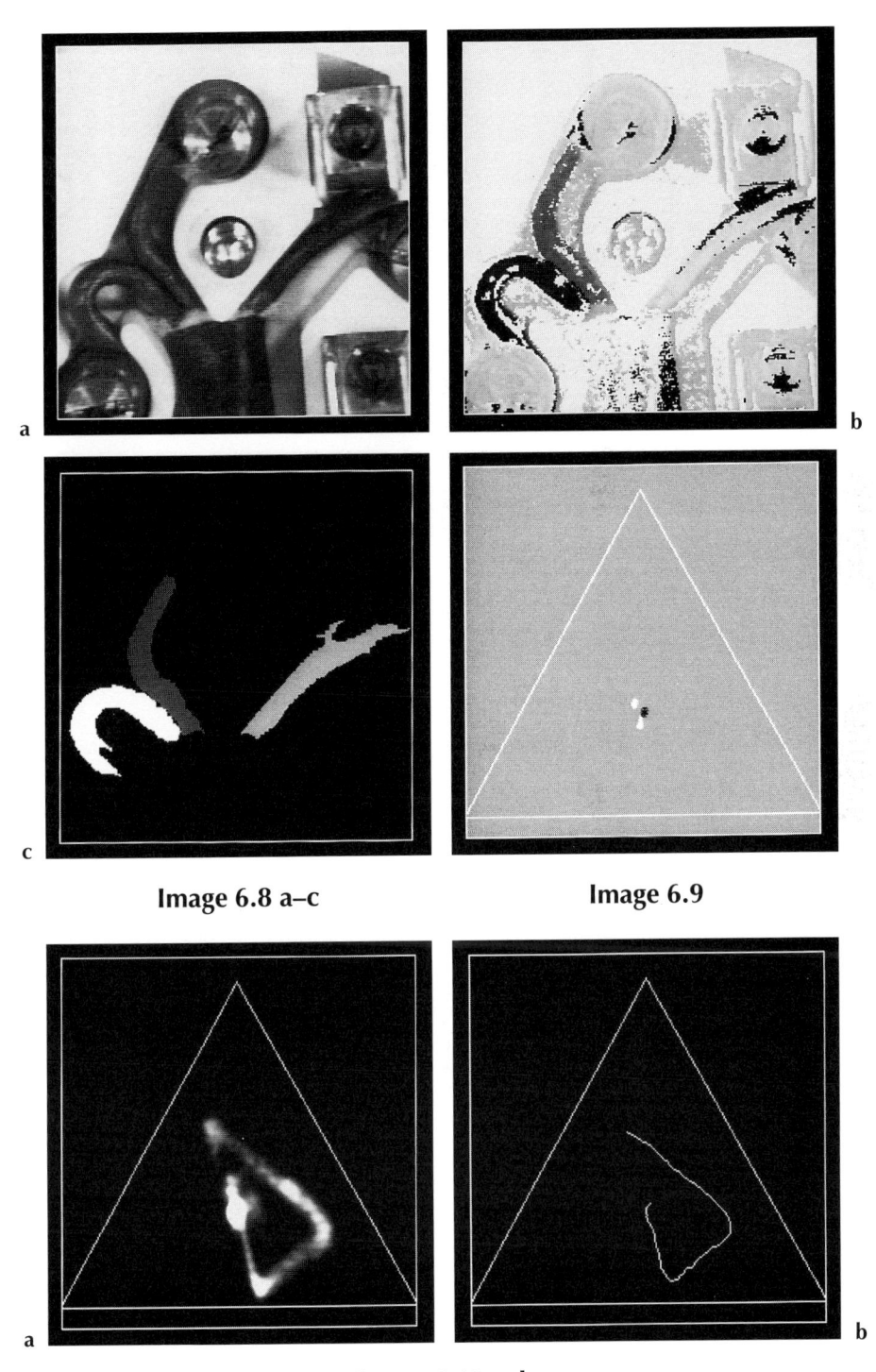

Image 6.8 a–c

Image 6.9

Image 6.10 a–b

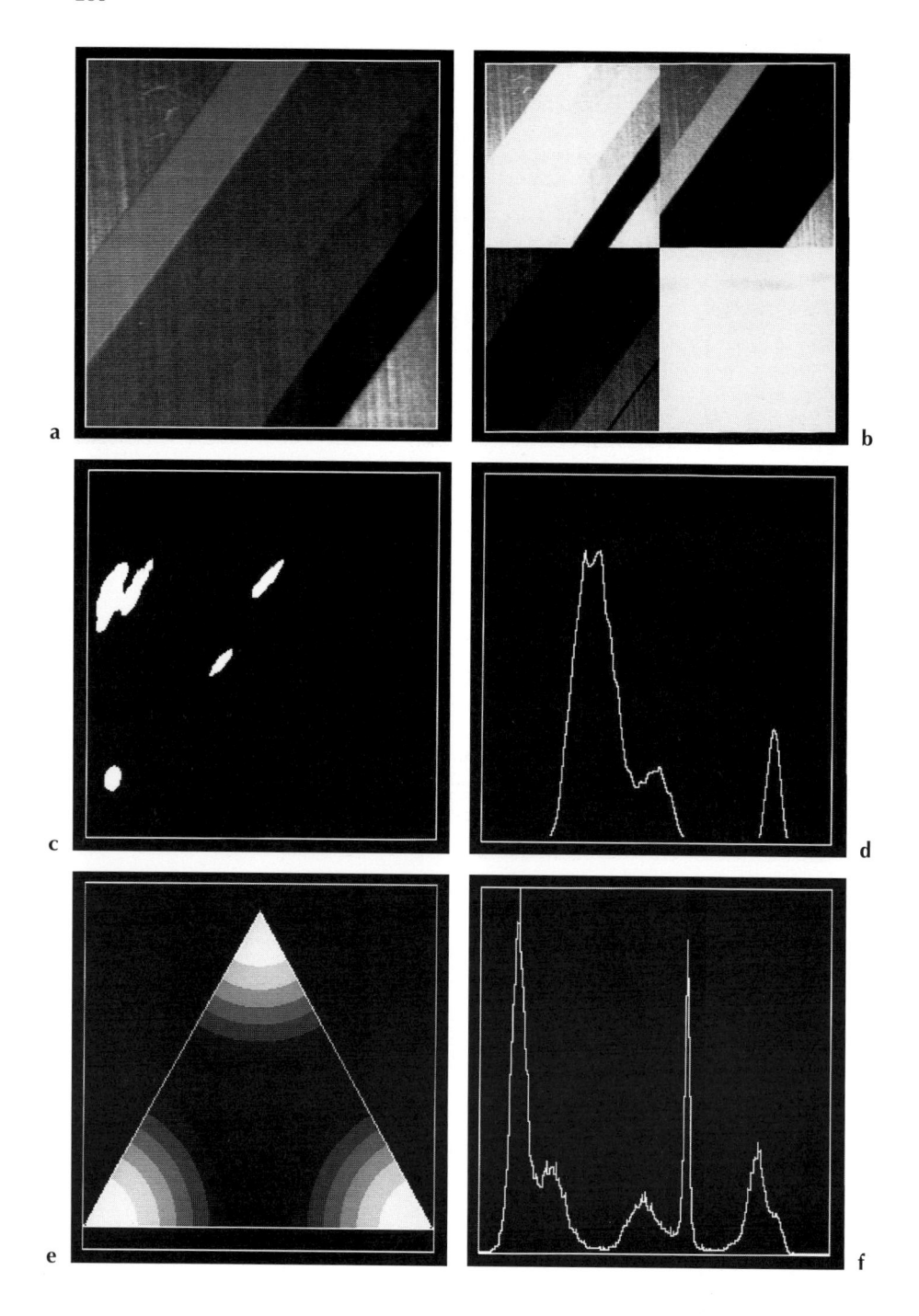

Image 6.11 a–f

Image 6.11 g–j

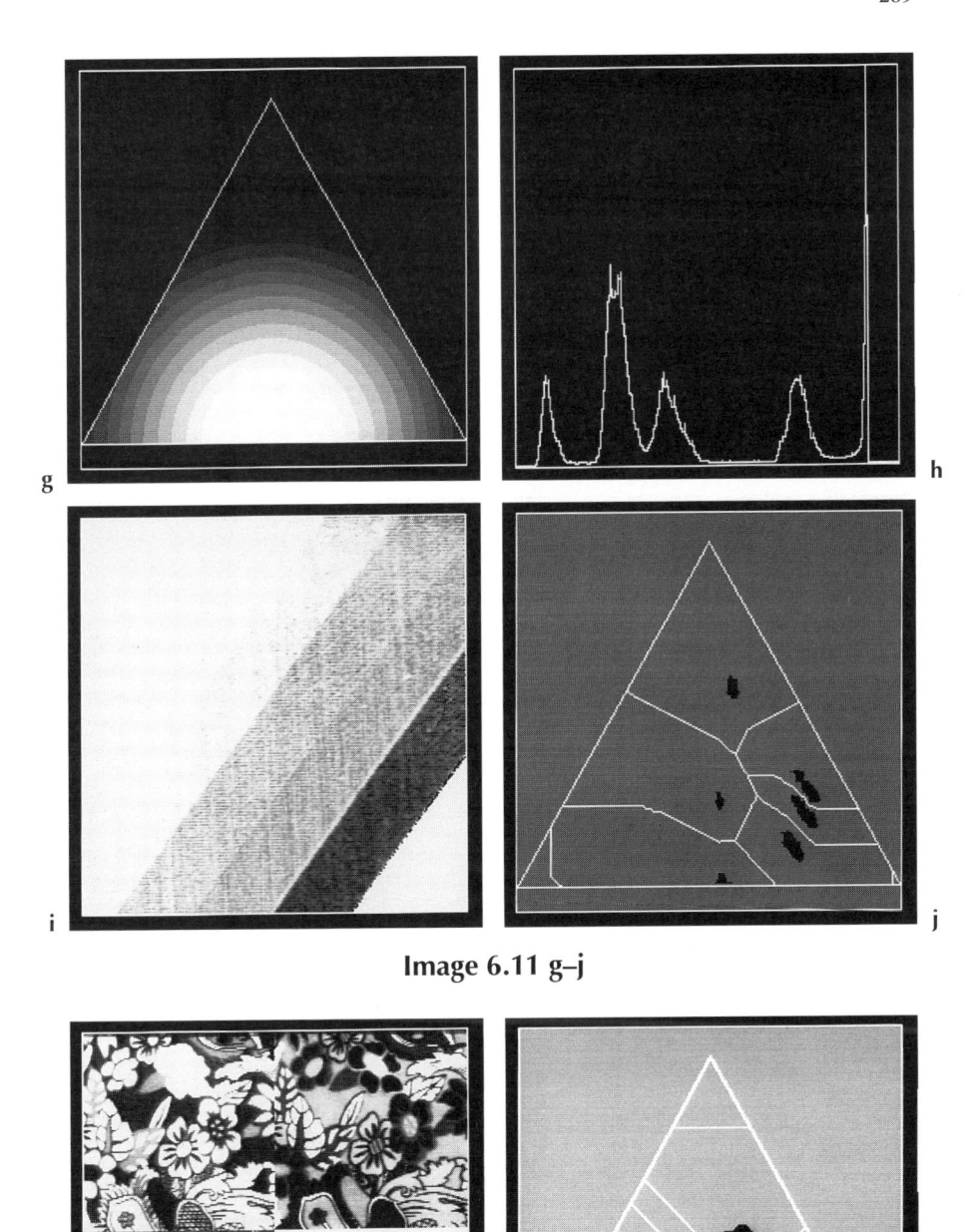

Image 6.12 a–b

COLOUR PLATES

Plate 1 (a) Pseudo-colour applied to: *top*, intensity stair-case (The black and white stripes correspond to levels 0 (zero) and 255, respectively.) *bottom*, intensity wedge (operator *wgx*). (b) Pseudo colour-triangle, formed by generating an intensity wedge and then using the Cartesian-to-polar transformation. (*ctp*) The radius of the central white disc (24 pixels here) can be varied at will. (c) Pseudo colour-triangle, forming the basis of a filter for measuring saturation. (d) Analysing the image of a set of 6 pencil erasers within a transparent "bubble" pack. (Unprocessed video image) (e) Colour histogram. The peaks in this histogram correspond to the following objects / surfaces (from left to right): blue background, blue erasers, green eraser, yellow eraser, orange-red background stripe, red eraser. (f) Output of the hue colour filter. (White disc radius = 24.)

●

Plate 2 (a) Analysing a scene containing three similar shades of yellow that are just distinguishable by the human eye. (Photographic copy of the original artwork) (b) Colour scattergram. (Pseudo-colour display on.) The general concept "yellow" would be represented by a blob that encloses all three of these small spots. (c) Output of the colour filter derived from (b). (d) Analysing the image of a small electronics component (black with shiny *silver* printing), in a transparent plastic bag. The bag has red printing on it and is resting on a white background. (Unprocessed video image) (e) Output of the hue colour filter. (White disc radius = 24.) (f) Binary image, obtained by thresholding (e). (Pseudo-colour display on.)

●

Plate 3 (a) Coloured stripes from the packaging of a well-known domestic product. (Unprocessed video image.) (b) Output of the hue colour filter. (c) Colour scattergram, superimposed on the pseudo colour-triangle. Notice that the white blob at the 3 o'clock position straddles the sharp transition between red and violet in the pseudo colour-triangle. This is the reason that the red stripe in the input image generates a noisy pattern in (b). (d) A new pseudo colour-triangle was created by flipping the *hue* pseudo colour-triangle about its vertical axis. A colour filter was then generated in the usual way. Since there is no sharp transition across the blob at 3 o'clock, the red stripe does not create a noisy pattern, as it did in (b). Although the colour filter outputs corresponding to the red and orange-red stripes are similar, they can be separated reliably, using simple thresholding. (e) Analysing a scene containing four children's building bricks. (Unprocessed video image) (f) Output of the hue colour filter. (White disc radius = 24.)

●

Plate 4 (a) Analysing colours on a piece of dress fabric. (Unprocessed video image.) (b) Output of the *hue* colour filter. (White disc radius = 16.) (c) Unprocessed video image of a quiche. (d) Output of the *hue* colour filter applied to (c). (e) Simulated product package. (f) Key lettering isolated from (e).

Plate 1

292

a

b

c

d

e

f

Plate 2

Plate 3

294

a

b

c

d

e High fat 250g e

f

Plate 4

7

Applications of Intelligent Vision

7.1 Recognition of Printed Patterns

Optical character recognition (OCR) is concerned with the reading of printed text by machine. The subject was first studied seriously in the 1960s, when it was regarded as a very expensive technology. OCR is now common-place; indeed, it is possible to buy a reliable software package that will read laser-quality printed text, for a few hundred dollars.

In this case study, we shall consider the recognition of printed patterns but not conventional text. The programs that we shall describe are capable of distinguishing printed letters and we shall discuss this particular topic. It should be understood, however, that our primary concern is not to re-invent OCR but to demonstrate the power of Prolog+. Our research in this area has been motivated by the observation that an industrial machine vision system is sometimes required to recognise members of just a few well-formed printed patterns. Figure 7.1.1 illustrates several tasks typical of this type and we shall discuss these in turn.

7.1.1 Non-picture Playing Cards

Recognising the playing card suit symbols is straightforward; we simply count the number of bays (i.e. blobs in the convex deficiency). That is, we construct the convex hull (Prolog+ operator *chu*), then apply the blob-fill operator (*blb*), followed by the exclusive OR operator (*xor*). Finally, we count the blobs. If there are four blobs, the card belongs to the *"club"* suit (♣). If there are 2, the card is a *"spade"* (♠). If there is only one, the card is a *"heart"* (♥) and if there are none, the card is a *"diamond"* (♦). Any other value indicates an error. Here is a Prolog+ program for recognising the suit and value of a *non-picture* playing card.

```
playing_card :-
   loa, enc, wri(temp1), thr, neg, big, wri(temp2), cwp(A),
   B is A//2, swi, ndo, hgi(C), list_elements_greater(C,B,D),
   length(D,E), V is E - 1, rea(temp2), cvd, cbl(F), suit(F,G),
   rea(temp1), writeseqnl(['The card is the',V,'of', G]).
```

```
suit(4,clubs).
suit(2,spades).
suit(1,hearts).
suit(0,diamonds).
suit(_, 'unknown suit').
```

The task of recognising the *picture* cards will be solved by recognising the letters A, J, Q and K, and is discussed in Section 7.1.4.

♣ ♠ ♥ ♦

❀ ❁ ✳ ✵

☺ ☻ ☹

A J Q K

Figure 7.1.1 Four typical discrimination tasks,. Each requires choosing among a small number of well-defined printed patterns.

7.1.2 "Stars"

The "stars" in Figure 7.1.1 can be distinguished almost as easily, as the suit of a playing card. The following table shows how this can be achieved.

Character	No. of bays	No. of lakes
❀	5	5
❁	8	9
✳	6	6
✵	8	8

Notice that this time, however, we must use another measure, the number of *lakes*. The program can be made more robust and we can, of course, distinguish more classes of patterns, if we derive more measurements simultaneously.

7.1.3 "Smiley Faces"

Although, the "smiley faces" cannot be distinguished quite so easily, the task is nevertheless quite straightforward. Here is the program:

```
face(A) :-
   pre_process_face,      % Convert to binary form & save image on disc
   test_face,             % Is this a valid face image ?
   find_smile(A).         % Find facial expression

face('Face not found').   % Image was not a face

pre_process_face :-
   grb,                   % Digitise image
   enc,                   % Enhance contrast
   thr,                   % Threshold
   neg,                   % Negate image
   blb,                   % Fill lakes
   xor,                   % Exclusive OR - to isolate "lakes"
   wri,                   % Save image for use later
   blb,                   % Fill lakes
   xor,                   % Exclusive OR - to isolate "lakes"
   biggest,               % Isolate largest blob i.e. mouth
   yxt,                   % Interchange X and Y axes
   keep.                  % keep image for further analysis

% Is the image likely to be a face ?
test_face :-
   rea,                   % Read image saved earlier
   blb,                   % Fill lakes
   shape_factor(A),       % Calculate (Area/Perimeter^2)
   A > 0.8,               % Is shape factor large enough for
                          % approximate circle?
   rea,                   % Read image again
   eul(B),                % Euler number
   B is -2.               % Face has one blob and three lakes

% Calculate the shape factor
shape_factor(C) :-
   cwp(A),                % Area
   perimeter(B),          % Perimeter
   pi(Pi),                % Pi = 3.14...
   C is (4*Pi*A)/(B*B).   % 4*pi*Area/Perimeter

% Find expression of mouth
find_smile(happy) :-      % Mouth turned up - happy
   fetch,                 % Recover image saved earlier
   rox,                   % Row maximum
   chu,                   % Convex hull
   max,                   % Logical OR
   blb,                   % Fill lakes
   xor,                   % Exclusive OR
   cwp(N),                % Area
   N ≥ 100.               % Is mouth turned up enough?

% Same as previous clause, except for line 2
find_smile(sad) :-
   fetch,
   lrt,                   % Invert X-axis
   rox, chu, max, blb,
   xor, cwp(N),
```

```
writenl(N),
N ≥ 100.

find_smile(neutral).        % Mouth is neither turned up nor down
```

While the above program was originally intended for recognising the printed faces in Figure 7.1.1, it is robust enough to cope with hand-drawn faces, provided the outline is a closed contour. (Figure 7.1.2) The reader may like to ponder about the changes needed to cope with drawings where this condition is not met.

Figure 7.1.2 Four hand-drawn face images.

7.1.4 Alphanumeric Characters

The task of distinguishing between the following *sans serif* characters (Arial font) can be easily achieved using the numbers of lakes, bays, skeleton limb-ends and joints.

A, J, Q, K

For such a simple discrimination task, a person can easily write down the feature values, without the use of an image processing system, and then verify that they are unique:

Character	[Lakes, Bays, Limb-ends, Joints]
A	[1,1,2,2]
J	[0,1,2,0]
Q	[1,2,2,1]
K	[0,3,4,2]

For more complex recognition tasks, requiring the discrimination of more character classes, it may well be necessary to employ more measurements. Table 7.1.1 lists the values of 8 parameters which can distinguish the numerals 0 - 9 in Times Roman font. The task of choosing a feature set which can distinguish the

26 lower-case, 26 upper-case, numerals 0 - 9, punctuation and special symbols (+, £, $, @, &, *, ?, /, etc.) is difficult and is almost certainly beyond the capability of a person working unaided. The situation is made worse by the fact that some patterns produce measurement vectors which are unstable. For example, the number of bays may change, depending upon noise.

Pattern	Measurements							
Times Roman Font	1	2	3	4	5	6	7	8
0	0	1	0	X	X	X	X	X
1	1	0	2	X	X	X	X	1
2	1	0	2	1	X	X	X	2
3	1	0	2	0	2	3	3	3
4	0	1	3	X	X	X	X	X
5	1	0	2	0	X	X	X	2
6	0	1	1	0	X	X	X	X
7	1	0	1	X	1	2	2	X
8	X	2	X	X	X	X	X	X
9	0	1	1	1	X	X	X	X
10	2	1	2	X	X	1	X	1
11	2	0	3	X	X	0	X	1
12	2	0	3	1	X	2	X	1

Table 7.1.1 Recognising the digits 0 - 9 and the compound patterns '10' - '12', in Times Roman font. X indicates *"Don't Care"*. Measurements 1 - 8 are defined as follows: *1.* Euler number. *2.* Number of holes (lakes). *3.* Number of indentations (bays). *4.* Equal to 1 if the largest blob in the convex deficiency (i.e. bay or lake) is above the second largest. Equal to 0 otherwise. *5.* The number of times the vertical line L1 intersects the character. (See Figure 7.1.3) *6.* The number of times the vertical line L2 intersects the character. (See Figure 7.1.3) *7.* The number of times the vertical line L3 intersects the character. (See Figure 7.1.3) *8.* The number of blobs generated by the following sequence: *rox, xor, skw, exw*.

When we try to accommodate a mixture of fonts, the situation quickly becomes quite unmanageable. This is why we need to employ learning techniques. Before we consider this topic in detail, here is a Prolog+ program which can discriminate the numerals 0 - 9, in Times Roman font. (See Figure 7.1.3.)

Program
```
% Top level predicate: recognising well formed printed alpha-
% numeric characters.
recognise_alpha_numeric(X) :-
   alpha_numeric_features(A,B,C,D,E,F,G,H),
```

```
      stored_features(A,B,C,D,E,F,G,H,Q),
      writeseqnl([[A,B,C,D,E,F,G,H], ' was classified as ',X]).

% Database: Times Roman font. Other fonts may not be recognised
% correctly
stored_features(0,1,0,_,_,_,_,_,zero).
stored_features(1,0,2,_,_,_,_,1,one).
stored_features(1,0,2, 1, _,_,_,2,two).
stored_features(1, 0, 2, 0, 2, 3, 3,3,three).
stored_features(0, 1, 3, _,_,_,_,_,four).
stored_features(1, 0,2, 0, _,_,_,2,five).
stored_features(0, 1, 1, 0, _,_,_,_,six).
stored_features(1, 0, 1, _, 1, 2, 2,_,seven).
stored_features(_,2,_,_,_,_,_,_,eight).
                       % Only one feature needed to find '8'
stored_features(0, 1, 1, 1, _,_,_,_,nine).
stored_features(_,_,_,_,_,_,_,_,not_known).
                       % Character is not recognised

% Eight "logical" shape features. (i.e. these must match the stored
% values exactly)
alpha_numeric_features(A,B,C,D,E,F,G,H) :-
    wri,                  % Save image for future reference
    eul(A),               % Measurement 1: Euler number
    holes,                % Isolate lakes (holes)
    cbl(B),               % Measurement 2. Number of lakes (holes)
    rea,                  % Get stored image back again
    bays,                 % Isolate bays (indentations)
    cbl(C),               % Measurement 3. Number of bays
    rea,                  % Get stored image back again
    biggest_bay_top(D),   % Measurement 4: Biggest bay above/below 2nd
                          % largest
    rea,                  % Get stored image back again
    rox,                  % Row maximum - form "shadow"
    xor,                  % Exclusive OR - by shadow by removing
                          % original figure
    skw,                  % Shrink white to eliminate any very small
                          % regions present
    cbl(H),               % Measurement 8: Number of blobs in "shadow"
    rea,                  % Get stored image back again
    vertical_scan_count(E,F,G). % Measurements 5-7. Vertical slicing

% Count number of times 3 equally spaced vertical slices cut the
% figure
vertical_scan_count(P,Q,R) :-
    normalise,            % Normalise position in centre of image
    wri,                  % Store image
    dim(A,B,_,_),         % Minimum and maximum X and Y
    C is (B - A)//4,
    E is 128 - C, F is 128 + C,
    zer,                  % Black image
    vpl(E,1,E,255,64),    % Draw vertical line left of centre of image
    vpl(D,1,D,255,128),   % Draw vertical line through centre of image
    vpl(F,1,F,255,192),   % Draw vertical line right of centre of image
    rea,                  % Read normalised image
    min,                  % Mask figure and 3 lines
    wri, thr(64,64),
    count(blobs,P),       % No. of chords formed by left-hand vertical
                          % line
    rea, thr(128,128),
    count(blobs,Q),       % No. of chords formed by central vertical
                          % line
    rea, thr(192,192),
```

```
    ccount(blobs,R).     % No. of chords formed by right-hand vertical
                         % line
% Is the biggest blob in the convex deficiency above below the
% second largest. (Similar to predicate bbt.)
biggest_bay_top(0) :-
    chu,              % Convex hull
    max,              % Maximum - superimpose convex hull on figure
    blb,              % Fill holes
    xor,              % Exclusive OR: isolate lakes and bays and lakes
    biggest,          % Largest blob
    cgr(_,Y1),        % Find its vertical position
    xor,              % Remove biggest blob in convex deficiency
    biggest,          % Gets second largest blob in convex deficiency
    cgr(_,Y2)         % Find its vertical position
    (Y1 > Y2, A is 0); A is 1).    % Fix "output" value
```

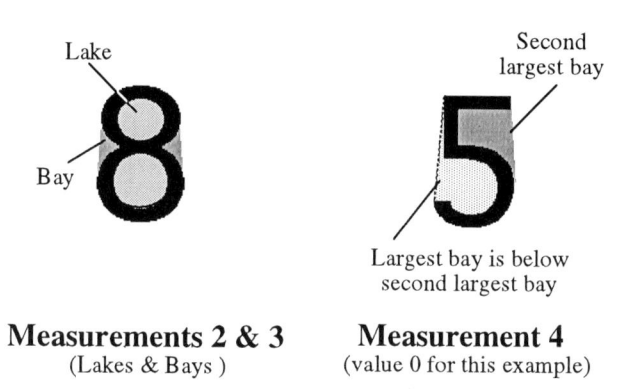

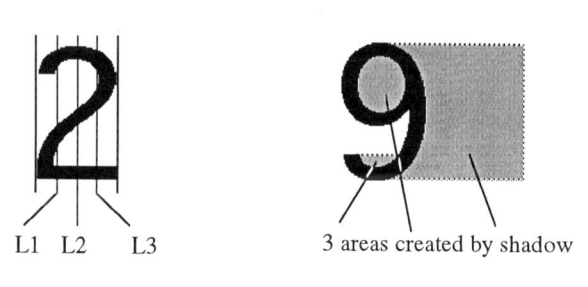

Measurements 2 & 3
(Lakes & Bays)

Lake
Bay

Measurement 4
(value 0 for this example)

Second largest bay

Largest bay is below second largest bay

Measurements 5 - 7
(Scan lines L1-L3 intersect the numeral (3,3,3) times)

L1 L2 L3

Measurement 8
("Shadow" produces 3 blobs)

3 areas created by shadow

Figure 7.1.3 Measurements 2 - 8 for recognising the digits 0 - 9 and the compound patterns '10' - '12', in Times Roman font.

Comments

An observant reader will have spotted the fact that the set of measurements created by *alpha_numeric_features* is redundant; the number of lakes is equal to

(1 - E), where E is the Euler number. However, there are some occasions when it is useful to know the Euler number. Two lower-case letters, **i** and **j** consist of more than one component and the Euler number provides an easy way to distinguish these letters from **1** and **l**. In its present form, however, the program could not distinguish **i** from **j**, nor **1** from **l**. The inclusion of the Euler number also enables the program to cope with compound patterns, such as **10**, **11** and **12**.

Logical and Analogue Shape Measurements

The predicate *recognise_alpha_numeric* suffers from a serious problem: the measured and stored feature lists must match one another perfectly. Hence, the set of features generated by *alpha_numeric_features* must be exactly the same as one of the lists held in *stored_features*. For this reason, the predicate *recognise_alpha_numeric* behaves in a logical manner, since it requires an exact match between measured and stored parameter values. (Of course, *"don't care"* conditions are allowed under this scheme.)

In Figure 7.1.4, three *analogue* measurements are defined for describing the alpha-numeric pattern **'5W'**. By the term *analogue*, we mean that the measurements are continuously variable and a perfect match with stored values may not be possible

Lake

Bays

Recognition Criteria (logical)

One lake
Five bays
Two limb ends

Recognition Criteria (analogue)

Aspect ratio (width : height)
Ratios of areas of lakes (e.g. biggest : second largest)
Ratio of area of convex hull : area of original figure

Figure 7.1.4 *Logical* and *analogue* measurements for describing the compound pattern **'5W'**. Notice that this consists of a singe blob.

7.2 Manipulation of Planar Objects

Our objective in what follows is to pick up only those flat laminate objects that are known to the vision system, using a visually guided robot. "Unknown" objects are to be located and the user warned that moving them would be unsafe. It is possible to use any robot which is able to manipulate objects lying on a plane table. We shall explain how the Flexible Inspection Cell (FIC) can be used for this purpose. However, a SCARA or gantry robot, fitted with an overhead camera could be used instead. There are three phases in the operation of such a system:

(a) *Calibration.* The automatic calibration of a robot vision system which uses an overhead camera has already been explained in detail (Section 5.7) and so will not be discussed again here.
(b) *Learning.* The vision system learns typical values for a set of size and shape parameters characterising each class of objects that is to be moved by the robot.
(c) *Recognition.* The vision system guides the robot a it picks up objects that are similar to those encountered during the learning phase.

7.2.1 Assumptions

As usual, a series of assumptions is imposed, to make the problem tractable.

(i) The Flexible Inspection Cell (Section 5.5) will be used to demonstrate the ideas outlined below. Remember that the FIC incorporates an (X,Y,θ)-table, a pneumatic pick-and-place arm, computer-controlled lighting and an overhead camera, which looks vertically downwards, onto the table top.
(ii) The top surface of the (X,Y,θ)-table is matt black.
(iii) A set of thin, nominally white laminate objects are placed haphazardly on the table top.
(iv) The objects lying on the table top do not touch, or overlap.
(v) During the *learning* phase, a person is able to name each object that the system sees.
(vi) During the *recognition* phase, the system is expected to work autonomously, without any human intervention. However, the user is to be informed about unknown objects and those that are considered to be unsafe to handle using the robot.
(vii) Objects that are similar to those seen during training are picked up by the robot.
(viii) Objects that are unlike any seen during training are identified but are not picked up by the robot.
(ix) Some objects would be unsafe to lift because the suction gripper is too large and overlaps their sides. Such objects are to be identified but not lifted by the robot.
(x) The FIC has already been calibrated. (See Section 5.7.1.)

(xi) A simple goal of the form *pick(X,Y,Theta,Q)* is used to order the FIC to pick up an object of type *Q*, located at [X,Y] and with orientation *Theta*. It is assumed that the label *Q* indicates where the object is to be placed. Thus, the system is able to sort objects, placing them in bins according to their type.

(xii) To minimise errors when calculating the positions of an object on the table-top, the lights should be placed close to the overhead camera. However, care must be taken to avoid glinting. It will be assumed that glinting does not occur.

Compared to our naïve expectations, the task that we have just described is surprisingly complicated; there are many different aspects of the system behaviour which do not immediately come to mind.

7.2.2 Significance

Many industrial assembly, and sorting tasks can be solved using the same basic arrangement, consisting of an overhead camera to guide a robot that moves an object on a flat table top. Consider the task of sorting components made using a stamping, moulding, or die-casting machine. It is assumed that the sprue has been removed automatically and that a series of different components have fallen in random position and orientation onto the table. The visually guided robot could be used to sort them, placing each type of product into a separate bin. Objects that touch or overlap can be accommodated within the constraints imposed by the assumptions listed above. This process consists of two stages:

(i) Two or more touching / overlapping objects together form a single "unknown object" that the robot will not try to lift. Thus, identifying them is a necessary prelude to step (ii). (It is, of course, a good idea to pick up "known" objects first.)

(ii) The robot can nudge an "unknown object" from the side, to try and separate it into objects that it can recognise and handle individually.

The task of identifying shapes is an important prelude to packing. (See Section 7.3 for a detailed discussion on the issues relating to automated packing systems.)

7.2.3 Simple Shape Measurements

In Section 7.1, we described a set of shape measurements, which could be used/adopted for the present task. However, the six parameters calculated by the predicate *measurements,* as used in our experiments, are slightly different and are defined below. (Also see Figure 7.2.1.)

1. Area of the object silhouette. (Variable A)

2. Length of the object, measured along the principal axis (i.e. axis of minimum second moment). (Variable D1)
3. Width of the object measured in a direction normal to the principal axis. (Variable D2)
4. The ratio D1/D2. (Variable R)
5. The area of the convex hull divided by the area of the silhouette. (Variable S)
6. The area of the minimum enclosing rectangle, divided by the area of the silhouette. (Variable T)

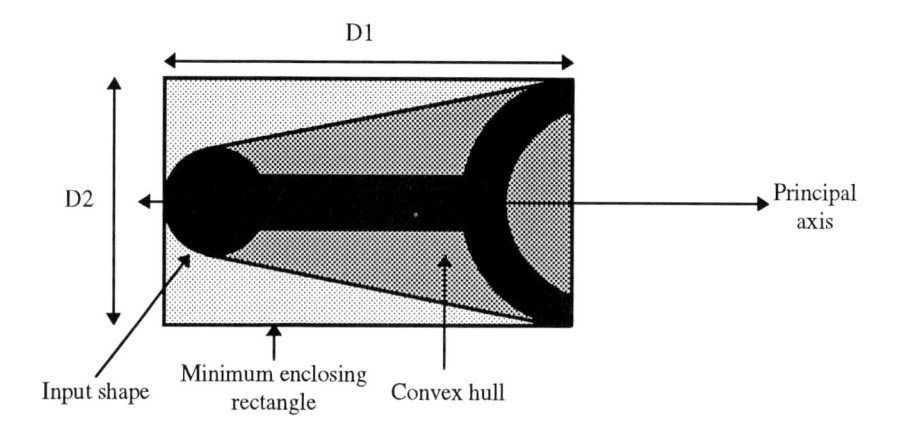

Figure 7.2.1 Parameters calculated by the predicate *measurements*.

A crude linear rescaling is included in the definitions of A, R, S and T, to make sure that each of the measured parameters lies in roughly the same range. Notice that variables R, S and T are all size independent, while D1 and D2 all vary with object size and optical magnification.

```
measurements(V)  :-
   rea,                    % Read "input" image saved earlier
   cwp(B),                 % Area  of the object silhouette
   A is B//40,             % Rescaling
   rea,                    % Read "input" image
   normalise,              % Normalise both position and orientation
   dim(X1,X2,Y1,Y2),       % Dimensions of minimum enclosing rectangle
   rea,                    % Read "input" image
   chu,                    % Convex hull
   blb,                    % Fill it
   cwp(C),                 % Area of convex hull
   S is (100*C)//B,        % Simple arithmetic, including rescaling
   D1 is X2 - X1,          % Length along principal axis
   D2 is Y2 - Y1,          % Width normal to principal axis
   R is (100*D2)//D1,      % Simple arithmetic, including rescaling
   T is (100*D1*D2)//C,    % Simple arithmetic, including rescaling
   V = [A,D1,D2,R,S,T],    % Fix "output" list
   rea.                    % Read "input" image
```

Figure 7.2.2 shows four objects used to demonstrate the ideas we are discussing. The learning program *(learn)* defined in Section 7.2.5 calls *measurements* and when applied to this image asserts the following facts into the database.

```
object_data([100, 141, 78, 55, 208, 131], 'Conrod').
object_data([67, 94, 89, 94, 197, 157], 'Y shape').
object_data([97, 123, 110, 89, 190, 182], '3-pointed star').
object_data([97, 158, 50, 31, 139, 144], 'Spanner').
```

It is important to note that *learn* has been written in such a way that it is a trivial matter to compute different or additional shape / size parameters: we simply redefine the predicate *measurements*. No other changes to the program are necessary.

Figure 7.2.2 Four objects for use with *learn* and *recognise*.

7.2.4 Learning and Recognition

Before we present the learning program, we must spend a little time discussing how decisions can be made when a set of imprecise measurements is available. When a program such as *measurements* is applied to a set of objects of the same nominal type, the parameter values so obtained are liable to vary. Even if they were all identical, quantisation noise would cause some fluctuation of the measured values. When we apply *measurements* in its present form, each object is described by a set of six numbers. In more general terms, an object (Q) can be described by a vector, X, containing n numbers $(X_1, X_2, ..., X_n)$. A set of m reference vectors, describing archetypal objects of each class, will be stored by our program. These will be denoted by $Y_1, Y_2 ... , Y_m$, where $Y_i = (Y_{i,1}, Y_{i,2}, ..., Y_{i,n})$

The Y_i ($i = 1, ..., m$) describe objects which have been examined by a human inspector and are known to be "good" and are typical of their class. (Later, we shall see that it is possible to have more than one archetype representing each class.) The *similarity* between two objects represented by vectors X and Y_i can be assessed by measuring the *Euclidean distance* ($D(X,Y_i)$) between them:

$$D(X,Y_i) = \sqrt{(X_1 - Y_{i,1})^2 + (X_2 - Y_{i,2})^2 + \ldots + (X_n - Y_{i,n})^2}$$

The larger $D(X,Y_i)$ is, the smaller the similarity is between the shapes they represent. Thus, an object of unknown type and which is represented by a vector X can be attributed to an appropriate class by finding which of the $Y_1, Y_2 \ldots, Y_m$ is closest to X. This leads us to the idea of a *Maximum Similarity Classifier*. The more usual name for this is a *Nearest Neighbour Classifier (NNC)*, and is explained in diagrammatic terms in Figure 7.2.3. (Also see Figure 6.22.) Each of the Y_i (i = 1, ..., m) is associated with some class label C_i (i = 1, ..., m). The NNC attributes X to class C_i if $D(X,Y_i) \le D(X,Y_j)$, (j $\ne$ i, 1, ..., m).

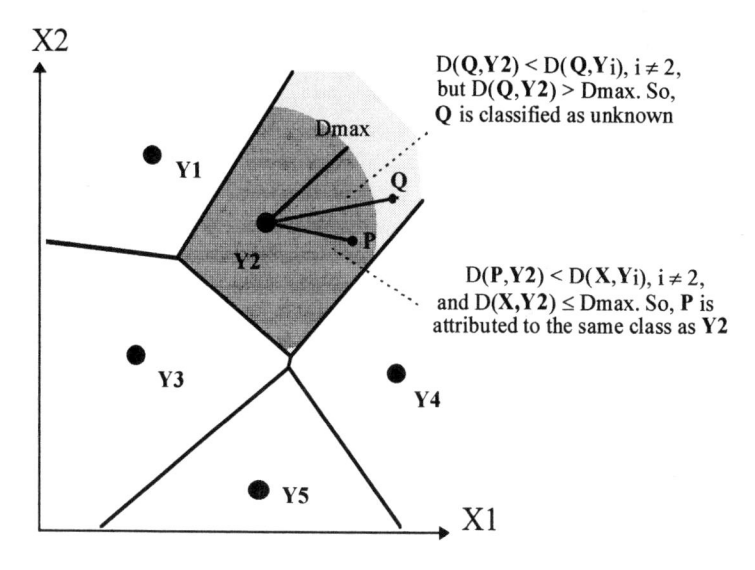

Figure 7.2.3 Nearest Neighbour Classifier (NNC).

In our simple exercise there is probably no need to store more than one reference vector Y_i for a given value of C_i. We are considering shapes that are very similar to their respective archetypes but are very different from members of any other class. However, the NNC does permit this option. In order to select suitable values for the Y_i in this straightforward application, we simply apply *measurements* to one "good" example of each category of shape that we wish to recognise in future. The parameters so obtained are then stored in the Prolog database. This is the basis of the predicate *learn,* whose listing we are about to present. The shape recognition process is accomplished by *recognise,* which implements a simple Nearest Neighbour Classifier.

7.2.5 Program Listing

```
% Learning phase
learn :-
   retractall(object_data(_,_)),   % Clear the database
   preprocess,                      % Simple image processing
   analyse_binary_image1.           % Measure each blob in the image

% Recognition phase
recognise :-
   preprocess,                      % Simple image processing
   analyse_binary_image2.           % Analyse each object/decide what to do

% Generate an image in which each blob is given a different
% intensity
preprocess :-
   grb,                             % Digitise image
   enc,                             % Enhance contrast
   thr,                             % Threshold
   neg,                             % Negate
   ndo,                             % Shade objects
   enc,                             % Enhance contrast
   wri(temp1).                      % Save image

% Analysing blobs during the learning phase
analyse_binary_image1 :-
   select_one_blob,                 % Choose one blob for analysis
   prompt_read(['Object name'],Q),
                                    % Ask user for name of that object
   measurements(V),                 % Now measure it
   assert(object_data(V,Q)),        % Save measurements in the database
   analyse_binary_image1.           % Repeat until all blobs have been
                                    % analysed

analyse_binary_image1 :- writenl('FINISHED').
                                    % No more blobs to be processed

% Analysing blobs during the recognition phase
analyse_binary_image2 :-
   select_one_blob,                 % Choose one blob for analysis
   measurements(V),                 % Now measure it
   writeseqnl(['Vector: ', V]),
                                    % Message for the user
   nnc(V,S,Q),                      % Nearest neighbour classifier
   pick_up(Q,_,_,_),                % Robot now picks up the object
   message(['Object class: ', Q, 'Distance =',S]),
   nl, nl,                          % Message for the user
   analyse_binary_image2.           % Repeat until all blobs have been
                                    % analysed

analyse_binary_image2 :- writenl('FINISHED').
                                    % No more blobs to be processed

% Selecting one blob from a stored image
select_one_blob :-
   rea(temp2),                      % Read stored image
   gli(_,X),                        % Find highest intensity in it
   !,                               % Force failure if next sub-goal fails
   X > 0,                           % Any more objects to be analysed?
   hil(X,X,0),                      % Eliminate selected blob from further
                                    % consideration
   rea(temp2),                      % Read stored image again
```

```prolog
    swi,                    % Switch current and alternate images
    wri(temp2),             % Save depleted image
    swi,                    % Switch current and alternate images
    thr(X,X),               % Select the next blob to be analysed
    wri(temp1).             % Save it

% Normalise position and orientation of blob in binary image. Robot
% is not moved
normalise :-
    lmi(X,Y,Z),             % Centroid, [X,Y]. Orientation, Z
    X1 is 128 - X, Y1 is 128 - Y, Z1 is -Z,
    psh(X1,Y1),             % Shift so centroid is at centre of image
    tur(Z1).                % Rotate so principal axis is horizontal

% Nearest Neighbour Classifier (slight variation compared to page
% 263). The first clause initialises the MacProlog property "nnc"
nnc(_,_,_) :-
    remember(nnc,[1000000,'Not known']),
                            % ≥ largest possible distance
    fail.

% Finding the stored point that is closest to the "input vector" X
nnc(X,_,_) :-
    object_data(Y,Z),       % Consult database of stored vectors
    euclidean_distance(X,Y,0,D),
                            % D= Euclidean dist. between X and Y
    recall(nnc,[E,_]),      % Find previous minimum
    E > D,                  % Is new values smaller?
    remember(nnc,[D,Z]),    % It is! So, store it for use later
    fail.                   % Backtrack to "object_data"

%   Terminating recursion
nnc(_,D,X) :-
    recall(nnc,[D,X]),      % Find minimum distance to stored point
    D ≤ 20,
    writeseqnl(['Class: ',X,'Distance to nearest neighbour =',D]),
    !.

nnc(_,'too large','not known').

% Euclidean distance (slightly different from page 255).
euclidean_distance([],_,A,B) :-
    sqrt(A,B),              % B is square root of A
    writeseqnl(['Euclidean distance =',B]).

% Second terminating clause - second "input" list is empty
euclidean_distance(_,[],A,B) :-
    sqrt(A,B),              % B is square root of A
    writeseqnl(['Euclidean distance =',B]).

% Find sum of squares of differences between corresponding elements
% of 2 input lists
euclidean_distance([A|B],[C|D],E,F) :-
    G is (A - C)*(A - C) + E,   % Sum of squares of differences
    !,
    euclidean_distance(B,D,G,F).  % Repeat until all list elements
                                  % done

% Virtual robot. We move the image not the real object
% What to do when the NNC cannot classify this object.
pick_up('not known',_,_,_) :-
    rea(temp1),             % Read saved image of blob
    normalise,              % Normalise position and orientation
```

```
        vpl(1,128,256,128,128),    % Horizontal line through image centre
        vpl(128,1,128,256,128),    % Vertical line through image centre
        writenl('Robot will not try to pick up an unknown object').

% Object is of known type and it is safe to lift it with suction
% gripper
pick_up(Q,X,Y,Z) :-
        rea(temp1),                % Read saved image of blob
        lmi(X,Y,Z),                % Position and orientation
        normalise,                 % Normalise position and orientation
        cwp(A),                    % Count white points
        draw_sucker,               % Draw "footprint" of the gripper
        cwp(B),                    % Count white points again
        A is B,                    % Is white area same?
        vpl(1,128,256,128,128),    % Horizontal line through image centre
        vpl(128,1,128,256,128),    % Vertical line through image centre
        writeseqnl(['Located at:', [X,Y],'Orientation:',Z]),
        writeseqnl(['Robot will pick up the',Q]). % Message for user

% Safe lifting is not possible because suction gripper overlaps
% edge of silhouette
pick_up(Q,X,Y,Z) :-
        rea(temp1),                % Read saved image of blob
        normalise,                 % Normalise position and orientation
        vpl(1,128,256,128,128),    % Horizontal line through image centre
        vpl(128,1,128,256,128),    % Vertical line through image centre
        hin,                       % Halve int. to make sucker-disc visible
        draw_sucker,               % Draw "footprint" of the gripper
        writeseqnl(['Sucker is too large for safe lifting - robot will
not pick up the',Q]).

% Draw a white disc to represent the "footprint" of the suction
% gripper
draw_sucker :-
        draw_disc(128,128,6,255). % Draw white disc, radius 6 pixels.
```

7.2.6 Sample Output of Recognition Phase

```
Vector:   [100, 141, 78, 55, 208,    Vector:   [97, 122, 108, 88, 190,
130]                                 177]
Euclidean distance = 2               Euclidean distance = 70.29
Euclidean distance = 74.12           Euclidean distance = 50.21
Euclidean distance = 81.51           Euclidean distance = 108.04
Class:  Conrod                       Robot will not try to pick up an
Distance to nearest neighbour = 2    unknown object
Located at: [190, 178]
Orientation: 132                     Vector:   [98, 159, 50, 31, 139,
Robot will pick up the Conrod        144]
                                     Euclidean distance = 81.70
Vector:   [67, 95, 89, 93, 197,      Euclidean distance = 116.68
157]                                 Euclidean distance = 0
Euclidean distance = 75.89           Class:  Spanner
Euclidean distance = 2               Distance to nearest neighbour = 0
Euclidean distance = 118.13          Located at: [138, 67]
Class:  Y shape                      Orientation: 24
Distance to nearest neighbour = 2    Robot will pick up the Spanner
Sucker is too large for safe
lifting - robot will not pick up     FINISHED
the Y shape                          Nº1                      yes
```

7.3 Packing and Depletion

The ability to manipulate previously unseen objects under visual control is one of the key tasks in the successful implementation of robotic, automated assembly and adaptive material handling systems. It is within the context of this framework that an industrial vision packing strategy has been developed [WHE-93, WHE-96]. Its two main components are a *geometric packer*, based on the principles of mathematical morphology [WHE-91], which takes an arbitrary shape in a given orientation and puts the shape into place, in that orientation. The second component, a *heuristic packer*, is concerned with the ordering and alignment of shapes prior to applying them to the geometric packer. This component also deals with other general considerations, such as the conflict in problem constraints and the measurement of packing performance. In addition, it deals with practical constraints, such as the effects of the robot gripper on the packing strategy, packing in the presence of defective regions, and anisotropy ("grain" in the material being handled) and pattern matching considerations.

Together, these form a flexible strategy that allows the packing of arbitrary two-dimensional shapes. While the technique about to be described will pack any set of shapes presented to it, the efficiency is critically dependent on the application. Therefore, we need to use any clues we may glean from the context information, to ensure that we obtain an efficient packing strategy for that application. (See Figure 7.3.1.)

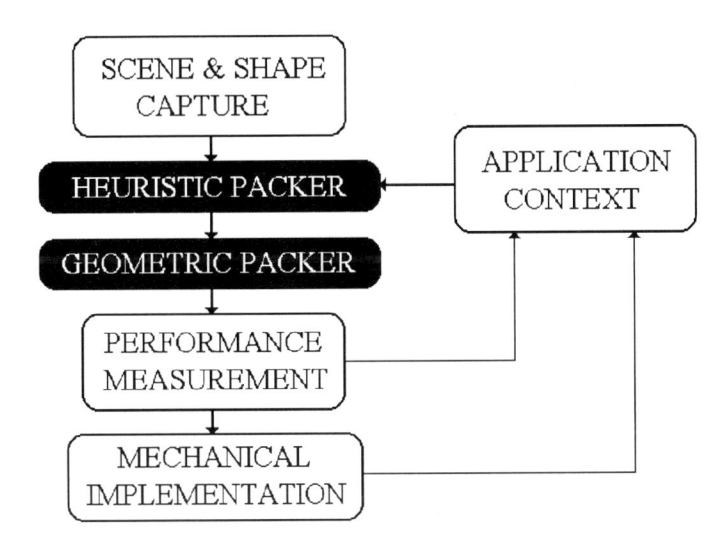

Figure 7.3.1 General packing strategy.

Since simpler packing problems, such as palletising [DOW-85], have been shown to be NP-complete [GAR-79], it is clearly impossible to guarantee that we will reach an optimal procedure for the more general problem. Hence, our aim has been to produce an efficient packing strategy (but not necessarily an optimal solution), that is flexible enough for industrial use.

7.3.1 Geometric Packer Implementation

The following section outlines the intermediate steps involved in the morphological packing of an arbitrary shape. This is denoted by the structuring element B. The image scene is denoted by the image set A. The morphological operations are also summarised in the image flowchart illustrated in Figure 7.3.2. A detailed description this procedure can be found in [WHE-91]. (Also see Section 2.4 for details on morphological imaging techniques.)

Step 1
Digitise the scene, image set A, and the shape to be packed, B.

Step 2
Erode the image scene A, by the structuring element B, to produce the erosion residue image $C = A \ominus B$. Every white pixel in this residue image represents a valid packing location. This step will be valid for any choice of structuring element origin point. (Conventionally, and to be consistent with the ideas discussed in the previous section, the origin will be taken to be the structuring element centroid.)

Step 3
Scan the erosion residue image, in a raster fashion, for the location of the first (top-left) white pixel. This location is denoted by (fitx, fity) in the program that follows and corresponds to the first possible packing location of B in the scene A, when it is scanned in this way. It has been suggested by Haralick [HAR-92] that further erosion of the residue image C by a standard 3x3 square structuring element, prior to searching for the first packing location, would enable control of the spacing between the packed shapes. Clearly, the number of pixel stripping operations, on the erosion residue, would be related to the spacing between the packed shapes.

Step 4
Translate (shift) the shape to be packed, B, to the location (fitx, fity). This effectively places B at the co-ordinate of the first possible packing location (when the residue image is scanned in a raster fashion). The resultant image will be denoted by $B_{(fitx, fity)}$.

Step 5
The resultant image from step 4 is subtracted from the original image set A to produce a new value for the image set A, therefore effectively packing B into the scene. (See Figure 7.3.2.) This can be represented algebraically as replacing A by $A - B_{(fitx, fity)}$.

This procedure is applied recursively to the image set A until an attempt to pack all the input shapes has been made or no more shapes will fit in the remaining space. The reapplication of the transform has no effect on the shapes already packed, due to the idempotent nature of this operation.

7.3.2 Heuristic Packing Techniques

The heuristic packer determines the orientation and order in which the shapes are applied to the geometric packer and operates upon two classes of shapes: simple polygons and (irregular) blobs. It is necessary to consider both these general shape classes separately, since no single scheme exists for all cases. While, the geometric packer is independent of the shape class and application context, the heuristic packer is not.

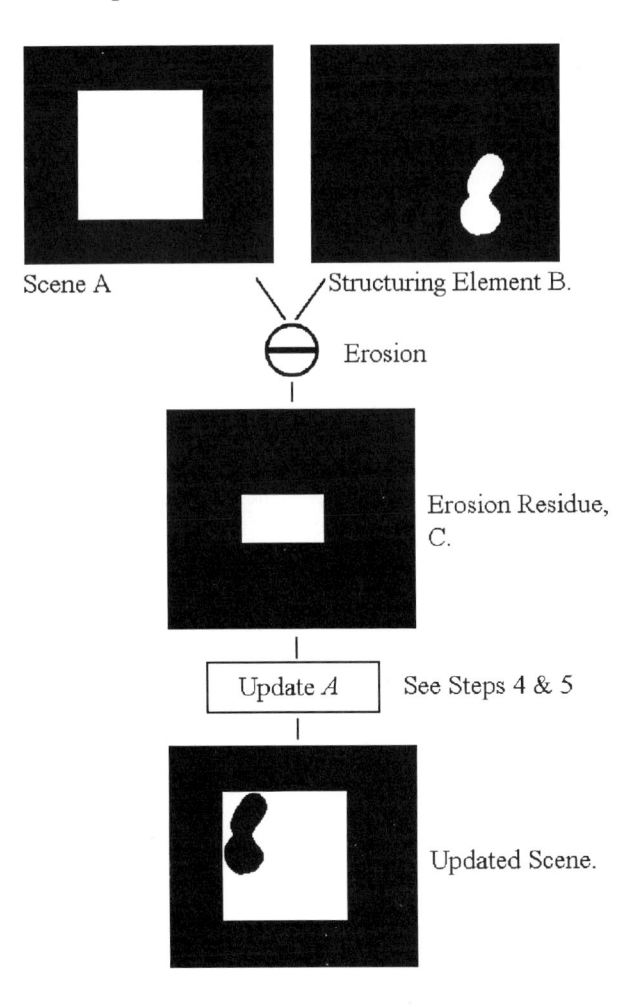

Figure 7.3.2 Morphological operations image flowchart.

Blob Packing

This section outlines some of the possible heuristics that have been devised to deal with two-dimensional binary images of random shape and size, prior to the application of the geometric packer. The approach outlined was designed specifically for off-line packing but the techniques developed could equally well be applied to an on-line packing application.

All the shapes to be packed are presented simultaneously to the vision system. The shapes are then ranked according to their bay sizes; the shape with the largest bay is the first to be applied to the geometric packer. Once the shape ordering has been decided, it is necessary to orientate each shape so that an efficient local packing strategy can be implemented. Four orientation rules are used to align the shape to be packed in the scene.

The order in which the shapes will be placed by the packer is determined by the *sort_by_bay* predicate defined below. If the area of the largest bay is significant compared to the area of the current shape, then the shape is sorted by its largest bay size (largest first). Otherwise the shapes are sorted by their size (largest first). The *bay_rot* predicate rotates a shape such that the largest bay is aligned with the scene's angle of least moment of inertia. This predicate also ensures that the biggest bay is facing into the scene (that is facing to the right and upwards). The operation of this predicate is summarised below:

- If *object_Y_coordinate* > *bay_Y_coordinate* then rotate shape by 180°
- If *object_Y_coordinate* = *bay_Y_coordinate* and *object_X_coordinate* > *bay_X_coordinate* then rotate shape by 180°
- If *object_Y_coordinate* = *bay_Y_coordinate* and *object_X_coordinate* ≤ *bay_X_coordinate* then no action required as in correct orientation
- If *object_Y_coordinate* < *bay_Y_coordinate* then no action required as in correct orientation

In the following program, the undefined predicate *main_pack* finds the valid packing location in the erosion residue. The appropriate shape is then placed at this location.

```
packbay:-
    get_all_shapes,     % Get all shapes to be packed.
    sort_by_bay,        % Sort shapes by bay size.
                        % Sorted data is stored in new_blob_db db.
    !,
    pack_bay_1.         % Pack shapes - largest bay first.

% Main packing co-ordination predicate
pack_bay_1:-
    big_bay_first,      % Pack the shapes by bay size.
    read_shapes,        % View remaining shapes.
    cwp(N),             % Count image pixels.
    N > 0,              % If no pixels to process - fail.
    centre_screen_se,
                        % Place shape in centre of FOV prior to
                        % rotation
```

```
    cwp(SHAPESIZE),      % Current shape area.
    get_shape_parameters(ROUNDNESS,BAYSIZE,BAY_ROUNDNESS),
                         % Current shape parameters.
    bay_rotate_options(SHAPESIZE,ROUNDNESS,BAYSIZE,
    BAY_ROUNDNESS),      % Choose rotation procedure.
    !, pack_bay_1.

pack_bay_1:-
    performance_meas.   % Calculate performance values.

% Shape alignment and rotation after sorting

/* Orientation rule 1: Use this sorting option if the bay size is
zero and the shape has a roundness <= 1. If the above conditions
occur then the object is classified as 'round' (i.e. a disk shape)
and therefore it need not be rotated or aligned. */
bay_rotate_options(_,ROUNDNESS,BAYSIZE,_):-
    BAYSIZE = 0, ROUNDNESS < 1,
    pack_bay_main.           % Morphological packing.

/* Orientation rule 2: Use this sorting option if the largest bay
size > a quarter of the shape area. Therefore if the bay is large
the shape is rotated such that the largest bay is at the angle of
the least MOI of the shape to be packed, ensuring that the bay
region always points into the main body of the shape to be packed.
By checking the bay roundness we ensure that we do not rotate the
image by its bay if the bay is elongated. */
bay_rotate_options(SHAPESIZE,_,BAYSIZE,BAY_ROUNDNESS):-
    BAYSIZE > SHAPESIZE/4, BAY_ROUNDNESS =< 2,
    bay_rot,
    pack_bay_main.           % Morphological packing.

/* Orientation rule 3: Use this sorting option if the bay size > a
quarter of the shape area. By checking the bay roundness we ensure
that we do not rotate the image by its bay if the bay is elongated.
We align the shape with respect to the least MOI of the scene. */
bay_rotate_options(SHAPESIZE,_,BAYSIZE,BAY_ROUNDNESS):-
    BAYSIZE > SHAPESIZE/4, BAY_ROUNDNESS > 2,
    shape_rot,               % Rotate shape such that it is at
                             % scenes angle of the least MOI.
    pack_bay_main.           % Morphological packing.

/* Orientation rule 4: Use this sorting option if the bay size <= a
quarter of the shape area. Therefore if the bay is considered small
we align the shape with respect to the scenes least MOI. */
bay_rotate_options(SHAPESIZE,_,BAYSIZE,_):-
    shape_rot,
    pack_bay_main.           % Morphological packing.

% Morphological packing predicate: Using system macros installed in
% the Intelligent Camera.
pack_bay_main:-
    read_structuring_element,
    $'PACK',         % Erosion residue using system macro call
                     % (indicated by the symbol '$').
    space_out,       % Dilate image by single pixel to make packing
                     % clearer.
    $'TRES',         % Pack original SE.
    main_pack.       % Place SE onto erosion residue.
```

Figure 7.3.3 shows the result of packing hand tools into a rectangular tray. The shapes were initially presented directly to the geometric packer, without the aid of the heuristic packer. (Figure 7.3.3(a).) This has the effect of packing each tool at whatever orientation it happened to be in when it was presented to the vision system. Figure 7.3.3(b) shows the resultant packing configuration when the heuristic packer precedes the geometric packer. Each shape is aligned and ordered, before it is applied to the geometric packer. Figure 7.3.3(c) shows the packing of the tools into a "random" blob region. The full packing strategy was used again here, as in Figure 7.3.3(b).

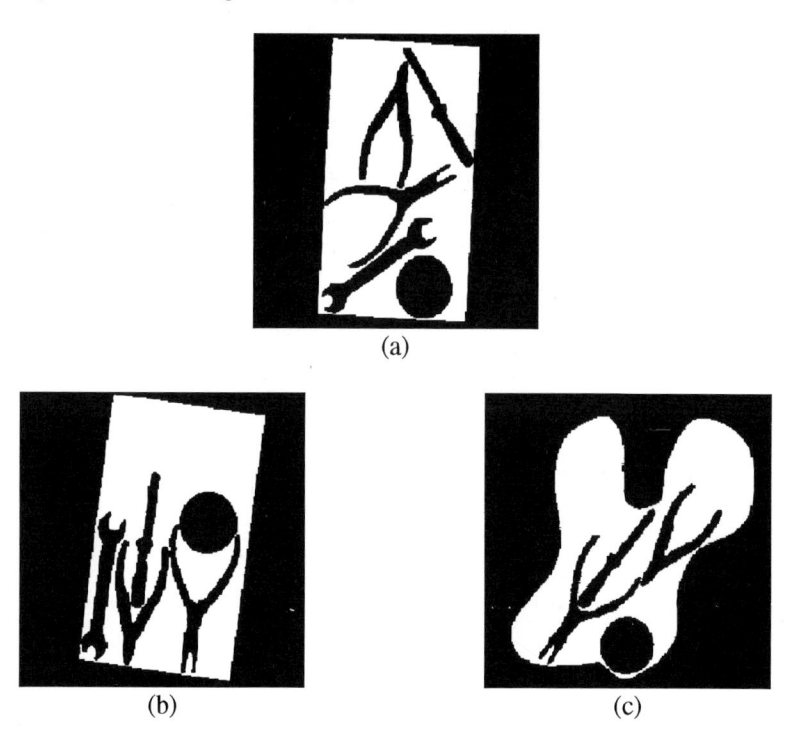

Figure 7.3.3 Automated packing of tool shapes. (a) Tools packed in their current orientation, (b) tools reorientated for better efficiency, (c) tools packed in an irregular scene.

Polygon Packing

The previous approach is not efficient, when packing shapes which do not contain bays of significant area. Hence, a different packing procedure is used to pack simple polygons which do not possess large bays. As before, this procedure was designed to work within an off-line packing system but could also be applied to on-line packing applications. Unlike the previous approach, however, this second procedure has the ability to determine the local packing efficiency for each shape and will reorientate it, if necessary, to ensure a more efficient configuration. (This local efficiency check could also be applied to the blob packing strategy.) In our second sample application, we chose to pack non-uniform box shapes (squares and rectangles) into a square scene. (Figure 7.3.4) Once all the shapes have been presented to the packing system, they are ordered according to size, with the largest shape being packed first. The shapes must then be orientated, prior to the application of the geometric packer.

In the initial versions of this packing procedure, each shape was aligned in such a way that its axis of least moment of inertia was matched to that of the scene under investigation. However, this method proved unreliable for packing squares, because quantisation effects produce a digital image, with a jagged edge. (A resolution of 256x256 pixels was used.) Furthermore, a square has no well-defined axis of minimum second moment. This can cause errors in the calculation of the moment of inertia. The problem was overcome by aligning the longest straight edge of the shape to be packed with the longest straight edge of the scene. The edge angles for the shape and scene were found by applying an edge detection operator, followed by the Hough transform. The latter was used, because it is tolerant of local variations in edge straightness. Once the peaks in the Hough transform image were enhanced and separated from the background, the largest peak was found [BAT-91c]. This peak corresponds, of course, to the longest straight edge within the image under investigation, whether it be the shape or the scene. Since the position of the peak in Hough space defines the radial and the angular position of the longest straight edge, aligning the shape and the scene is straightforward.

Once a polygonal shape has been packed, a local packing efficiency check is carried out. This ensures that the number of unpacked regions within the scene is kept to a minimum. The shape to be packed is rotated through a number of predefined angular positions. After each rotation, the number of unpacked regions in the scene is checked. If a single unpacked region is found, then a local optimum has been reached. In this case, the local packing efficiency routine is terminated and the next shape is examined. Otherwise, the local packing efficiency check is continued, ensuring that, when a shape is packed, a minimum number of unpacked regions exists. This reduces the chance of producing large voids in the packed scene, and improves its overall efficiency of packing.

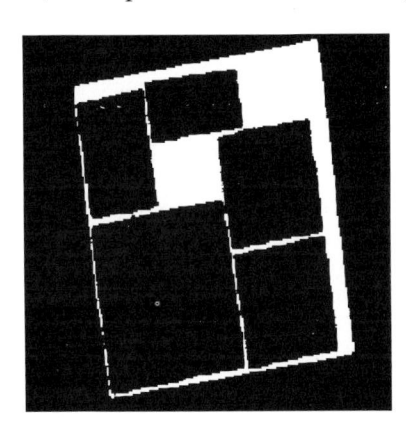

Figure 7.3.4 Automated packing of non-uniform boxes in a square tray.

The packing order is determined by the sizes of the shapes to be packed (largest first). The rotation of the shapes by the packer is based on the angle of the largest face (longest straight side of the polygon) of the unpacked region. The predicate *shape_face_angles* finds the largest face angle and stores it in the face angle database. This database also contains a selection of rotational variations for the current shape. The face angles are sorted such that the angle of the largest face appears at the top of the database. The other entries are modified (by a fixed angle rotation factor) versions of this value. The predicate *blob_cnt* counts the number of "free space blobs", that is the number of blocks of free space available to the packer. The polygon packer operates according to the following rules:

- If *blob count* is 1 then the best fit has occurred, so exit and view the next shape.
- If *blob count* is 0 then read the new angle from face angles database and retry.
- If *blob count* < *local optimum* then update blob count and update the local optimal storage buffer before trying the next angle in the face angles database.
- If *blob count* ≥ *local optimum* then try the next angle in database.

```
polypack:-
    scene_capture,           % Capture scene to be packed.
    get_all_shapes,          % Find all the input shapes.
    poly_pack_1.             % Pack shapes - largest first sorted.

% Main packing co-ordination predicate
poly_pack_1:-
    big_shape_first,         % Pack the shapes by SE size.
    centre_screen_se,        % Centre SE in the field of view.
    shape_face_angles,       % Find the angle of the largest face.
    face_angles_db(LONGEST),
    retract(face_angles_db(LONGEST)),
                             % Recover a face angle database value.
    turn(LONGEST,A,B),       % Rotates the structuring element about
                             % its centre of gravity, by the angle
                             % recovered from the face angle
                             % database.
    poly_pack_main,!,
    poly_pack_1.             % Get the next shape to pack.

poly_pack_1:-
    performance_meas.        % Calculate performance values.

poly_pack_main:-
    morph_pack,              % Morphological packing.
    read_updated_image, blob_cnt.

% Morphological packing predicate
morph_pack:-
    updated_image,
    $'PACK',                 % Erosion residue using system macro
                             % call indicated by the symbol '$'.
    space_out,               % Dilate image by single pixel to
                             % make packing clearer.
```

```
        residue_check,          % No residue exists then try next
                                % angle. No residue indicates that the
                                % shape cannot be packed in current
                                % orientation so quit.
        $'TRES',                % Pack original SE using a system macro
        main_pack.              % Place SE onto erosion residue.
```

7.3.3 Performance Measures

To ensure that we have confidence in the global efficiency of any packing strategy, there must be some way of measuring its performance. Traditionally, packing performance has been measured by a single number, called the *packing density* [STE-91]. This is the ratio of the total area of all the packed shapes to that of the total area of the scene. This is referred to as the *worst case analysis* packing measure. A number of other performance measures have been developed in the field of *Operational Research*, particularly for comparing different heuristics for packing rectangular bins by odd-sized boxes. (See [DOW-85] for a review of packing procedures used in Operational Research.) These alternative performance measurements can be quite useful in well-constrained packing problems, but they are of little use in dealing with the packing of arbitrary shapes. [WHE-96].

Predicates

The following predicates evaluate the 'goodness-of-fit' of a given packing procedure. The *Packing Density* is defined as the ratio of the optimal packing area (which is the sum of the area of the individual shapes to be packed) to that of the area of the convex hull of the packed shapes (minus the area of any scene defects). This is a standard packing measurement, and has a maximum value of one. We have defined a parameter, called *Performance Index,* which is a modified version of the *Packing Density* and accounts for unpacked shapes. The *Performance Index* equals *Packing Density* times the *Count Ratio*, and has a maximum value of one. The *Count Ratio* is defined as the ratio of the total number of shapes packed to the number of original shapes presented to the packing procedure. Another parameter that we have defined is *Space Usage,* and this equals the ratio of the area of the shapes packed to that of the unpacked original shape. This gives us an idea of the amount of space unpacked in the original scene. The *Space Usage* ratio has a maximum value of one. This occurs when no space remains unpacked.

Figure 7.3.5 shows the results of placing (using blob packing techniques) some standard household items, such as scissors, keys and pens, into a rectangular tray (Figure 7.3.5(a)) and into an irregular scene (Figure 7.3.5(b)). Figure 7.3.6 shows the automated packing of simple polygon shapes, i.e. assembling a simple block jigsaw. The performance measures for the packing examples discussed in this section are shown in Table 7.3.1.

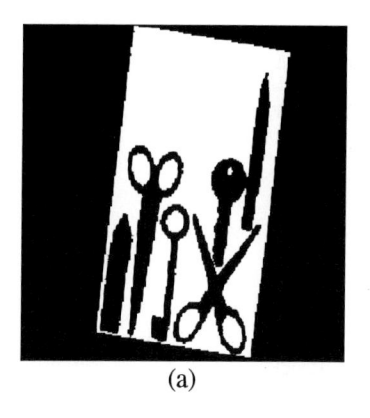

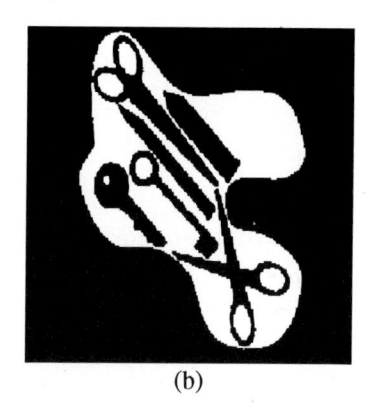

Figure 7.3.5 Packing items into: (a) a rectangular tray, (b) an irregular scene.

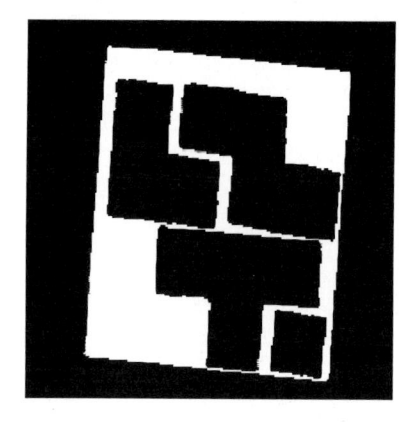

Figure 7.3.6 Automated packing of simple polygons in a rectangular scene.

Figure Number	Packing Density	Shapes Presented	Shapes Packed	Performance Index	Space Usage
7.3.3(a)	0.36	5	5	0.36	0.25
7.3.3(b)	0.435	5	5	0.435	0.25
7.3.3(c)	0.381	5	4	0.305	0.191
7.3.4	0.86	6	5	0.72	0.735
7.3.5(a)	0.38	6	6	0.38	0.261
7.3.5(b)	0.44	6	6	0.44	0.317
7.3.6	0.81	5	4	0.64	0.566
7.3.8	0.41	5	3	0.246	0.23
7.3.9(a)	0.36	5	5	0.36	0.249
7.3.9(b)	0.511	8	5	0.319	0.381

Table 7.3.1 Comparison of packing configurations using the performance measures defined in Section 7.3.3.

7.3.4 Robot Gripper Considerations

Any supposedly general purpose strategy for packing must be robust enough to cope with a range of different type of material handling systems. For all of the applications considered above, we have tacitly assumed that some form of suction or magnetic gripper could be used to lift and place the objects during packing. In this case, the "foot-print" of the gripper is assumed to lie within the outer edge of the shapes being manipulated. (See Section 5.9.)

Automated material handling systems frequently make use of robotic grippers which have two three or more "fingers". This complicates the problem of packing, since the gripper requires access to objects within a partially packed scene. Therefore, any packing strategy must make allowances for the gripper. The worse case position usually (but not always) occurs when the gripper is fully open, just after placing an object in position. The problem of gripper access can be dealt with very effectively, by the simple expedient of overlaying a gripper template on the shape to be packed prior to the application of the geometric packer. (Figure 7.3.7) The gripper "foot-print" is based on the positions of the fingers in both the open and closed positions. In fact, the convex hull of each of the finger tips in the open and closed positions is formed when computing the composite "foot-print". This convex hull is indicated by the shaded region in Figure 7.3.7.

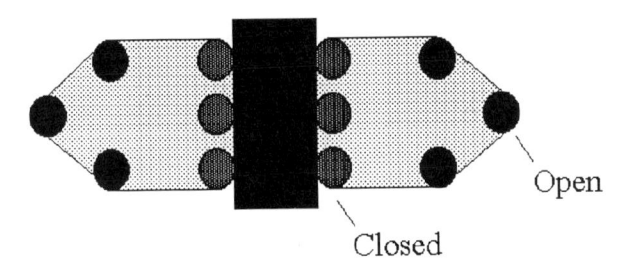

Figure 7.3.7 Generation of the gripper "footprint" based on the fully open and closed positions of a multi-fingered robot gripper.

Figure 7.3.8, shows the result of packing tools into a rectangular tray, taking the gripper "foot-print" into account. Although the general blob packing remains the same, the procedure's performance is inevitably weakened when allowance is made for the robot gripper. For example, compare Figures 7.3.3(b) and 7.3.8. (See Table 7.3.1 for a comparison of the packing performance measures.) Clearly, the gripper "foot-prints" indicated in Figure 7.3.8 are not those for the ideal gripping positions for these objects. They are used merely to indicate how our packing strategy can cope with multi-finger grippers.

In a practical situation, care must be taken to ensure that any change in the shapes of the objects to be packed, due to squeezing by the robot gripper, does not adversely effect the packing. The same is true of articulated and other hinged objects, such as scissors or pliers, which can change their shape during handling.

Again, this type of application constraint, could also be dealt with by the introduction of suitable heuristic packing rules, and may also be used as a factor when calculating the gripping position.

The strategy outlined above for working with multi-finger grippers does have the advantage of allowing the shapes to be *unpacked* from the scene in any order. One possible modification to the approach outlined above, results in a denser configuration that, in general, can only be unpacked safely in reverse order. This modification consists of packing each shape, taking the robot "foot-print" into account, but removing the "foot-print" from the scene prior to the application of the next shape.

Figure 7.3.8 Tool packing with worse case gripper "foot-print".

7.3.5 Packing Scenes with Defective Regions

Any practical automated packing system for use in such industries as leather or timber processing must be able to pack "objects" into a scene which may contain defective regions. (This is an alternative way of representing the stock cutting problem.) The importance of good packing procedures in the leather industry is obvious, since the raw material is both expensive and non-recyclable. Our packer can readily accommodate defects like these; we simply define the initial scene to contain a number of holes. Figure 7.3.9(a) illustrates the effect of packing tools into a rectangular tray which contains four small blob-like "defects". By comparing the packing configuration shown in Figures 7.3.9(a) and 7.3.3(b) (see Table 7.3.1 also), it is clear that the packing is not as tight when defects are taken into account. Figure 7.3.9(b) shows the packing of leather templates onto a hide. The small blob-like regions indicate the defective areas of the hide. These defective regions are not to be included in the leather pieces to be cut. Both of the results shown in Figure 7.3.9 indicate the flexibility of the packing strategies described.

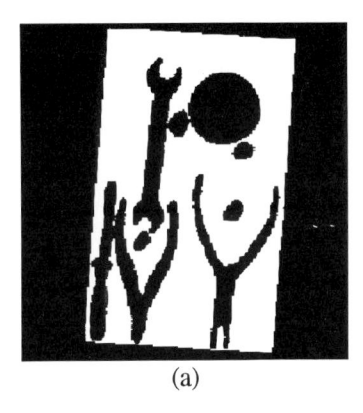

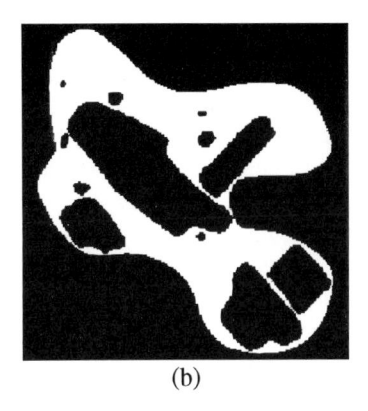

Figure 7.3.9 Packing items into defective regions: (a) Tools into a defective tray, (b) Leather template pieces into a leather hide which contains defects.

7.3.6 Discussion

The results presented here show the power of the heuristic approach when presented with a wide range of problems, including the packing of shapes into materials with defective regions. We have attempted to maximise the use of application specific information, to produce an efficient packing strategy. The example application also outlines a technique that will allow a range of performance measures to be computed, so that different packing procedures can be compared. This is necessary due to the fact that the heuristic approach we have taken does not guarantee an optimal result. For a more detailed discussion on automated packing techniques and systems, the reader is directed to [WHE-96, WHE-93 and WHE-91].

7.4 Handedness of Mirror-Image Components

7.4.1 Handedness and Chirality

We explain how Prolog+ can be used to determine the "handedness" of planar objects, such as leather, plastic and fabric components for clothing, bags, upholstery, gloves and shoes, using so-called *Concavity Trees (CTs)* and a number of other heuristic techniques. A large number of industrial products are either symmetrical, or are sold in mirror-image pairs. In such a case, for every "right handed" component, there will be another "left handed" one. Objects which are mirror images of each other, but are in all other respects identical, are said to differ in their *chirality*. This quantity is closely related *"handedness"* and the relationship between them will be defined later. Mirror-image pairs of objects are found in many other situations, most notably where they are intended to fit some other symmetrical object, most notably the human body. The list of products existing as mirror-image pairs is not simply limited to clothing. Many of the components of a bicycle, for example, exist in mirror-image pairs. Another use for a chirality test is to determine whether a planar object, such as a metal stamping is *"face up"* or *"face down"*. A safety-critical application of this kind was encountered by the authors several years ago and was concerned with components for automobile brakes.

Our prime objective in this case study is to present techniques for finding the chirality of objects that can be viewed in silhouette. An important secondary goal is to introduce the reader to the concepts of concavity trees, which have a range of other applications, some of which are explained elsewhere in this chapter. A recursive program for generating concavity trees is listed below and provides a good demonstration of Prolog+.

Relating Chirality and Handedness

Chirality is an abstract property and does not indicate whether a given piece of leather forms part of a left or right shoe, or glove. Chirality has the value *left* or *right*, and this value is inverted when we pick up an object and turn it over. A piece of leather may form part of a *right* shoe, yet have a chirality value equal to *left*. When that piece of leather is turned over, its chirality values becomes *right* but, of course, it is still a component of a right shoe. (Figure 7.4.1) Thus, chirality is a function of posture, not an indicator of intended use. We shall define handedness in terms of chirality, in the following way:

```
% If the object is laying "face up" then the handedness is equal to
% the chirality
handedness(U) :-
   face(up), chirality(U), !.

% If the object is laying "face down", then the handedness is the
% inverse of the chirality
handedness(U) :-
   face(down), chirality(V),
   (U = left, V = right); (U = right, V = left)), !.
```

Handedness, defined in this way, is still not sufficient to tell us whether a given component forms part of a left or right shoe. We also need a parts list having the following general form.

Component identity number	Handedness of component to make	
	right shoe	*left shoe*
1	right	left
2	left	right
...	...	...
10	left	right

Such a table can be represented using a set of simple Prolog statements :

```
% Component A with handedness value of "right" is part of left shoe
component(left, A,right) :- component(right, A,left), !.
% Component A with handedness value of "left" is part of left shoe
component(left, A,left) :- component(right, A,right), !.

% If handedness = right, comp. 1 is part of right shoe
component(right, 1,right).
% If handedness = left, comp. 2 is part of right shoe
component(right, 2,left).
......
% If handedness = left, comp. 10 is part of right shoe
component(right, 10,left).
```

(a)

(b)

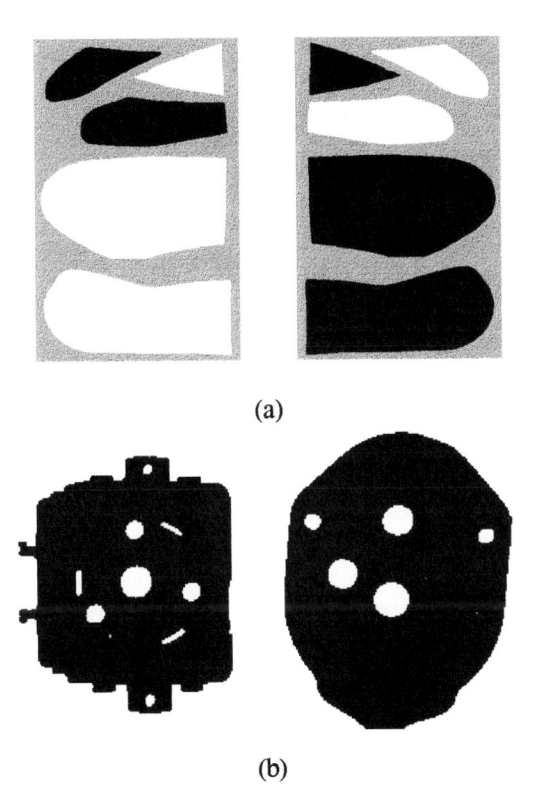

Figure 7.4.1 Chirality and its relationship to handedness and function. (a) Mirror image components of leather mittens, all viewed "face up". The group on the left, form the left glove and those on the right are parts of the right glove. Shading indicates chirality, as it is calculated by the second version of *chirality*. If an object is "turned over", its *chirality* is reversed but the *handedness* is not and, of course, it remains a component of the same glove, so its *function* is unaltered. (b) Typical objects considered in this case study.

Now that we have seen the significance of chirality, we shall describe how it can be calculated from a concavity tree. First, however, we shall digress briefly, to explain the significance and generation of concavity trees.

7.4.2 Concavity Trees

There is a general requirement for an inspection process which combines both local and global information about an object's shape, within a single integrated data structure. It is often necessary to pay detailed attention to a number of small regions of a large artefact and then to verify that its overall shape is correct. In essence, concavity trees provide a multi-level representation of an object's shape, which makes this type of analysis fairly simple. Concavity trees are ideally suited for a variety of industrial inspection tasks, such as inspecting stampings, pressings and mouldings, where severe but local defects can occur. Metal-forming tools, such as punches and dies are not subject to significant changes in shape during normal operation. Hence, the objects produced by them are likely to be dimensionally correct. However, abnormal operation can occur, as a result of failure to feed the metal stock properly, through broken or chipped tools or displaced tooling.

There is a specific need for a "general purpose" inspection method which can examine piercings (lakes) and indentations (bays), both of which are intended to mate with other parts. In many applications, inspection may be achieved with relatively low precision, since there is no common fault which can introduce small overall errors. (Tool wear can but this accurately predictable and, for this reason, need not concern us here.) In this type of situation, parts are either correctly made, or are badly malformed locally. Thus, for example, an inspection system might be asked to decide whether an object possesses a certain hole, indentation (or spur), needed to provide mechanical linkage to another component. A similar inspection technique is needed in any industry where a *"pastry cutter"* is used to form the products. Hence, we may expect to see applications of concavity trees in the plastics, leather, clothing and food industries, amongst others.

Concavity Trees have a variety of possible uses, including shape recognition, parts assembly and finding object orientation. The one feature of Concavity Trees that makes them particularly well suited for discussion in this book is that they require the use of *recursion*, which is, of course, an essential feature of Prolog. The use of such a sophisticated technique for the seemingly straightforward task of finding whether a 2D object forms part of a left- or right-handed shoe might seem to be unjustified. Certainly, it is possible to find the chirality in other ways, for example by shape matching, or comparing the sequences of left-right turns, taken as we traverse the perimeter of a blob-like object. These may well be faster than the approach based on concavity trees but are unlikely to be as versatile in the range of objects they will handle successfully. For the sake of completeness, we present alternative methods for finding chirality later. For the moment, however, we shall concentrate on concavity trees.

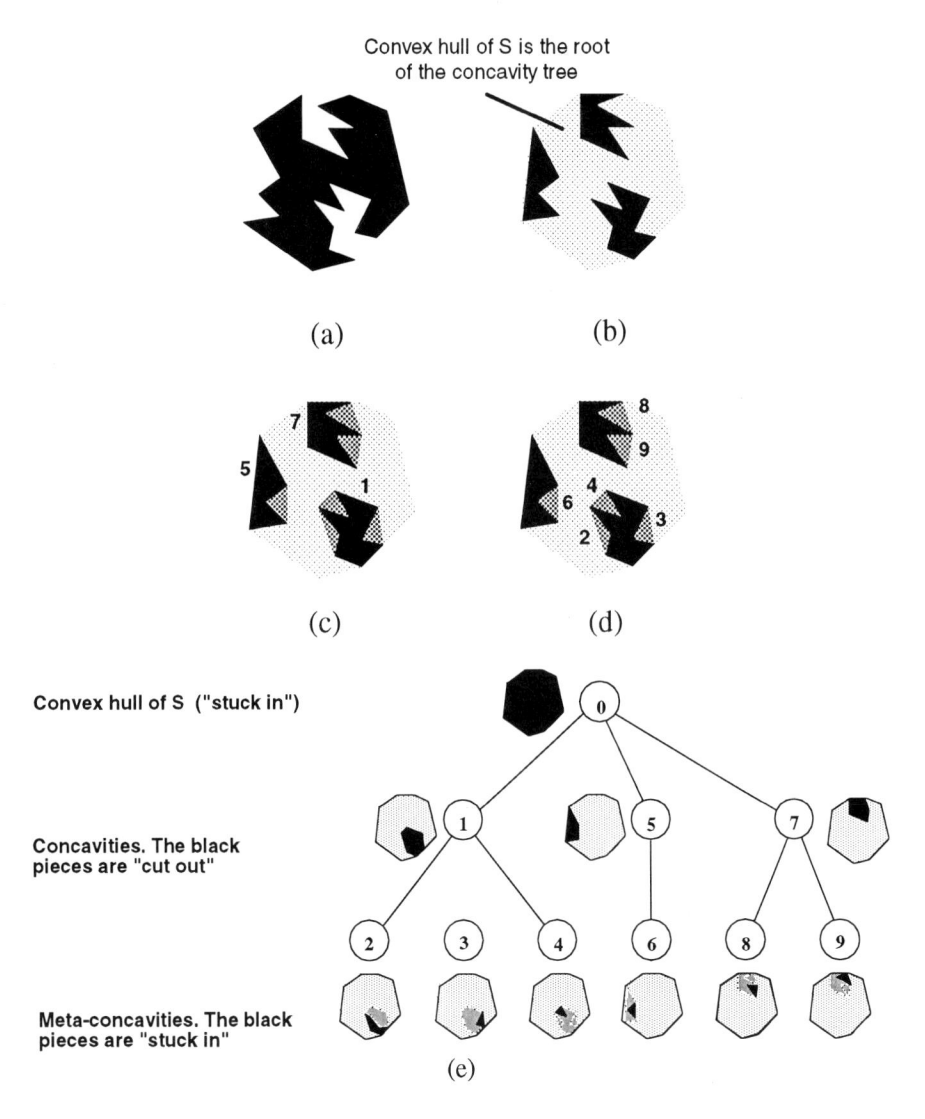

Figure 7.4.2 Generating concavity trees. (a) Original shape (S) to be analysed. (b) Convex hull of S. The three concavities are represented here as black regions. Together, they form the convex deficiency of S. (c) Convex hull of each concavity. The numerals indicate the node labels generated by the first program. (The nodes are analysed in this order, thereby implementing a depth-first search.) (d) Meta-concavities of S. These are the concavities of the concavities of S. (e) Concavity tree generated by the first program. The shape corresponding to each node in the tree is shaded black.

A Concavity Tree (CT) is an hierarchical tree-like structure, whose nodes represent convex polygons. These are of varying sizes and are either *"cut out"* or *"stuck in"*, beginning with the convex hull of whatever object is to be analysed. This cut-and-paste process enables us to approximate a given blob-like object, to varying degrees of precision. The tree-like representation of shape is based on the idea of a *concavity*, which is a term intended to encompass both lakes (holes) and bays (indentations). Concavities are defined in terms of the dual concepts of *convex hull* and *convex deficiency*. (Prolog+ operators *chu* and *cvd*). The convex deficiency is the difference between the *filled* convex hull of a blob and the blob itself. It consists of a number of distinct regions, each of which is a concavity. A CT combines both global and local information about shape, in a single integrated data structure and it is possible to label its nodes, using as many shape, size and position descriptors as we can conveniently measure. In this respect, a CT is much more general than any of the other shape description / representation techniques that we discuss elsewhere in this book. As we shall see later, an elegant, and very short, Prolog+ program can be used to generate a CT from a binary image. In fact, this particular program shows Prolog+ off to very good effect.

Formal Definition

In the following explanation of how to create concavity trees, we shall find it convenient to use the terminology of set theory. Recall that an object in a binary image is a connected set of white points. The idea of a concavity tree can be understood most easily with the help of a series of simple diagrams. In Figure 7.4.2, S denotes the initial blob that is to be represented by a CT. To generate the tree, we begin by computing the *convex deficiency* of S. (In general, the convex deficiency will consist of a number of large, disjoint blobs, representing the bays and lakes of S and numerous very small regions, created as artefacts of camera and quantisation noise.) Let Q^* denote the filled convex hull of any given set, Q. Furthermore, let $(X \otimes Y)$ be the set of white points formed by computing the difference *(exclusive OR)* between two given sets X and Y. We shall also assume that the convex deficiency of S, i.e. the set $(S^* \otimes S)$, consists of N distinct blobs, which will be denoted by $\{C_{S1}, C_{S2}, ..., C_{SN}\}$.

The C_{Si} (i = 1, , ..., N) are the *concavities* of S. We now apply the same type of analysis to each of the C_{Si} as we did to S. That is, we compute the convex deficiency of each member of the following set: $\{(C_{S1}^* \otimes C_{S1}), (C_{S2}^* \otimes C_{S2}), ..., (C_{SN}^* \otimes C_{SN}),\}$. This process of analysing the blobs within the convex deficiency of a given blob is repeated *recursively*. To terminate recursion, we simply impose a lower size (area) limit on the objects which we analyse in this way. The concavity tree is used to relate the concavities, concavities of concavities *(meta-concavities)*, concavities of concavities of concavities *(meta-meta-concavities)* etc. to each other and to S^*.

To understand the CT-generation process in physical terms, let us equip ourselves (mentally) with scissors and adhesive tape. To approximate S, we begin with a convex shape, S^*, and *cut out* N convex shapes, corresponding to the filled convex hulls of the concavities: $\{C_{S1}^*, C_{S2}^*, ..., C_{SN}^*\}$. We now *stick back* some smaller convex pieces corresponding to the filled convex hulls of each of the meta-concavities. Then, we *cut out* pieces corresponding to the filled convex hulls of all of the meta-meta-concavities. Next, we *stick back* pieces corresponding to the filled convex hulls of all of the meta-meta-meta-concavities. This process of alternately *cutting out* convex shapes and then *sticking back* other, smaller convex shapes continues indefinitely (i.e. until the objects are too small to make it worthwhile continuing). (See Figure 7.4.2)

Generating Concavity Trees

As we mentioned earlier, the CT is defined recursively. To an experienced programmer, the mere whisper of the word "recursion" immediately suggests the use of Prolog. However, the program *concavity_tree*, which is listed below, uses a somewhat unusual form of recursion; the predicate *analyse_node* does not contain a direct call to itself but instead contains the line:

```
eab_modified(E,analyse_node(E)),
```

This is perfectly legal Prolog programming practice and leads to a very compact program for calculating CTs. The reader should understand that *eab_modified(A,B)* tries to satisfy goal B on all blobs in the binary image held in file A. The first (recursive) clause of *analyse_node* measures various shape parameters of a given (single) blob, using the undefined "general purpose" predicate *shape_measurements*, and then applies *analyse_node* to each 8-connected set in the convex deficiency of that blob. We can use *shape_measurements* to obtain as many shape, size and / or position parameters as we wish. We might, for example, choose to derive such measurements as: X-co-ordinate of centroid; Y-co-ordinate of centroid; area; perimeter; shape factor; area of filled convex hull; ratio of area of filled convex hull to the area of the original blob; aspect ratio of minimum area rectangle ... etc. (See Section 7.2.3.)

Prolog+ contains three standard operators

gob	If the stored image is empty, *gob* fails.
	Otherwise, select (*get*) one blob from a stored image. The latter is modified by deleting that blob.
gob_init	Used to initialise *gob*.
eab(A)	Evaluate goal A for all blobs in the current image

These three predicates had to be modified slightly for the concavity tree program; one extra parameter has been added to each one, enabling us to pass the name of the "input" image file. The revised definitions are given below. Here is the program listing for the CT generator.

```
% Top level predicate for computing Concavity Trees
concavity_tree:-
  retractall(ct_node(_)),
                    % Clear database, ready for new CT
  psk,              % Push image onto the stack
  init_gensym(node), % Initialise the symbol generator
  analyse_node([]), % This bit does all of the hard work
  pop.              % Restore image

/* Define the minimum size of blob to be analysed. Adjust this
parameter to taste */
min_blob_size(50).

/* Analyse one node in the Concavity Tree. The first clause
contains the indirect recursive call to "analyse_node" */
analyse_node(A) :-
  cwp(C),           % Measure blob area
  min_blob_size(D), % Consult DB for minimum blob size
  C ≥ D,            % Is blob large enough to bother with?
  gensym(node,E),   % Name node in CT - standard symbol generator
  shape_measurements(F),  % General shape measurement - undefined
  cvd,              % Convex deficiency of "input" blob
  kgr(D),           % Ignore tiny blobs
  eab_modified(E,analyse_node(E)),
                    % Analyse convex deficiency - recursive
  writeseqnl(['Blob: ',E,'Parameters: ',F,'Parent: ', A]),
  assert(ct_node(E,A,F)). % Assert node data into Prolog DB

analyse_node(A).    % Ending recursion. Force goal to succeed

/* Initialise "gob_modified". This is the standard predicate,
"gob_init," modified slightly, by adding the parameter A, which
specifies the name of an image file. */
gob_modified_init(A) :-
  ndo,              % Shade image
  wri(A).           % Save image in file A

/* Get one blob. Standard Prolog+ predicate, "gob," modified
slightly to facilitate recursion. */
gob_modified(A,B) :-
  rea(A),           % Read image file named A
  gli(_,B),         % Upper intensity limit
  ((B = 0, !, fail) ; % Fail if no more objects left to analyse
  (hil(B,B,0),      % Remove next blob to be analysed
  wri(A),           % Save image with one blob deleted
  swi,              % Switch images
  thr(B,B))).       % Isolate next blob for analysis

gob_modified(A,B) :- gob_modified(A,B).

/* Evaluate a named goal, for all blobs in an image. This is the
standard predicate, "eab", modified slightly to facilitate
recursion. A is the name of a file containing a multi-level image
with several blobs. B is the predicate to be "applied" to all blobs
in the image in file A. */
eab_modified(A,B) :-
  psk,              % Push image onto the stack
  gob_modified_init(A)
                    % Initialise, ready for "gob_modified"
  ->                % Conditional evaluation
  (gob_modified(A,C), % Get one blob from image held in file A.
  call(B),          % Satisfy goal specified by B
  fail).            % Repeat for all blobs in file A
```

```
% Goal always succeeds and restores original image
eab_modified(_,_) :- pop.
```

Sample Concavity Trees

For the sake of illustration, three simple shape / size parameters were calculated by *shape_measurements*, viz *area, perimeter* and *shape factor*. In the experiments reported below, these parameters were calculated on the blob and its (meta)-concavities, not on their convex hulls. The following output was generated by the program, given the starting image shown in Figure 7.4.3(a).

(a)

(b)

(c)

(d)

(e)

(f)

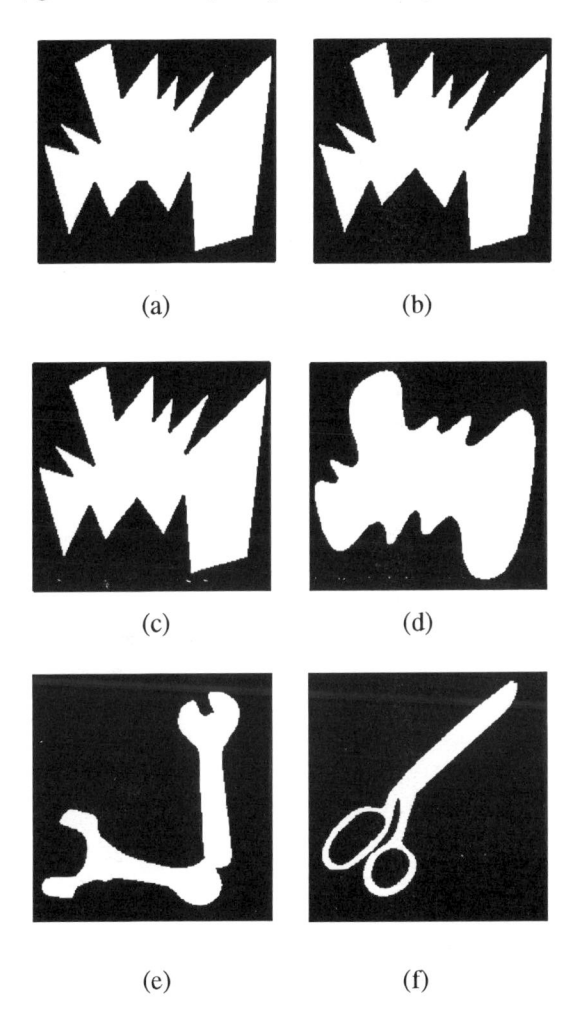

Figure 7.4.3 Six binary objects to be analysed. Notice that the three "peaks" in the bottom-most concavity in (a - c) are at slightly different heights.

```
Blob:  node2                    % Give arbitrary label to this node
Parameters:   [2651, 242, 0.571]
                                % [Area, Perimeter, Shape factor]
Parent:  node1                  % Parent of node2 is node1

Blob:  node3
Parameters:   [1755, 214, 0.485]
Parent:  node1                  % Parent of node3 is node1

Blob:  node1
Parameters:   [17196, 768, 0.367]
Parent:  node0                  % Parent of node1 is node0

Blob:  node5
Parameters:   [1261, 195, 0.420]
Parent:  node4                  % Parent of node5 is node4

Blob:  node4
Parameters:   [6095, 530, 0.274]
Parent:  node0                  % Parent of node4 is node0

Blob:  node9
Parameters:   [904, 154, 0.483]
Parent:  node8                  % Parent of node9 is node8

Blob:  node8
Parameters:   [4502, 408, 0.341]
Parent:  node7                  % Parent of node8 is node7

Blob:  node7
Parameters:   [7594, 671, 0.213]
Parent:  node6                  % Parent of node7 is node6

Blob:  node6
Parameters:   [16310, 1106, 0.168]
Parent:  node0                  % Parent of node6 is node0

Blob:  node0                    % node0 is the root of the CT
Parameters:   [73717, 2433, 0.070]
Parent:  []                     % There is no parent for the root

No.1 : yes                      % Goal always succeeds
```

In addition, a set of facts describing the tree was placed in the database. (This is achieved by *assert(ct_node(E,A,F))*, i.e. the last line of the first clause of *analyse_node*.)

```
ct_node(node2 , node1,   [2651, 242, 0.571] ).
ct_node(node3 , node3,   [1755, 214, 0.485]).
......
ct_node(node0 ,[], 73717, 2433, 0.070] ).
```

The corresponding tree is drawn in Figure 7.4.4(a).

333

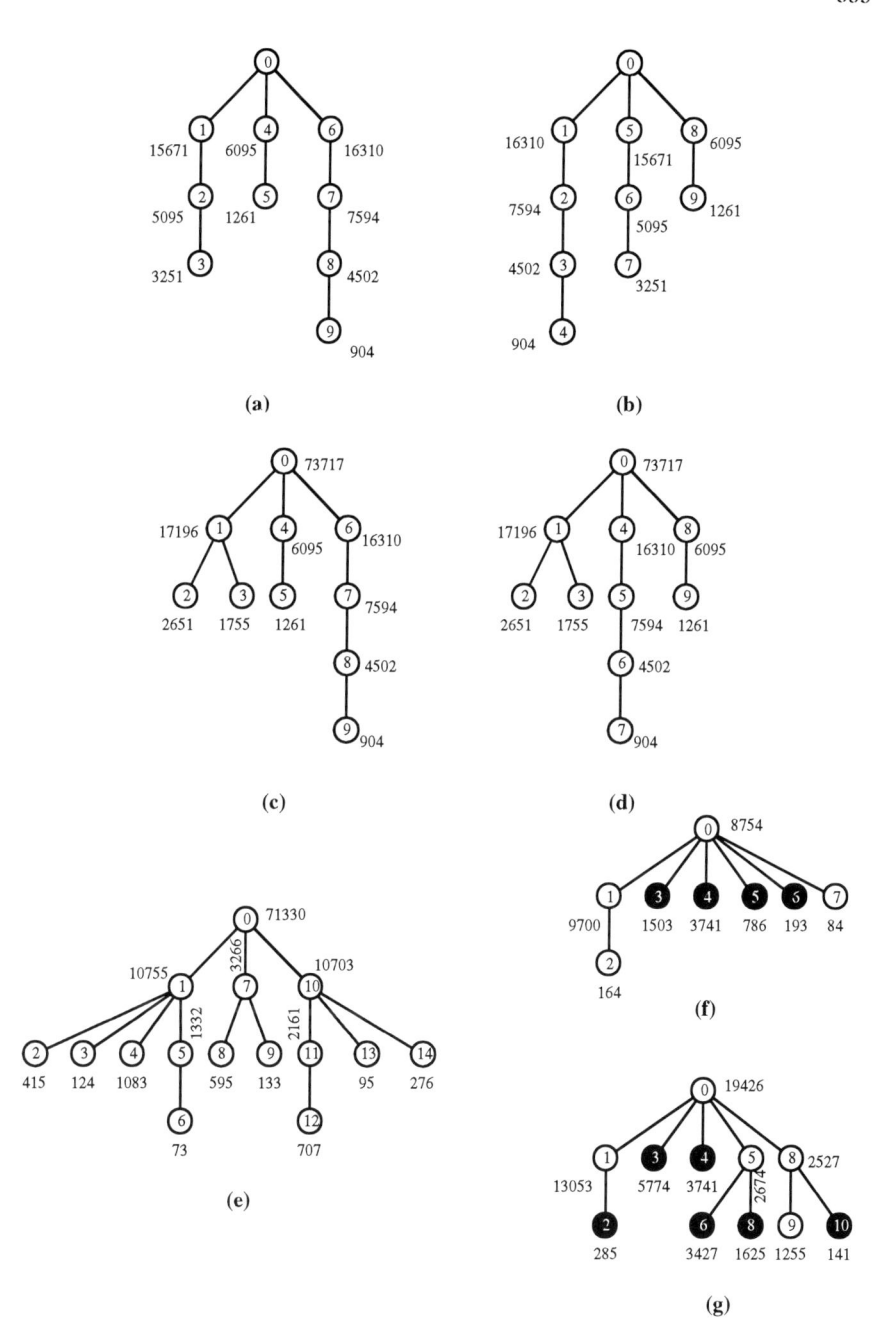

Figure 7.4.4 Concavity trees of various objects in Figure 7.4.3.

The CT results given in Figure 7.4.4 require some explanation, which is given by following table. It should be noted that the numbers inside the circles indicate the order in which the nodes of the CT were analysed, while the numbers printed beside them indicate the areas of the corresponding shapes.

Input shape	Concavity tree	Comments
Fig. 7.4.3(a)	Fig. 7.4.4(a)	Non-canonical form of CT
Fig. 7.4.3(a)	Fig. 7.4.4(b)	Canonical CT (described in next section).
Fig. 7.4.3(b)	Fig. 7.4.4(c)	Compare to Fig. 7.4.4(a). Major change in CT occurs even though there is a very small change in the input shape.
Fig. 7.4.3(c)	Fig. 7.4.4(b) or (c)	Result depends on noise; 3 "peaks" on bottom-most indentation of input are directly in line.
Fig. 7.4.3(b)	Fig. 7.4.4(d)	Canonical CT
Fig. 7.4.3(d)	Fig. 7.4.4(e)	Input is smoothed version of Fig. 7.4.3(a)
Fig. 7.4.3(e)	Fig. 7.4.4(f)	4 black nodes correspond to bays that do not change as a result of objects touching.
Fig. 7.4.3(f)	Fig. 7.4.4(g)	Black nodes remain unchanged whether scissors are open or closed.

Canonical Form of Concavity Trees

The order in which the nodes are added to the CT is indicated by simple identifiers: *node0, node1, node2, ...* (These names are generated by *gennsym*.) The reader will observe that this ordering reflects the fact that the CT is generated using a *depth-first* search. However, this is only part of the story, since detailed examination of the program reveals that the order in which the "children" of a given node are analysed is determined by two predicates: *gob_modified* and *eab_modified*. If we rotate the initial shape to be analysed, the program defined above may well generate a CT with its nodes labelled in a different order. So that we can make the calculation of chirality easier and to facilitate other important operations, such as shape matching, it is better to create CTs in some standard way that is independent of orientation. This can be achieved very simply, by redefining the predicates *gob_modified, eab_modified* and *gob_modified_init*. This incurs very little cost, in computational terms. The key is to use the predicate *big*, to select one blob at a time for analysis. Here are revised definitions of these three predicates, which use this idea:

```
% One line has been deleted from "gob_modified_init"
gob_modified_init(A) :- wri(A).

eab_modified(A,B) :- psk, wri(A)
  ->
  (gob_modified(A), call(B), fail).
```

```
eab_modified(_,_)  :- pop.

gob_modified(A)  :-
    rea(A),              % Read image file named A
    cwp(N),              % Count white pixels in image
    ((N = 0,!,fail);     % If there are none, then fail
    (big,                % Select biggest blob for analysis first
    xor,                 % Delete that blob from image to be stored
    wri(A),              % Save the depleted image
    swi)).               % Revert to the single blob image

gob_modified(A)  :- gob_modified(A).
```

The higher level predicates, *concavity_tree* and *analyse_node*, remain unchanged. CTs, generated by the second version of the program, are evaluated and the nodes identified in the same order, whatever the orientation of the original object; the biggest (meta-) concavities are analysed first. (See Figure 7.4.4(b) and (d).) A CT generated in this way is said to be in *Canonical Form*. There is no unique form for a canonical tree and in some situations, it may be preferable to use an alternative definition. For example, we might choose instead to list the nodes of the tree so that they are in cyclical order, following a clockwise tour around the object perimeter, beginning from some convenient starting point (e.g. largest (meta-)concavity). (Figure 7.4.5)

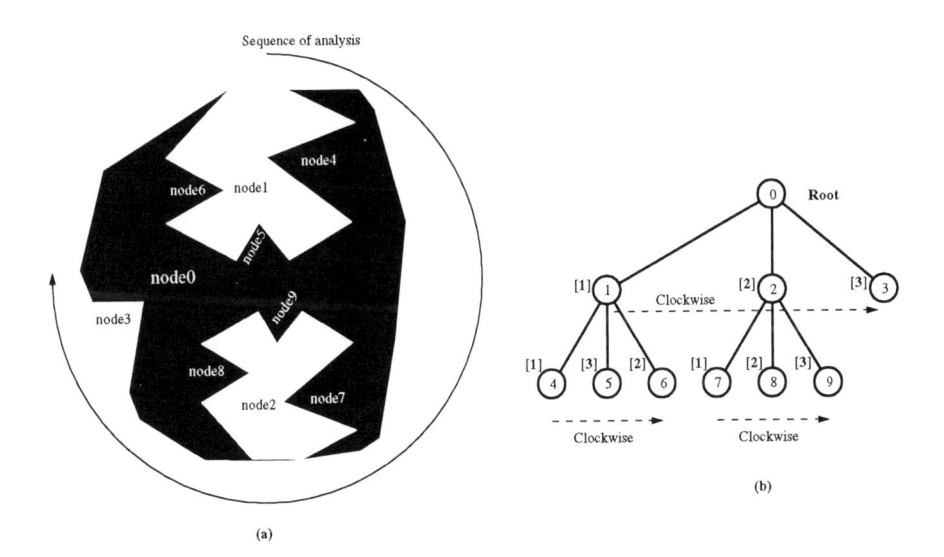

Figure 7.4.5 Alternative definition of Canonical Concavity Trees. (a) Input blob. (Meta-)concavities are analysed in cyclic (clockwise) order, beginning with the largest. (b) Concavity tree. Numbers in square brackets indicate the ranked sizes, *at that level* in the tree. Thus, *node4* [1] is bigger than *node6* [2], which is bigger than *node5* [3].

Program to find Chirality

Normally, we might assume that the concavities are all sufficiently different for us to base the test of chirality on them. If this is not the case, we can use the meta-concavities, meta-meta-concavities etc. instead. The first parameter of *chirality(A,B)* allows us to specify the node whose children are to be compared so that we can find the chirality. (Figure 7.4.6) The following program is based on the assumption that the first two parameters computed by *shape_measurements* are the [X,Y]-co-ordinates of the centroids of the (meta-)concavities.

```
% The chirality (Y) of a given shape is defined in terms of angular
% positions of the (meta-)concavities associated with the children
% of node X. The chirality is computed on the canonical CT.
chirality(X,Y) :-
    findall(M,(ct_node(_,X,M)),Z),   % Z is list of children of node X
    ct_node(X,_,[X0,Y0|_]),          % Where is node X?
    Z = [[X1,Y1|_], [X2,Y2|_], [X3,Y3|_] | _],
    angle(X1,Y1,X0,Y0,A1),           % Angle of line twixt [X1,Y] & [X0,Y0]
    angle(X2,Y2,X0,Y0,A2),
    angle(X3,Y3,X0,Y0,A3),
    chirality_database(Y,A1,A2,A3).  % Consult database

chirality_database(right,A1,A2,A3) :- rank_order(A1, A2, A3).
chirality_database(right,A1,A2,A3) :- rank_order(A2, A3, A1).
chirality_database(right,A1,A2,A3) :- rank_order(A3, A1, A2).
chirality_database(left,A1,A2,A3)  :- rank_order(A1, A3, A2).
chirality_database(left,A1,A2,A3)  :- rank_order(A2, A1, A3).
chirality_database(left,A1,A2,A3)  :- rank_order(A3, A2, A1).

rank_order(A1, A2, A3) :- A1 ≤ A2, A2 < A3.
```

Notice that *chirality(A,_)* fails if there are fewer than 3 children of node A. An important additional point to note is that the program defines chirality for an arbitrary node in the CT and that there is no reason why these values should be the same. (Figure 7.4.7)

7.4.3 Properties of Concavity Trees

In addition to their providing a test for chirality, concavity trees have some useful properties, which may be summarised as follows:

1. A concavity tree combines both global and local information in a single, integrated structure.
2. The accuracy of the representation of a given shape is under the control of the programmer, who can "prune" the tree to ignore small features, or retain them, in order to obtain improved precision.
3. The nodes all correspond to *convex* polygons, which are the (filled) convex hulls of the (meta-)concavities.
4. Nodes in *odd* numbered levels correspond to shapes which are *"stuck in"*. (The root, node0, is taken to be in level 1.)
5. Nodes in *even* numbered levels correspond to shapes which are *"cut out"*.

6. Each of the nodes in a CT can be labelled, by calculating a set of shape, size and position measurements for the corresponding blob and / or its convex hull. In the sample CTs given in Figures 7.4.4, the nodes have been labelled with the areas of the (meta-)concavities, not their convex hulls.

7. Repeated edge features, such as the bays formed between the legs of an integrated circuit, are obvious when we inspect the CT, which contains several similar sub-trees.

8. Semi-flexible shapes, such as a pair of scissors, often retain certain nodes / sub-trees unchanged when they are flexed. (Figure 7.4.4(g).)

9. Touching / overlapping objects will often lead to CTs in which certain nodes are identical to those generated by each object analysed individually. (Figure 7.4.4(f))

10. The CT can be used to determine the orientation of "difficult" shapes, for example, where there is no obvious "long axis". The angles used to test for chirality can also be used to determine orientation.

11. CTs suffer from two different types of instability, which are described below. However, this does not usually lead to serious problems.

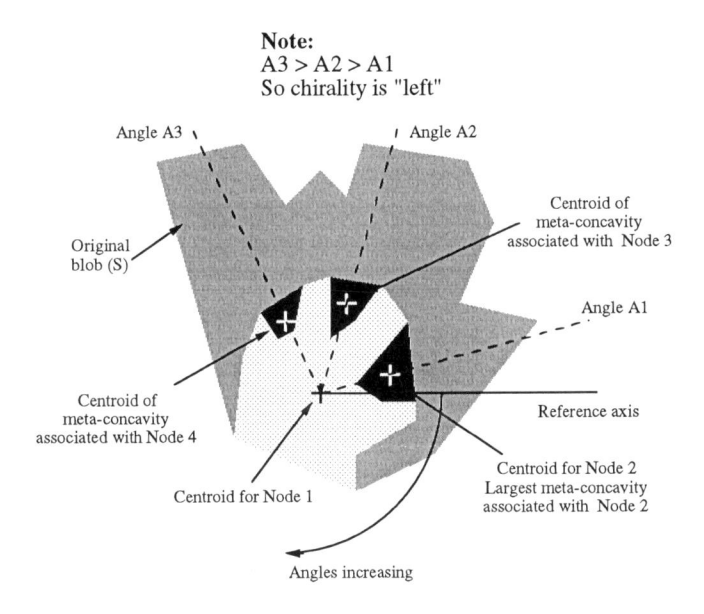

Figure 7.4.6 Chirality test based on the first program for *chirality*. The diagram shows the test being applied to Node 1. (Prolog+ goal *chirality(node1,X)*) The input blob is shaded mid-grey. Its largest concavity (Node 1) is light grey and its children (meta-concavities, labelled Node 2 - Node 4) are very dark grey. Crosses indicate the centroids of this concavity and its three meta-concavities. Since A3 < A2 < A1, the chirality is *"left"*. Notice however, that the chirality calculated on Node 0 is "right". Hence, the goals *chirality(node0,right)* and *chirality(node1,right)* both succeed.

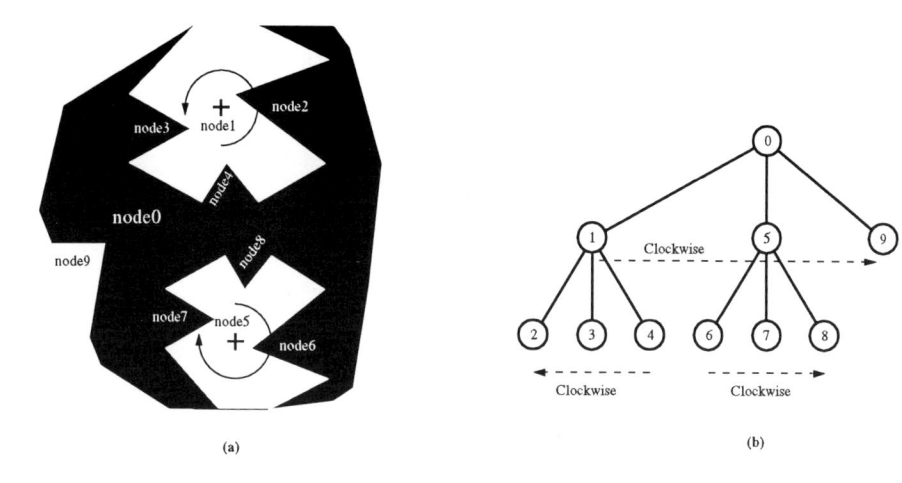

Figure 7.4.7 Chirality is defined in an arbitrary way, so calling *chirality* with different parameters does not necessarily yield the same results. In this example, *chirality(node0,right), chirality(node1,left)* and *chirality(node5, right)* are all satisfied. (a) "Input" blob, Arrows show the directions of decreasing meta-concavity area. (b) Concavity tree, Arrows show the clockwise direction.

Instability

The two blobs shown in Figure 7.4.3(a) and Figure 7.4.3(b) differ in only one small detail: the central "peak" of the bottom-most concavity is lower in Figure 7.4.3(a) than it is in Figure 7.4.3(b). Hence, the bottom-most concavity of this part of the object in Figure 7.4.3(a) gives rise to one meta-concavity and one meta-meta-concavity. The equivalent sub-tree for Figure 7.4.3(b) has two meta-concavities. The remaining features (i.e. the other two concavities) of these two objects are identical. The concavity trees in Figures 7.4.4(a) and 7.4.4(c) are quite different, even though the blobs that they represent are almost identical. This illustrates a fundamental feature of CTs, which has far-reaching implications for their practical application. In Figure 7.4.3(c), the three "peaks" at the top of the bottom-most concavity are aligned exactly. In this situation, it is impossible to predict which form the tree will take; it could resemble either Figure 7.4.4(a) or (c). In practice, camera and quantisation noise will determine which tree is actually generated by the program. For this reason, it is occasionally necessary to store and process more than one CT representing each class of "good" objects. Instability is caused, of course, whenever two (meta)-concavities are separated by only a very narrow channel between the edge of the convex hull and the edge of the blob itself. Thus, a simple, *sans serif* letter 'E' may be viewed as having either one or two concavities. (Figure 7.4.8.) If we were to use the CT approach for shape recognition, we would, clearly, need to store two or more CTs for each case of instability.

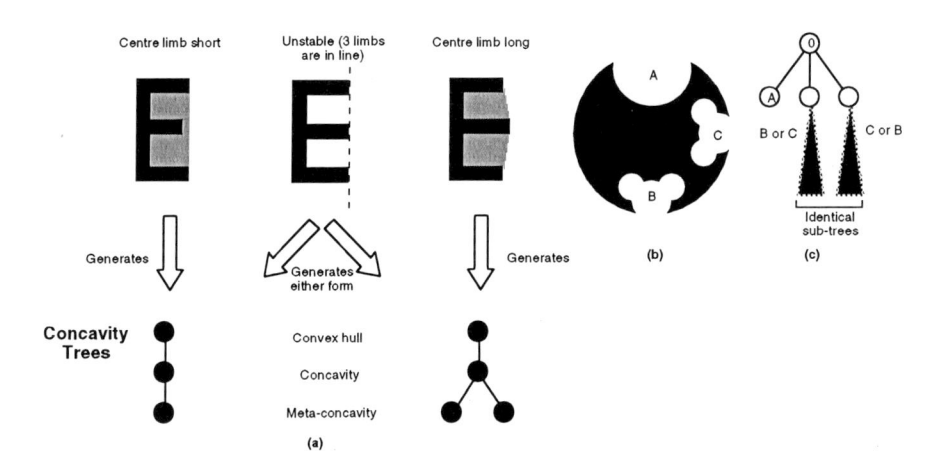

Figure 7.4.8 Instability in Concavity trees. (a) An **E**-shaped object in which the ends of the three limbs are in line exhibits the first type of instability. (b) An object with two (very nearly) identical concavities. Of course, concavity A is analysed first. However, either B or C may be analysed next, depending on camera and quantisation noise. (c) The canonical CT generated for the shape in (b). Notice that there is an ambiguity about the identity of the sub-trees corresponding to concavities B and C. This is the second type of instability.

Canonical CTs, generated by the second version of the program, are *apparently* evaluated and labelled in the same order, whatever the orientation of the input blob, since the biggest (meta-)concavities are always analysed first. (See Figure 7.4.4(b) & (d).) However, noise can again cause problems, if two concavities are very similar. In Figure 7.4.8(b), for example, there are three concavities, two of which are almost identical. In such a case, it is impossible to calculate a unique canonical form of the CT, in the way that we have described. We could, of course, redefine the concept of a canonical tree, based for example on meta-concavities. This second form of instability can be detected very easily by a Prolog+ program, which simply hunts for pairs of similar sub-trees in the CT.

7.4.4 Simpler Tests for Chirality

While Concavity Trees are very versatile, generating them can be quite time consuming. In order to determine chirality, much simpler computational techniques will often suffice and can be made to operate much faster. In this section, just a few of the many possibilities will be described.

Second Program

This simplified procedure to determine chirality, is based on a standard Polar-to-Cartesian co-ordinate mapping. Consider the following program:

```
chirality(right)  :-
    npo,            % Normalise position and orientation
    chf,            % Flip horiz. axis if longest vert. section is left
                    % of image centre
    yxt,            % Interchange X and Y co-ordinate axes
    ptc,            % Map from polar to Cartesian co-ordinates
    yxt,            % Interchange X and Y co-ordinate axes
    rin,            % Integrate intensities along rows
    csh,            % Make all columns the same as RHS of the image
    gli(_,Q),       % Peak intensity
    thr(Q,Q),       % Threshold at peak intensity
    big,            % Isolate biggest region at peak intensity, if more
                    % than one
    cgr(_,Y),       % Find its vertical position
    dgw(_,_,_,A),   % Get image size
    B is int(A/2),  % Find mid-point of image (vertical axis, only)
    Y < B,          % Is peak above centre of image?
    !.              % Do not allow back-tracking

chirality(left).
```

This procedure on which this based is illustrated in Figure 7.4.9. The task of determining chirality is reduced to testing whether or not the peak in the integrated intensity profile (Figure 7.4.9(c)) is above or below the centre of the image.

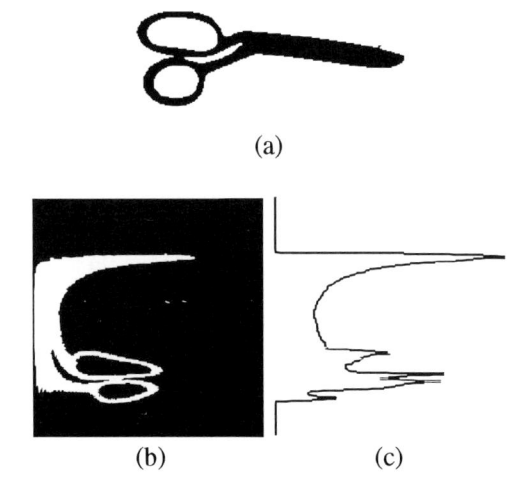

Figure 7.4.9 Chirality test based on the second program for *chirality*. (a) Input image, after *npo*. (b) After *[npo, yxt, ptc]*. (c) *[rin,plt]* applied to (b). The program finds the position of the peak integrated intensity. If the peak is above the centre of the image, the chirality is taken to be *"right"*. Otherwise it is *"left"*.

Third Program

This program fixes the orientation and position of the "input" shape using *[npo, chf]* and then tests whether the centroid of the largest concavity (i.e. either a lake or a bay) is above or below the middle of the image. (Figure 7.4.10) The operator *npo* has the effect of aligning a blob so that its principal axis lies along the horizontal axis, while *chf* makes sure that its longest vertical section is to the right of the centre of the image. N is a control parameter, which allows the user to select which concavity is to be used for finding the chirality.

```
chirality(N,right) :-
   psk, npo,
   chf,            % Flip horizontal axis if longest vertical section
                   % is left of image centre
   cvd, big(N), cgr(_,Y), dgw(_,_,_,A),
                   % Centroid of largest concavity
   B is int(A/2),  % Find mid-point of image (vertical axis, only)
   Y < B,          % Is peak in integrated intensity profile above
                   % centre of image?
   pop, !.

chirality(left).
```

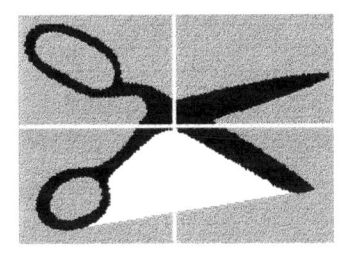

Figure 7.4.10 Chirality test based on the third program for *chirality*. The input image has been normalised, using *[npo,chf]*, so the centroid of the blob is at the centre of the image and its principal axis is horizontal. Since the centroid of the largest bay is below the centre of the image, the chirality is taken to be *"left"*. Notice that the chirality calculated in this way is different from that derived by the fourth method.

Fourth Program

In certain situations, it may be preferable to base the chirality test on lakes (holes), and ignore indentations. (Figure 7.4.11)

```
chirality(N,right) :-
   psk,            % Put input image onto stack
   npo,            % Normalise position and orientation
   chf,            % Flip horizontal axis if longest vertical section
                   % is left of image centre
   blb,            % Fill lakes (holes)
   xor,            % Exclusive OR - isolates lakes
   big(N),         % Isolate N-th biggest lake.
   cgr(_,Y),       % Find its vertical position
```

```
    dgw(_,_,_,A),     % Get image size
    B is int(A/),     % Find mid-point of image (vertical axis, only)
    Y < B,            % Is peak in integrated intensity profile above
                      % centre of image?
    pop,              % Restore input image
    !.                % Do not allow back-tracking

chirality(left).
```

The reader may like to contemplate how the program can be modified to allow indentations but not holes to be used as the basis for the chirality test.

Fifth Program

In some cases, it is impossible to obtain a reliable estimate of the orientation of a component from the principal axis (i.e. axis of minimum second moment), which forms the basis of the operators *npo* and *lmi*. In this case, it is possible to use two bays, two lakes or one lake and one bay to find the orientation first. The following program uses the two largest bays to fix the orientation (Figure 7.4.12) Here is the program, which in other respects resembles the second program.

```
chirality(right) :-
    nlk,                    % Normalise position/orientation using lakes
                            % ranked 1 & 2
    rin, gli(_,Q), thr(Q,Q),
    big,                    % Isolate biggest region at peak intensity,
                            % if more than one
    cgr(_,Y),               % Find its vertical position
    dgw(_,_,_,A),           % Get image size
    B is int(A/2),          % Find mid-point of image (vertical axis, only)
    Y < B,                  % Is peak in integrated intensity profile
                            % above centre of image?
    !.                      % Do not allow back-tracking

chirality(left).
```

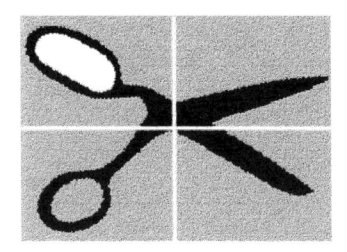

Figure 7.4.11 Chirality test based on the fourth program for *chirality*. The input image has been normalised, using *[npo,chf]*, so the centroid of the blob is at the centre of the image and its principal axis is horizontal. Since the centroid of the largest lake is above the centre of the image, the chirality is taken to be *"right"*. Notice that the chirality calculated in this way is different from that derived by the third method.

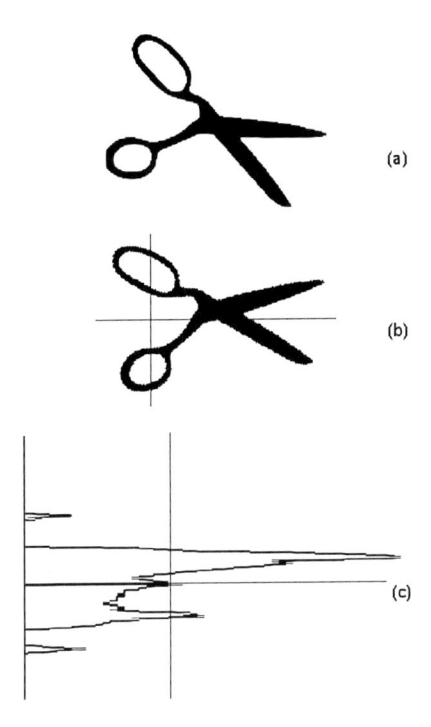

Figure 7.4.12 Chirality test based on the fifth program for *chirality*. (a) Input image. (b) After *nlk* has been applied. (c) *[rin,plt]* applied to (b). The program finds the position of the peak integrated intensity. If the peak is above the centre of the image, the chirality is taken to be *"right"*. Otherwise it is *"left"*.

7.5 Telling the Time

In this case study, the camera views a conventional "analogue" clock. The vision system interprets the image of the clock and calculates what time it is. The output is given in both digital and symbolic "casual" format. Examples of the latter are *"twenty past three"*, *"quarter past eleven"* and *"ten o'clock"*.

7.5.1 Significance

Analysing the image of a clock face serves as a model for a number of important tasks for manufacturing industry. Although less common than they once were, moving-needle meters are still widely used in automobiles, aircraft, industrial process control instruments, domestic electricity, water and gas meters. A machine vision system might appropriately be used to "read" instruments such as these, during calibration and inspection.

The task of telling the time by analysing the image of an analogue clock is less straightforward than might at first be thought. Since both hands move continuously and are linked by gears, the significance of the position of the hour hand can only be determined after first locating the minute hand. In addition, there are two "discontinuities", which occur when the minute and hour hands pass the 12 o'clock position. Furthermore, special provision has to be made for analysing the image when the minute and hour hands overlap. The task of telling the time can be broken down into several sub-tasks, which is, of course, well-suited to Prolog's multiple-clause structure.

7.5.2 Simplifying Assumptions

The following assumptions were made initially, in order to make the problem manageable.

(a) The clock face consists of a white circular disc, with a black annulus surrounding it.
(b) The hands have black tips but are white near the centre of the face.
(c) There is no second hand.
(d) Black numerals, minute / hour "tick" marks and lettering are tolerated on the clock face but they must be thin compared to the hour and minute hands.

Some of these conditions can be relaxed but the general effect is to make the program more complicated. Unless we make these, or some other similar simplifying assumptions, it seems unlikely that it would be possible to write an effective program for telling the time, given any type of clock face. Unusual and bizarre clock designs, such as those showing pictorial scenes, animals, cartoon characters, etc. are outside the scope of our present discussion, since our theme in this book is to describe techniques that are relevant in the context of manufacturing industry.

In certain industrial calibration tasks, it may be possible to move the hands / needle of an instrument quite quickly, under software control. This makes one particular type of vision algorithm (image subtraction) more attractive than it seems when we consider clocks. We shall therefore discuss how several images of a clock, obtained over a period of several hours, can be combined. While this seems an unreasonable approach when writing a program to tell the time, it is nevertheless quite attractive in certain industrial applications.

7.5.3 Lighting

It is clearly important to avoid producing shadows (of the rim and hands) and glinting (on the hands, front glass). Lighting the clock face with a broad illumination source is ideal. One possible way to do this is to use flood lamps to project light onto a board that has been painted matt white. The camera views the clock face through a small hole cut in the centre of the board.

7.5.4 First Program

The following program represents our first attempt at writing a program to tell the time. It returns the time in both digital and casual formats.

```
telling_time(H, M, C) :-
    grb,                            % Digitise image
    yxt,                            % Interchange X & Y axes
    tbt,                            % Flip vertical axis
    crack,                          % Non-linear filter & threshold
    blb,                            % Fill black holes
    cgr(U,V),                       % Centroid of clock face
    xor,                            % Isolate face
    blb,                            % Fill black holes
    xor,                            % Isolate hands
    big_blobs(200),                 % Eliminate small blobs (noise)
    count(blobs,N),                 % To check whether hands overlap
    biggest,                        % Select minute hand (larger)
    cgr(Xm,Xm),                     % Find its centroid
    angle(Xm,Xm,U,V, Mangle),       % Orientation of minute hand
    ( (N = 2,                       % Test whether hands overlap
    xor,                            % No - so select hour hand
    cgr(Xh,Yh),                     % Find its centroid
    angle(Xh,Yh,U,V,Hangle))        % Orientation of hour hand
    ;                               % Prolog OR operator
    (Hangle is Mangle) ),           % Do this if hands do overlap
    A is 60 - Mangle/6,             % Calculate M, floating pt.
    int(A,M),                       % Convert to integer
    calculate_hour(Hangle, M, H),   % Calculate hours
    casual_time_conv(H, M, C),      % Convert to casual format
    !.

% Interpret hour hand position, considering minute hand position.
calculate_hour(Hangle, M, H) :-
                                    % Calculate hours when minutes < 30.
    M < 30, A is (367 - Hangle)/30, int(A,H).

calculate_hour(Hangle, M, H) :-
                                    % Calculate hours when minutes ≥ 30
    M ≥ 30, A is (353 - Hangle)/30, int(A,H).

% Break task of converting to casual time format into twelve small
% units.

% 0 - 2 mins past hour. Example of output: [six, o_clock]
casual_time_conv(A, B, C) :-
    B ≤ 2,                          % Check minutes ≤ 2
    number_to_words(A,D),           % Convert hour to word format
    C = [D,o_clock].                % Create output list

% 3 - 7 mins past hour. Example of output: [five, past, four]
casual_time_conv(A,B,C) :-
    B ≥ 3, B ≤ 7,                   % Check minutes
    number_to_words(A,D),           % Convert hour to word format
    C = [five,past,D].              % Create output list
```

axis corresponds to distance, measured from the face centre. The minute hand is the longer of the two dark grey "fingers". The dark horizontal band is due to the rim of the clock face, while the hour "tick" marks are located beneath it. Notice that the operator *ptc* requires two parameters which define the centre of the polar-Cartesian axis transformation. This can conveniently and easily be derived from the centroid of the face disc. Slight errors arise as a result of quantisation noise, when calculating the centre of the face. This inevitably leads to a small "sinusoidal" wave in the image of the rim. Its effect is not important for our application. Image 7.5.3(c) illustrates the result of applying the command sequence [*yxt, rin, csh, wgx, sub, thr, bed, yxt*] applied to (b). This is equivalent to integrating the intensity in Image 7.5.3(a) along a series of radii. The major peaks here are easy to detect and locate, thereby enabling the clock to be "read". The minor peaks are, of course, due to the hour "tick" marks. Image 7.5.3(d) illustrates the original image from a real clock. Notice the fluorescent tips of the hands. It would, of course, be possible to increase the image contrast by illuminating with ultra-violet light. However, we chose not to do this, so that we could demonstrate the method on a more complex visual analysis task.

Image Subtraction One possible method of analysing the clock image that is potentially much more robust is to compare it with a face which has no hands. A similar effect can be obtained by comparing the clock image with another image, obtained some time (at least one hour) before. Then, simple image subtraction, allows the hands to be identified easily. (Image 7.5.4)

The following program allows an image of the face to be reconstructed, as if the clock did not have any hands. This program operates on the assumption that the face is bright compared to the hands, the camera and clock are both fixed rigidly and the lighting is constant over a 12 hour period.

```
build_face_image :- timer(0), grb, wri, fail.

build_face_image :-
   delay,                  % Don't do things too often - reduces noise
   grb, rea, max, wri, timer(X),
   X < 12.0,               % Observe clock for  12 hours
   !,                      % Needed to improve program efficiency
   build_face_image.       % Repeat

build_face_image :- rea.   % Get clock face image from disc
```

It is now a straighforward matter to locate the hands. To do this, we simply subtract the image created using *build_face_image* from the picture of the clock. Simple thresholding will then isolate the hands. (See Image 7.5.4)

7.5.6 Concluding Remarks

The purpose of this case study was to demonstrate that, what may seem to be a simple application, may be quite complicated and require the use of AI techniques. None of the methods described above is "complete", in being able to tell the time correctly for all designs of clock face. Indeed, it is quite to easy to contrive special cases where they all fail. While the image subtraction method is probably the most robust, it does require that the clock be stationary within the camera's field of view for a long period. As we have pointed out, this is not necessarily a serious problem in some industrial instrument calibration tasks. For example, suppose that we wish to calibrate an aircraft's barometric altimeter. During calibration, the signal from the transducer (measuring air pressure) going into the cockpit display instrument is replaced by one generated electronically. The test equipment then moves the instrument needle to a number of pre-set positions (e.g. 5000, 10000, 20000 feet, etc.) and the vision system locates the needle(s). In this type of application, the image subtraction method will probably be acceptable, whereas it would be far too slow for setting the time on clocks.

7.6 Food and Agricultural Products

7.6.1 Objective

In this section, we shall present a series of case studies, demonstrating how Prolog+ can be used to ensure that agricultural and food products are well formed, safe (i.e. have no large-body contaminants) and are attractive in appearance. Outline solutions will be presented, in the hope of convincing the reader that there is a large potential for machine vision in the agri-food industry, where there is inevitably a high degree of product variability.

7.6.2 Industrial Relevance

Since no physical contact is made with the objects being inspected, automated visual inspection is inherently and totally hygienic. Furthermore, machine vision can be used to examine soft and semi-liquid materials, such as purées of fruit and vegetables, tomatoes, uncooked dough, whipped cream, icing, butter etc., without any chance of deforming them. Since it is theoretically possible to examine a wide variety of features on food products, it would be reasonable to expect that we would find machine vision being used extensively throughout the food industry. In fact, this is not so. In the past, machine vision has been applied successfully in many widely different situations in the "hard" manufacturing industries (e.g. automobile, electronics, aircraft and consumer goods). These industries typically manufacture objects with close tolerances, often with micron-level accuracy on linear dimensions. By way of contrast, the food industry produces artefacts which are much more variable; tolerances are measured in terms of millimetres. This has, in the past, presented quite severe difficulties for standard (i.e. *"non intelligent"*) machine vision systems, which are much better suited to verifying that *well-defined* products are being made as they should be. It has been found to be much more difficult to build machine vision systems which are able to cope with the high degree of variability in shape, size and appearance that characterises food products. While there is no doubt that this high level variation has hindered the acceptance of machine vision technology in the food industry, there has been some notable work in the agri-food area. [CHA-95]

- Biscuits, inspecting biscuits.
- Bread, analysing texture.
- Carrots, controlling trimming.
- Cherries, detecting kernel shells embedded in flesh.
- Chicken / fish, detecting bones, using x-rays.
- Chips (french fries), inspecting for black spots.
- Confectionery, controlling decoration.
- Fish fingers, counting.
- Flat fish, controlling trimming.
- Flour, measuring bran content.
- Lettuce, harvesting.
- Loaves, measuring shape and volume.
- Measuring the thickness of chocolate on confectionery.
- Meat, measuring fat: lean ratio.
- Mushrooms, harvesting mushrooms.
- Potatoes inspecting.
- Sacramental wafers, inspection.
- Seeds (e.g. rice, wheat kernels, etc.), sorting.

Chan [CHA-95] provides an overall review of the achievements and potential for applying machine vision in the food processing industry. Another notable source of information about the need for improved food quality is to be found in newspapers. It is all too common to see headlines which report that a foreign body, such as a sliver of glass, metal bolt, dead mouse, bird's skull, or snail, has been found in food products, ranging from packaged peanuts and bottled milk to loaves and pizza. Numerous cases of foreign body contamination of food products are heard in the lower courts, but few cases ever reach the higher courts. It seems that in the UK and Ireland, most food manufacturers either settle out of court, or are fined in a lower court for selling products which contain dangerous / unpleasant foreign bodies.

We are left with several main conclusions:

(a) Machine vision is ideally suited to inspecting food products since it is totally hygienic.
(b) There remains a major problem of detecting foreign bodies in food.
(c) There is a continuing problem which manufacturers face, in making malformed products which lead to customer dissatisfaction, even though they are perfectly safe and nutritious. Such products are sometimes pulped and

recycled, or simply sold as scrap for animal feed. In both cases there is a loss of valuable product and hence profit.

(d) There are numerous potential (i.e. unsolved) applications of machine vision, where the existing technology has been unable to provide a cost effective solution.

(e) In both Europe and USA, even more stringent product safety laws are being imposed, particularly on the food and pharmaceutical industries.

(f) There has been a distinct reluctance to employ machine vision systems in the food industry, on account of the high capital cost of such systems and low profit margins in this industry.

The principal reasons why low-tolerance products, such as confectionery, loaves, meat pies, pizzas, etc. are difficult to inspect using machine vision are as follows:

(i) It is impossible to guarantee that viewing / lighting angles are optimal, if the product shape is changing drastically and unpredictably.
(ii) There are no firm points (e.g. corners, straight sides, drilled holes, etc.) on which to "anchor" measurements.
(iii) The interpretation of measurements and other data derived from highly variable artefacts requires subtle (i.e. rule-based) analysis.

For these reasons, we need to employ more "intelligent" machine vision systems in the food industry than have been used in the "hard" manufacturing industries.

7.6.3 Product Shape, Two-dimensions

Many mass-produced food products, such as pies, tarts and certain types of loaf are made in moulds, while others are extruded and then cut off to a given length. The latter is especially popular for making confectionery, since it employs a highly reliable continuous manufacturing process, which lends itself very well to full automation. Another popular manufacturing technique is to stamp out (complicated 2D) shapes from a sheet of soft material, such as dough, using a specially shaped knife. In all of these cases, the shape of at least part of the product is quite well controlled. However, there are numerous instances of food products being made without any constraints, except that they lie on a flat tray. For example, (American-style) cookies, macaroons, meringues, Welsh cakes, and scones are formed by depositing a preformed (stamped) shape of dough-like consistency, or an amorphous "dollop" of a semi-liquid material, onto a flat surface where it is baked. As it cooks, the "dollop" takes on a new shape, which develops in an unpredictable and uncontrolled way. Croissants, Cornish Pasties, and filled pies all have widely varying shapes for this reason. Checking product shape, given a binary image representing its silhouette, is one of the most important tasks in food inspection. We shall therefore consider this task in some detail, beginning with very simple shape checks and progressing to more sophisticated methods.

Image Acquisition

In order to generate a high-contrast image, which can then be thresholded to create a binary image, we can use one of several lighting-viewing techniques. [BAT85]

(i) Back lighting, with a light source located behind the object being examined.
(ii) Front lighting, using an ultra-violet light source and a fluorescent background. A UV-blocking filter is placed in front of the camera lens.
(iii) Front lighting using an ultra-violet light and a UV-absorbent, non-fluorescent background. This works only if the object being examined is fluorescent. Again, a UV-blocking filter is placed in front of the camera lens.
(iv) The CONSIGHT structured-lighting system. [HOL-79] This requires that the objects being inspected are carried on a smooth conveyor belt. (A chain belt would not be appropriate.)
(v) Off-axis front illumination, using carefully collimated (parallel-beam) light sources which shine on the object but not on the background. The latter must be placed some considerable distance behind the object, to avoid light falling on its surface.
(vi) Front lighting using coaxial illumination and viewing, and a retro-reflective background surface.
(vii) Front lighting and a highly coloured background. A programmable colour filter or a simple optical filter can improve the contrast between the object and its background. This technique only works with objects which are not strongly coloured.
(viii) Thermal imaging camera. The object being examined must be hot and no external light is required.

The final choice of lighting-viewing method can only be made when the full application requirements are known. Let us turn our attention now to algorithmic and computational techniques for inspecting silhouettes of food products, assuming that a binary image has already been created

Rectangular and Circular Biscuits

First, let us discuss how we can inspect rectangular objects such as biscuits. The following program will do this quickly and reliably, provided that the straight sides of the biscuit lie parallel to the image border.

```
rectangular_biscuit :-
  grab_and_process_image,   % Generate binary image
  mar,            % Draw minimum  area rectangle
  max,            % Superimpose MAR onto original (binary) image
  blb,            % Fill holes
  xor,            % Isolate differences between MAR & original image
```

```
3•skw,          % Shrink white areas.
cwp(0).         % Have all white regions disappeared?
```

rectangular_biscuit succeeds if the object being viewed is nearly rectangular *and* is aligned to the image axes. If the biscuit is rotated, relative to the camera, or the biscuit is broken, *rectangular_biscuit* fails. In many food manufacturing applications, it is perfectly reasonable to expect that the objects being inspected will arrive in front of the camera in known orientation. (The material cutting and mechanical handling arrangements ensure this.) However, if this condition is not satisfied, we simply add one extra step, *npo*, before the second sub-goal in *rectangular_biscuit*.

Another method of aligning the silhouette is needed if the alignment assumption is invalid and the biscuit is square. One possible way to do this is explained in Figure 7.6.1. We identify the top- and left-most points of the biscuit (A and B respectively) and then draw two vertical lines to intersect the biscuit at points C and D. Notice that the vertical lines through points A, B, C and D are equally spaced. The orientation of the biscuit is then determined by that of the line CD. We shall use a similar technique when we consider the inspection of slices of bread and will describe the program to do this, in detail then.

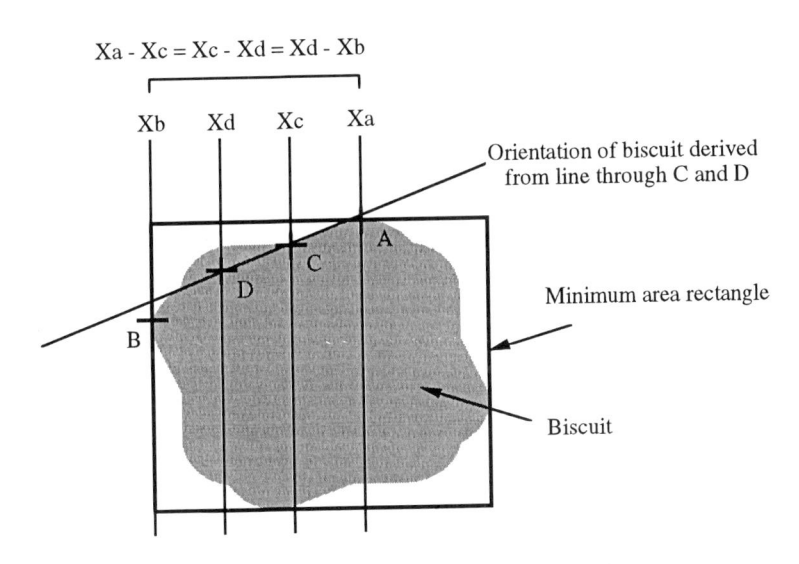

Figure 7.6.1 Alignment of a biscuit.

It is a straightforward task to verify that a biscuit, cake, pizza, or pie is nearly circular. Here is a program to do this.

```
circular :-
    grab_and_process_image,  % Generate binary image
    bed,          % Derive edge
    cwp(A),       % Count edge points
    wri,          % Save image for later
```

```
        ccc,         % Draw circumcircle
        3.exw,       % Make it thicker. Adjust loop parameter to taste
        rea,         % Recover edge image
        min,         % Apply "thick circumcircle" as a mask on edge image
        cwp(A).      % Do all edge pixels lie within 3 pixels of
                     % circumcircle
```

A simple alternative to the above scheme will be defined in a little while. This second procedure fits an ellipse into the *minimum area rectangle* (MAR) surrounding the blob which is to be tested for circularity. The aspect ratio of this ellipse and the areas of protuberances and indentations, defined by this ellipse, are computed. If the aspect ratio is outside defined limits, or the areas of the protuberances or indentations are too large, *circular* fails.

```
    circular :-
        psk,                % Push image onto stack
        dim(L,T,R,B),       % Find limits of minimum area rectangle (MAR)
        zer,                % Draw black image
        cir(L,T,R,B,255),   % Draw ellipse within MAR
        tsk,                % See image at top of stack
        sub,                % Subtract images
        thr(200),           % Isolate areas outside fitted circle
        cwp(U),             % Count points
        swi,                % Area of protuberances
        thr(0,100),         % Isolate areas inside fitted circle
        cwp(V),             % Area of indentations
        pop,                % Restore input image whatever the result
        S is ((R - L)*(R - L)) / ((B - T)*(B - T)),
                            % Squared aspect ratio of MAR
        circularity_tolerance(U1,V1,W1,W2),
                            % Consult DB for tolerance limits
        S ≥ W2,     % Fails if object is an ellipse, rather than circle
        S ≤ W1,     % Fails if object is an ellipse, rather than circle
        U ≤ U1,     % Fails if area of protuberances is too large
        V ≥ V1.     % Fails if area of indentations is too large

    circularity_tolerance(200,202,0.95,1.05).
                            % Database for shape parameters
```

This second definition of *circular* could, of course, be extended easily to inspect elliptical objects. We simply add *npo* after the first sub-goal and adjust the aspect-ratio limits stored in *circularity_tolerance* (third and fourth parameters). There are, of course, many other possible techniques for examining "geometric" shapes such as squares, rectangles, circles and ellipses. It must be borne in mind, however, that the shapes found in food products bear only a very loose resemblance to the mathematical entities which bear the same names. For example, a "circular biscuit" is not truly circular in the mathematical sense. Nevertheless, inspecting food products with "geometric" shapes is rather simpler than some of the other applications that we shall consider later. We could use any of the techniques listed below to examine nominally circular food products, which have a wide range of variability.

(a) Shape ratio (*shf*, which is based on the ratio of the area of a blob to the square of its perimeter) to examine circular objects.

(b) Ratio of the area of the convex deficiency [*cvd*] of a blob to that of the blob itself. This detects indentations quickly and easily. Convex shapes, such as squares, rectangles, circles and ellipses, all produce a small value for this ratio (close to zero).

(c) Find three edge points and then fit a circle to intersect all of them. Suppose that the radius of the fitted circle is R and its centre is at [X,Y]. We then draw a circular annulus, centred on [X,Y] with radii R ± K.R, where K is much smaller than 1.0 and then test, to make sure that all edge points lie within this annulus.

(d) Compare the minimum bounding circle [*mbc*] with a nominally circular object.

(e) Use the modified version of the Hough transform which can locate circles.

(f) Compute the differences between the minimum area rectangle [*mar*] and the silhouette. This is suitable for inspecting rectangular objects, such as biscuits, with rounded corners. Here is the program code:

```
rounded_rectangular_biscuits :-
    grab_and_process_image,   % Generate binary image
    npo,                      % Normalise position and orientation
    cwp(A),                   % Area
    mar, max, blb, xor,       % Isolate corners
    skw,                      % Ignore minor indentations
    blp(B),                   % Measure parameters of all blobs
    length(B,),               % Check there are 4 rounded corners
    compare(A,B).             % Simple Prolog test for corners
```

Slices of Bread

Examining the silhouette of a slice of bread is rather more complicated than inspecting the simple moulded / pressed shapes found in biscuits (Images 7.6.1 and 7.6.2). The ideal shape of slices from a lidded tin loaf is square, so that the filling of a sandwich does not ooze out at the edges. It is, of course, a simple matter to modify the rules described above to inspect the square slices taken from a lidded tin loaf. Hence, we shall not discuss this type of loaf in much detail here. On the other hand, non-lidded tin loaves are only partially constrained during baking and hence are more variable in the shape of the top surface. This makes inspecting them a more challenging and interesting task. Of course, there are many other types of loaf. The most difficult ones to inspect are those which are baked on a flat tray and are therefore totally unconstrained as the dough rises during baking. We shall concentrate for the moment upon the non-lidded tin loaf and merely illustrate the processing of images from a lidded tin loaf in passing.

The most complicated part of the task of inspecting a slice taken from a non-lidded tin loaf is the identification and measurement of appropriate features. Once a set of measurements has been derived from a slice, it is a relatively straightforward matter to compare them with stored values, using learning and recognition rules similar to those outlined in Section 7.2.4. There are several steps in the process of deriving suitable measurements:

(a) Locating the base of the slice, calculating and then normalising its orientation. (This will provide a reference for subsequent angle measurements.)
(b) Isolating the straight parts of the sides and calculating their orientations relative to each other and to the base.
(c) Locating the top of the slice and measuring its radius of curvature.
(d) Isolating and measuring the "overspill". (This is the upper part of the loaf which spreads out as the dough rises, overspilling the rim of the open-top baking tin.)

Locating the Base and Determining Orientation

One possible way to determine the orientation of a slice of bread, prior to analysing its shape, is to use the *Hough transform* [*huf*]. Image 7.6.2 shows the Hough transform image derived from the slice silhouette, after the binary edge detector operator [*bed*] has been applied. The location of the brightest point in the Hough transform output image (Image 7.6.2(d)) indicates both the position and orientation of the bottom edge of the slice. We can, of course, use this information to good effect, to normalise the orientation of the image of the slice, since this is the longest linear edge segment. Our experience has shown that the Hough transform is more accurate in this application than the faster operator, *lmi*. (Image 7.6.2(c)) The reason is that the loaf is almost square and has no obvious "long axis". This remark applies particularly to the *lidded tin* loaf (Image 7.6.1), which ideally has a square cross section.

Another possible technique for determining the orientation of a slice is to choose two points on the bottom (nearly straight) edge of the slice. (Image 7.6.2(k)) We then find the orientation of the line joining those points, using the operator *ang* (c.f. Figure 7.6.1). This technique is rather faster than that based on the Hough transform and is sufficiently accurate and reliable, provided that the slices are always orientated approximately. Assuming that the orientation of the bottom edge of slice does not vary by more than about ±30°, relative to the horizontal axis of the image, this method will work reliably. In many industrial applications of machine vision, an assumption of this general type is perfectly reasonable and reflects reality, where approximate, but not precise, positioning and alignment can be guaranteed by simple mechanical means, such as guide rails, deflector plates, etc. Assumptions like this often allow much simpler / faster computational methods to be used in machine vision. Despite its great popularity among image processing specialists, the Hough transform is often far less attractive for industrial applications than an alternative heuristic procedure, which is often simpler and faster and hence more likely to find its way into an industrial machine vision system.

It often happens, when studying industrial applications of machine vision, that problem-specific knowledge allows a simpler and faster *heuristic* procedure to be used, in preference to an algorithmic technique, even though the latter has received the benefit of detailed mathematical analysis. We very often find that implementation details preclude the use of certain "mathematically proven" computational techniques.[1] The authors would prefer to use some computationally convenient heuristic methods in this particular application, rather than the Hough transform, reflecting their conviction that Industrial machine vision should properly be regarded as a Systems Engineering discipline and not as part of the *science* of computer vision.

However, for the sake of completeness and to demonstrate that Prolog+ is sufficiently versatile to accommodate both approaches, we shall present two programs with which we can calculate and normalise the orientation of a slice of bread taken from a non-lidded tin loaf, beginning with one based on the Hough transform:

```
normalise_loaf_orientation1 :-
    psk, skw, xor,            % Edges are left behind
    huf,                      % Hough Transform
    gli(_,A), thr(A,A), big, cgr(X,Y),
    dgw(L,T,R,B),             % Image size
    Z is int(-90*X/(R-L+1)),  % Rescale to calculate angle
    tsk,                      % See image at top of the stack
    tur(Z),  % Rotate so strongest linear feature (base) is horizontal
    pop,     % Recover original image
    swi.     % Switch images - normalised image is in current image
```

The simpler alternative, in which we determine the orientation using two edge points may be implemented thus (c.f. Figure 7.6.1):

```
normalise_loaf_orientation2 :-
    psk,                      % Push image onto stack
    dim(A,_,B,_),             % Dimensions of loaf
    X1 is int(0.7*A + 0.3*B), % Vert. scan line towards left of base
    X2 is int(0.3*A + 0.7*B), % Vert. scan line towards right of base
    scan(X1,X2,Y1,Y2),        % Intersections of 2 scan lines & edge
                              % of base
    ang(X1,Y1,X2,Y2,_,Z),     % Angle of base w.r.t. horizontal
    Z1 is int(-Z),            % Negate angle
    tur(Z1),                  % Turn loaf image
    pop,                      % Restore input image
    swi.                      % Switch images
```

Locating Straight Sides

A program will be described which fits a straight line to the linear section of the side of a non-lidded tin loaf. Refer again to Image 7.6.2(d), which shows the Hough transform derived from the silhouette of a slice. Notice that there are three strong, well-defined peaks. We have already located and analysed one of these, to find the orientation of the base. The two other peaks correspond to the straight sides of the slice and can be analysed in a similar way. Notice that these two

---

[1] This is, of course, a dynamic situation, since a procedure that is too expensive / slow to implement now may become more attractive in a few years time, after computational and electronic hardware techniques have improved.

peaks have approximately equal co-ordinates along the horizontal axis, indicating that the lines they represent are almost parallel.

A procedure can be defined in terms of the Hough transform and which uses a set of edge points located on one side of the slice. Of course, this uses a similar process to that embodied in *normalise_loaf_orientation2* and hence will not be described in detail. Instead, we shall concentrate on a third method which uses *lmi* in an unusual way. Although we dismissed using this operator earlier, it can be used to good effect provided that we apply it to a section of the edge contour and not to the whole slice silhouette.

```
% Fit a straight line to the linear part of the left-hand-side of
% the loaf
loaf_left_side(Z) :-
    psk,                            % Push image onto stack
    dim(L,T,R,B),                   % Dimensions of loaf
    Y1 is int(T + (B-T)*0.5),       % Upper part of linear section
    Y2 is int(T + (B-T)*0.8),       % Centre of linear section
    tsk,                            % See image at top of stack
    bve(L,Y1,R,Y1,X1,_,_,_),        % Intersection of top scan line with
                                    % edge of loaf
    tsk,                            % See image at top of stack, again
    bve(L,Y2,R,Y2,X2,_,_,_),        % Intersection of middle scan with edge
    zer,                            % Black image
    fld(X1,Y1,X2,Y2),               % Fit straight line to linear part of
                                    % left of loaf
    angle(X1,Y1,X2,Y2,Z),           % Calculate angle of this line
    pop,                            % Restore original image
    swi.                            % Switch images

% Fit a straight line to the linear part of the right-hand-side of
% the loaf
loaf_right_side(A) :-
    psk,                            % Push image onto stack
    lrt,                            % Flip horizontal axis
    loaf_left_side(B),              % Fit straight line to edge on LHS
    A is -B,                        % Invert angle as calculated for LHS
    lrt,                            % Flip horizontal axis again
    pop,                            % Restore original image
    swi.                            % Switch images
```

Measuring Overspill

The lines formed by *loaf_left_side* and *loaf_right_side* allow us to identify the *overspill*. (Image 7.6.2(g)(h)). We explained earlier that this is the upper part of the slice which overhangs the sides of the baking tin.) Thus, the overspill can be reduced to two blobs, which can in turn be represented by two sets of simple shape parameters, such as area, perimeter, dimensions and aspect ratio.

Radius of Curvature of Top Edge

Let us assume, for the sake of simplicity, that the orientation of a slice from a non-lidded tin loaf has already been normalised (i.e. the bottom edge is nearly parallel to the image border). Using the following program, we can measure the radius of the top edge. (Figure 7.6.2)

Figure 7.6.2 Fitting a circle to the top surface of a non-lidded tin loaf, using three sample points.

```
loaf_top(X0,Y0,Rad) :-
   psk,                       % Push image onto stack
   dim(A,_,B,_),              % Dimensions of blob
   X1 is int(0.8*A + 0.2*B),  % X-position: first vertical scan line
   X2 is int(0.5* (A + B)),   % X-position: second vertical scan line
   X3 is int(0.2*A + 0.8*B),  % X-position: third vertical scan line
   scan_3_lines(X1,X2,X3,Y1,Y2,Y3),  % Three vertical scan lines
   fit_circle(X1,Y1,X2,Y2,X3,Y3, X0,Y0,Rad),
                              % Fit circle to intersections
   pop,                       % Pop image stack
   swi.                       % Switch images

% Find Y intersections of blob with three vertical scan lines given
% X-positions
scan_3_lines(X1,X2,X3,Y1,Y2,Y3) :-
   psk,                 % Push image onto stack
   dgw(L,T,R,B),        % Image dimensions
   bve(X1,T,X1,B,_,Y1,_,_),
                        % Find intersections with first scan line
   tsk,                 % See image at top of stack (Do not POP)
   bve(X2,T,X2,B,_,Y2,_,_),
                        % Find intersections with second scan line
   tsk,                 % See image at top of stack (Do not POP)
   bve(X3,T,X3,B,_,Y3,_,_),
                        % Find intersections with third scan line
   pop.                 % Recover "input" image

% Fit a circle to three points
fit_circle(X1,Y1,X2,Y2,X3,Y3, X0,Y0,Rad) :-
   circle(X1,Y1,X2,Y2,X3,Y3, X0,Y0,Rad),    % Defined below
   L is int(X0 - Rad), R is int(X0+ Rad),
   T is int(Y0 - Rad), B is int(Y0 + Rad), zer,
   cir(L,T,R,B,255).           % Draw circle inside rectangle

% Radius R & centre [Px,YPy] of circle through points [Ax,Ay],
% [Bx,By] & [Cx, Cy]
circle(Ax,Ay,Bx,By,Cx,Cy,Px,Py,R):-
   D is 2*(Ay*Cx + By*Ax -By*Cx - Ay*Bx  - Cy*Ax + Cy*Bx),
   Px is ( By*Ax^2 -  Cy*Ax^2 -  Ay*By^2 + Ay*Cy^2 +  Cy*Bx^2 +
   By*Ay^2 + Ay*Cx^2 -  By*Cy^2 -  By*Cx^2 -  Ay*Bx^2 +
   Cy*By^2 -  Cy*Ay^2 )/ D,
```

```
Py is ( Cx*Ax^2 + Cx*Ay^2 + Ax*Bx^2 -  Cx*Bx^2 + Ax*By^2 -
Cx*By^2 -  Bx*Ax^2 -  Bx*Ay^2 -  Ax*Cx^2 + Bx*Cx^2 -
Ax*Cy^2 + Bx*Cy^2 ) / D,
R is sqrt((Ax - Px)^2  + (Ay - Py)^2).
```

7.6.4 Analysing the 3D Structure of an Uncut Loaf

To conclude this section, we briefly discuss the analysis of the 3-dimensional shape of the top surface of a loaf using a so-called *Depth Map*. This is an image in which the "intensity" indicates the height of a surface, not the amount of light coming from it. An optical arrangement for generating depth maps is shown in Figure 7.6.3. This technique is variously called *light stripe sectioning, structured lighting* and *triangulation*, and relies on the fact that the loaf is moved progressively past the camera and light-stripe generator. The latter can very conveniently be built, using a diode laser, fitted with a cylindrical lens. At each X-position for the loaf, a vector of height-measurement values is created. This vector describes the height of that curve formed by the intersection of the top surface of the loaf with a vertical plane. Of course, some parts of a loaf, such as indentations in its sides and ends, are obscured. Occlusion can cause difficulties for subsequent analysis of the range map. To overcome this, more sophisticated range-measurement techniques are needed. For example, it is possible to use three, or more, laser light-stripe generators and cameras to obtain an "all-round" view of loaf sides and top surface. (We cannot, of course, measure the base.) As we shall see, minor occlusions can result in black spots appearing in depth maps and these require the use of special processing techniques.

An important point to note is that in Figure 7.6.3(a), the light-stripe generator is located directly above the loaf, while the camera views it obliquely. Many articles and books, even a wall-poster produced by a learned society show the camera placed above the sample and the laser off-set to one side. However, this makes the analysis of the data very much more difficult and it is far easier to use the arrangement shown in Figure 7.6.3(a). The reason is simple: the light stripe effectively forms a section of the object being measured. If a ray produced by the laser does not lie in the vertical plane, its point of intersection with the surface is not fixed, along the X axis. The reader is therefore strongly urged to use the arrangement shown in Figure 7.6.3(a).

Image 7.6.3(a) and (b) show the light stripe falling on the top surface of a round-top bread roll. Of course, we would normally choose to operate any optical rig in the dark, in order to avoid interference from highly-variable ambient light. However, it is possible to make the light-stripe sectioning technique very robust, since the laser generates high-intensity monochromatic light, whereas ordinary (i.e. pan-chromatic) room lighting typically has very little energy within the narrow pass band of a notch filter placed in front of the camera lens. The image of the light-stripe as detected by the camera can easily be processed and the list of surface-height values created. The program to do this is straightforward and is based on the "crack detector" [*crk*] and skeletonisation [*ske*] operators:

```
% Find list of surface height values for one X-position of sample.
one_row_height_data(L) :-
    grb,              % Digitise an image
    neg,              % Negate image, so stripe is dark
    crk,              % Crack detector finds thin dark features
    thr(32),          % Threshold - adjust parameter to taste
    3*exw, 3*skw,     % Noise removal
    ske,              % Skeletonisation
    rin,              % Integrate intensities along row - finds
                      % where light stripe is.
    vgt(L).           % Instantiate L to list of height values for
                      % this X-position
```

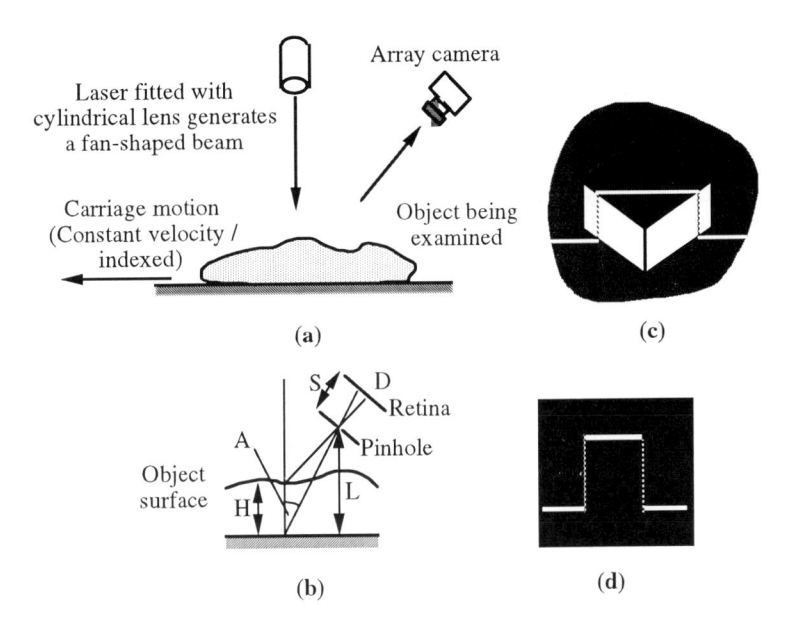

Figure 7.6.3 Generating depth maps using structured lighting. (a) Optical-scanning arrangement. (b) Ray geometry. The surface height, H, is given by the formula: $H = D.L.(1 + \tan^2(A))/(S.\tan(A) + D)$. (c) Projecting a light stripe onto a block on a plane surface. (d) What the camera "sees". The height of the "pulse" is D. Using the formula just given, the height of the block (H) can be determined.

Another particularly important point to note is that the technique for generating depth maps outlined in Figure 7.6.3 is inherently slow; a depth map with N rows can be created in N video frame-scan periods. If N = 512 and the frame-scan period is 0.040 seconds (PAL / CCIR standard), the depth map can be formed in 10.24 seconds. This may well be unacceptably long for many industrial applications. For example, it would not be possible to perform 100% inspection of loaves in a commercial bakery with this method. To overcome this difficulty, various other techniques are have been devised, including simultaneously projecting a series of (monochrome) light stripes (Image 7.6.4) and multi-coloured bands, from which a series of cross-section profiles can be created from a single image (Image 7.6.5). However, we shall ignore the problems caused by the slower arrangement shown in Figure 7.6.3, since we merely want to illustrate the basic principles of depth map generation and analysis.

Image 7.6.6 shows depth maps derived from a croissant, Cornish pastie (a "parcel" of meat and vegetables, wrapped in a pastry case) and a non-lidded tin loaf with a "split" at the top. It is possible to calculate intensity contours (isophotes) using the following program (also see page 104):

```
contours :-
        psk,            % Push image onto the stack for display later
        raf,            % Filter to make contours a bit smoother
        sca(3),         % Reduce number of grey levels to 8. Adjust
                        % parameter to taste
        sed,            % Edge detector
        thr(1),         % Threshold - result is binary image
        pop,            % Recover original image
        swi.            % Switch images
```

Since "intensity" in a depth map indicates surface height, the isophotes are also height contours, just like the contours of elevation drawn on a map. It is also possible to draw both horizontal and vertical height profiles using *plotit*.

```
plotit(A) :-
        psk,            % Push image onto the stack for display later
        lrt,            % Flip image about horizontal axis
        psh(A),         % Shift to right by amount A
        csh,            % Copy RHS to all other columns of image
        wgx,            % Intensity wedge
        sub,            % Subtract
        thr,            % Threshold at mid grey;
        bed,            % Contour is desired intensity profile
        pop,            % Recover original image
        swi.            % Switch images
```

The observant reader will have notice some black spots in Image 7.6.6(d). These are due to occlusion, which occurs when there are steep-sided pits or "cliffs" in the surface being measured. It is possible to "fill" these, if they are very small, using a fairly simple filter:

```
% Fill occlusions in depth maps if they are less than 2.A pixels
% wide / diameter

fill_occulsions(A) :-
        psk,            % Push image onto the stack for display later
        thr(0,0),       % Find occlusions
        wri(temp),      % Save binary image of occlusions
        swi,            % Revert to grey-scale image
        A*lnb,          % Expand bright regions to "fill" occlusions
        rea(temp),      % Recover binary image, showing occlusions
        min,            % Occlusions appear as shaded islands
        pop,            % Restore original image
        max.            % Fill holes with the "shaded islands"
```

From the depth map generated in this way, it is a straightforward matter to derive a range of measurements, of which the following are perhaps the most obvious ones:

(i) Maximum height, H_0.
(ii) Area, A(H), above a given height, H. The parameter H is allowed to vary in a step-wise manner over a range defined by taking the expected variations in height into account. For example, we might measure $A(H_1)$, $A(H_1-h)$, $A(H_1-2.h)$, $A(H_1-3.h)$, ..., where H_1 is the maximum allowed value for H_0 and h is a small increment, typically 1 - 5 mm. Each of the A(.) can then be compared individually to previously measured tolerance limits. Alternatively, we might choose to consider the vector { $A(H_1)$, $A(H_1-h)$, $A(H_1-2.h)$, $A(H_1-3.h)$, ... } as a complete entity and apply it as the input to some suitable learning process. (Image 7.6.6(b) and Section 7.2.4)
(iii) For some suitable value of the height parameter H, we "slice" the depth map. (This is a similar process as that defined for step (ii).) This results in a binary image, where white indicates points on the loaf surface that are higher than H, while black areas indicate points that are lower. For an ideal non-lidded tin loaf, which has a single well-rounded dome-like top "slicing" the depth map at say 90% of the maximum height, yields one nearly elliptical blob-like figure. We might then compute the position of the centroid of this figure, its aspect ratio and the orientation of its longest axis. These parameters can then be compared to "ideal" values, either by conventional statistical analysis, or by using some multi-parameter learning procedure.
(iv) The same "slicing" process might be applied to a split-top tin loaf. This type of loaf is expected to possess a top with a valley, lying between two elongated ridges. Ideally, the result of "slicing" should be a binary image, containing two elongated ellipse-like figures. Various tests for symmetry might be applied to these contours. Individually, they should be approximately symmetrical, about both their long and short axes. Their long axes should be parallel and the distance between them should lie within a defined range. Moreover, "ellipses" created by "slicing" at different height values, should be concentric.
(v) The "slicing" process applied to a loaf with a more complex shape might well result in a number of distinct blob-like features. For example, a certain type of bread roll is made by tying a simple knot in a rope of dough, while another kind of roll consists of a short plait. Again, "slicing" the depth map will result in the generation of a number of blobs, whose areas, shape and positions can all be analysed individually and collectively.

Of course, these rules can all be represented readily in Prolog+. In fact, Prolog+ is an ideal language for the task of analysing loaf shape, since the rules for recognising an acceptable loaf are likely to be expressed in terms of abstract, ill-defined quantities, such as "ellipse", "concentric", "symmetrical". While these words are associated with precisely defined mathematical entities, they are used here in the informal sense that a non-mathematician would use them.

In conclusion, we show that depth maps can be precise enough to be useful for engineering components. Image 7.6.7 shows a zinc die-cast component, which contains several step-like edges. These are visible as sharp intensity gradients in the depth map. Notice however, that occlusions occurs, appearing as black shadows.

IMAGES

Image 7.5.1 Using filtering and thesholding to analyse a simple clock, which produces a high contrast between the hands and a plain face. (a) Original image. (b) The image processing sequence [*wri, 2•(3•lnb, neg)), rea, sub, thr(140)*] was applied to (a). (c) [*blb,xor*] was applied to (b). (d) [*blb, xor*] was applied to (c). (e) Only blobs with an area in excess of 200 pixels have been retained. (f) [*2•skw,2•exw*] was applied to (e).

●

Image 7.5.2 The Hough transform applied to simulated and real clock faces. (a) Original image of a simple clock (simulated). (b) [*enc, thr*] applied to (a). (c) [*huf, neg, sqr*] applied to (a). The subsequence [*neg, sqr*] was used simply to improve visibility of some of the minor detail. (d) The two major peaks in (c) were detected automatically and the corresponding lines reconstructed. This shows that these peaks correspond to the minute and hour hands. (e) The medial axis transformation [*ske, mdl*] applied to (b). (f) [*huf, neg, sqr*] applied to (e). Notice that the peaks are better defined than in (c). (g) Same processing as in (d) but this time applied to (f). (h) The Hough transform method was applied to locate the minute hand on the real clock. Although the minute hand has been correctly located, the fluorescent tip of the hour hand makes this method unreliable.

●

Image 7.5.3 The Polar-to-Cartesian axis transformation [*ptc*] applied to a simulated and real clock faces. (a) Original (grey-scale) image of a simple clock face. (b) The operator *ptc* applied to (a). (c) The command sequence [*yxt, rin, csh, wgx, sub, thr, bed, yxt*] applied to (b). (d) Original image from a real clock. (e) *ptc* applied to (d). (f) As in (c), processing applied to (d). (g) Another clock design. (h) *ptc* applied to (g). The minute and hour hands are responsible for the two major peaks. (i) As in (c), processing applied to (h).

●

Image 7.5.4 (a) Original clock face. (b) Circular scan which intersects the hands but avoids the printing on the face. (c) Intensity plotted against angular position around the circular scan in (b). (d) The hour "tick" marks can be identified easily, using a circular scan, followed by simple filtering and thresholding. The image shown here was obtained using the Cartesian-to-polar co-ordinate axis transformation. (e) Difference image obtained by subtracting two images of the clock, taken at [9:10] and at [9:37]. (f) Multi-level thresholding produces a "cleaner" picture which is easy to analyse. (g) Image of the face of the clock, without the hands (obtained using the *max* operator, applied to two images, taken at [9:10] and [9:37]). (h) Difference image, obtained by subtracting the image in (g) from the (unprocessed) image of the clock taken, at [9:41]. (i) Simple thresholding applied to (h).

●

Image 7.6.1 Inspecting bread slices from a *lidded* tin loaf. (a) Silhouette of the slice. (b) Axis of minimum second moment [*dpa*]. (c) Slice after reorientation, so that the principal axis is vertical. (d) Difference between the minimum enclosing rectangle and (c). (e) A similar result to (d) can be obtained by computing the convex deficiency of (a). [*cvd*]

Image 7.6.2 Analysing the silhouette of a slice of bread from a *non-lidded* tin loaf. (a) Silhouette. (b) Outer edge. *[bed]* (c) Centroid and principal axis (axis of minimum second moment) *[dpa]*. (d) Hough transform *[bed,huf]*. (e) Hough transform image enhanced for easier viewing *[bed,huf,sqr,neg]*. (f) Line corresponding to brightest point in (d). Inverse Hough transform applied. (g) Three lines obtain by applying the inverse Hough transform to the three principal peaks in (d). (h) Overspill found from lines drawn in (g). (i) Points of high curvature (corners) are highlighted. (j) Corners have been used to "cut" the edge contour. (k) After removing very small blobs in (j). The three straight segments and curved top of the slice have been isolated.

●

Image 7.6.3 Projecting a single light stripe onto a bread roll, using a diode laser.(a) Image obtained in ambient light. (b) Improved image obtained in darkened room. (c) Light stripe has been reduced to one-pixel wide arc. (Note discontinuities)

●

Image 7.6.4 Projecting many (white) light stripes onto a bread roll.

●

Image 7.6.5 Projecting coloured light stripes onto a bread roll. (a) Original image. There were 4 light stripes in a repeating pattern. (b) Edge contours derived from (a).

●

Image 7.6.6 Depth maps. (a) Depth map of a croissant. (b) *sca(3)* applied to (a) reveals the height contours. (c) Intensity plot. (d) Cornish pastie. (e) Intensity plot (horizontal section). (f) Intensity plot (vertical section). (g) Depth map of a loaf. (h) Height contours of (g). (i) Intensity plot of (g).

●

Image 7.6.7 Structured lighting applied to an engineering component (zinc die-casting). (a) Depth map. (b) Height profile, across horizontal line in (a). (c) Height contours.

Image 7.5.1 a–f

Image 7.5.2 a–f

Image 7.5.2 g–h

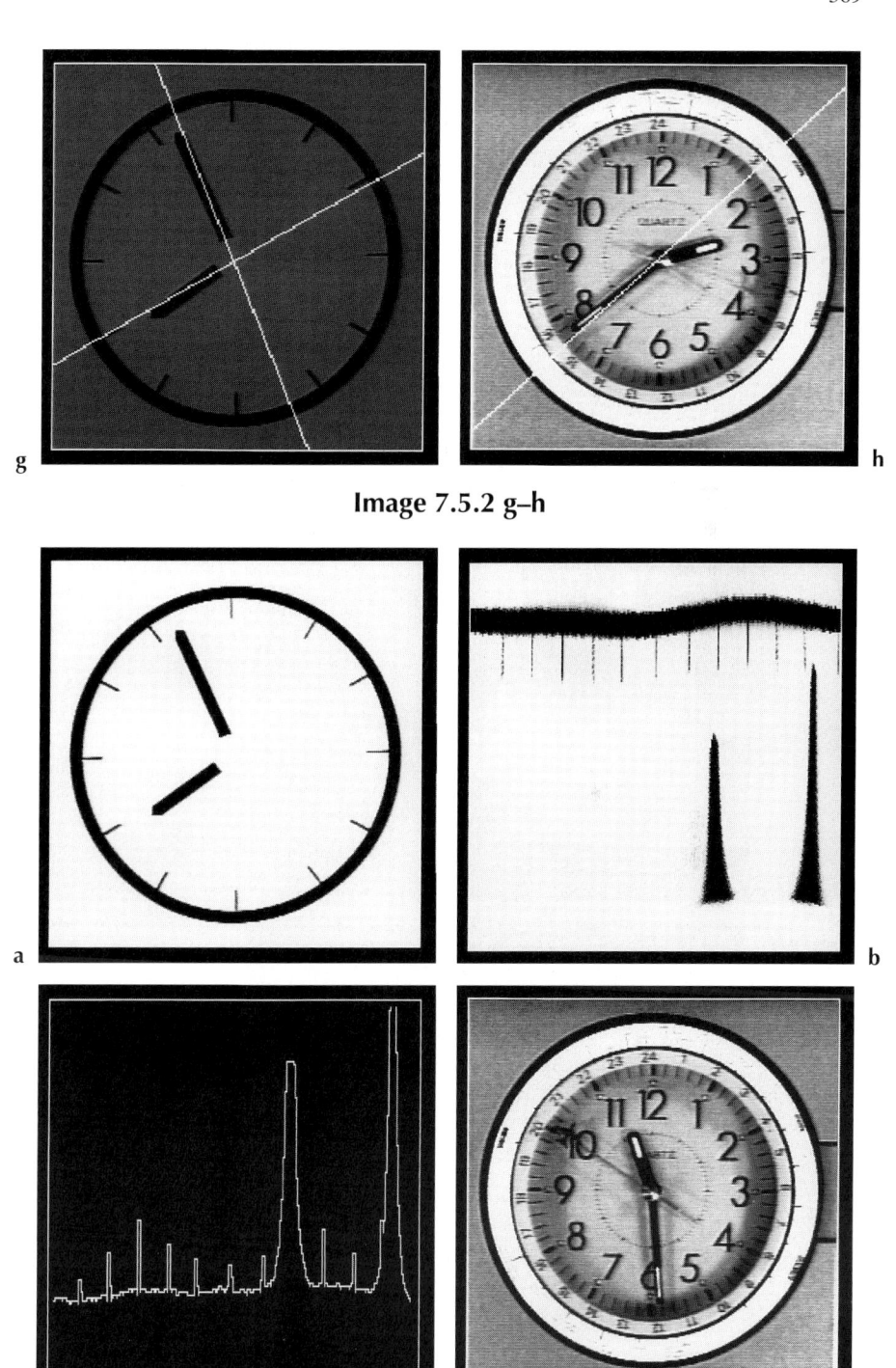

Image 7.5.3 a–d (continued on next page)

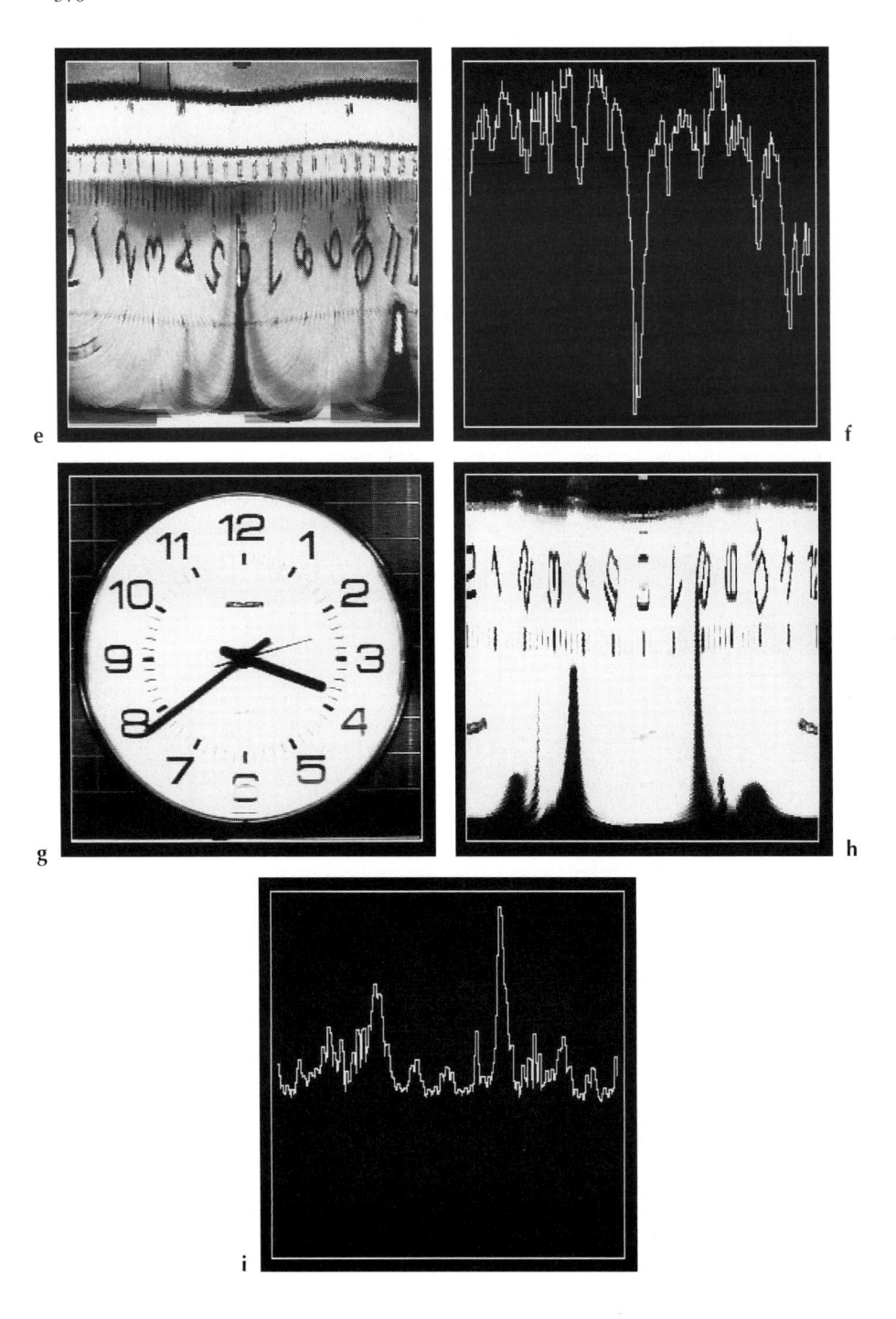

Image 7.5.3 e– i (continued)

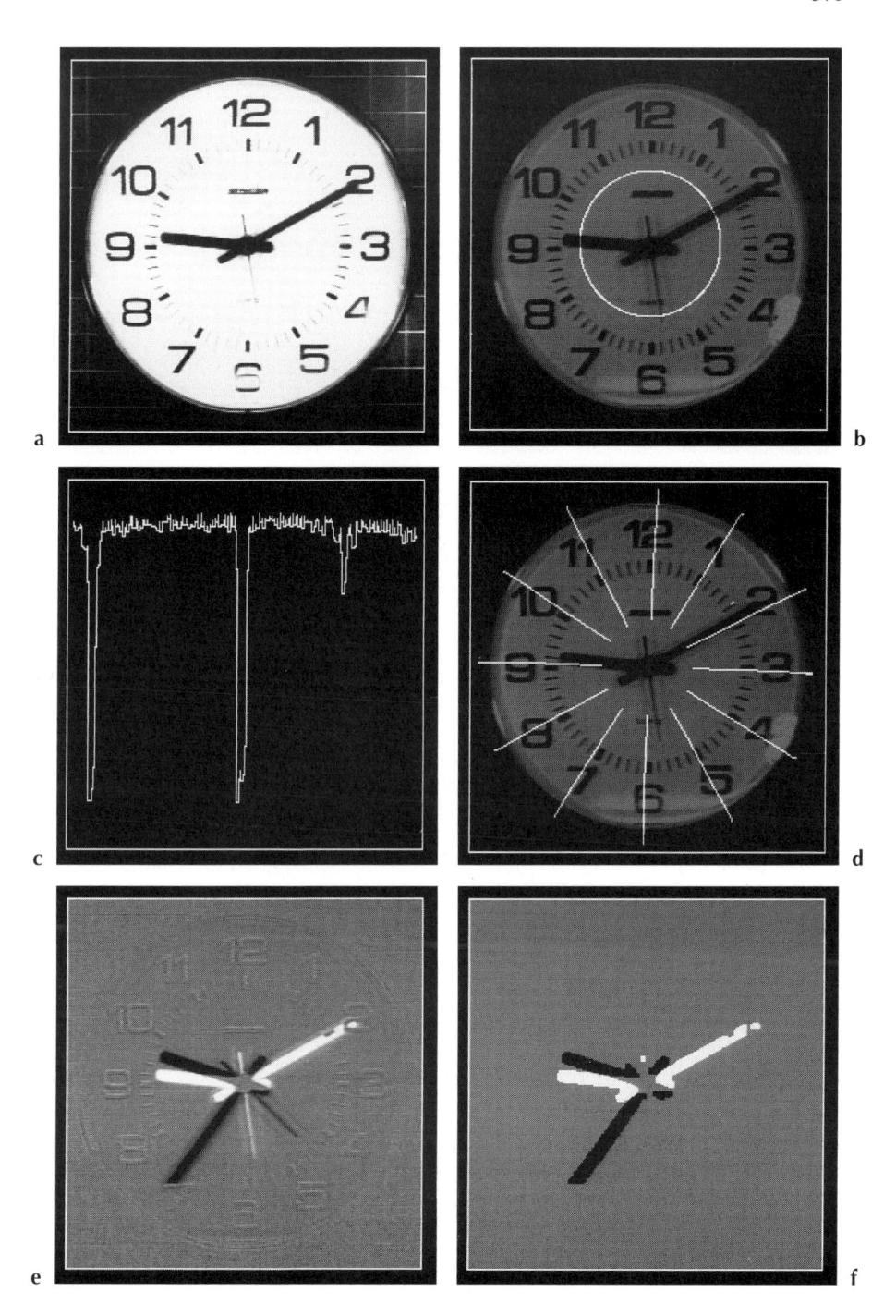

Image 7.5.4 a–f (continued on next page)

Image 7.5.4 g–i (continued)

Image 7.6.1 a–e

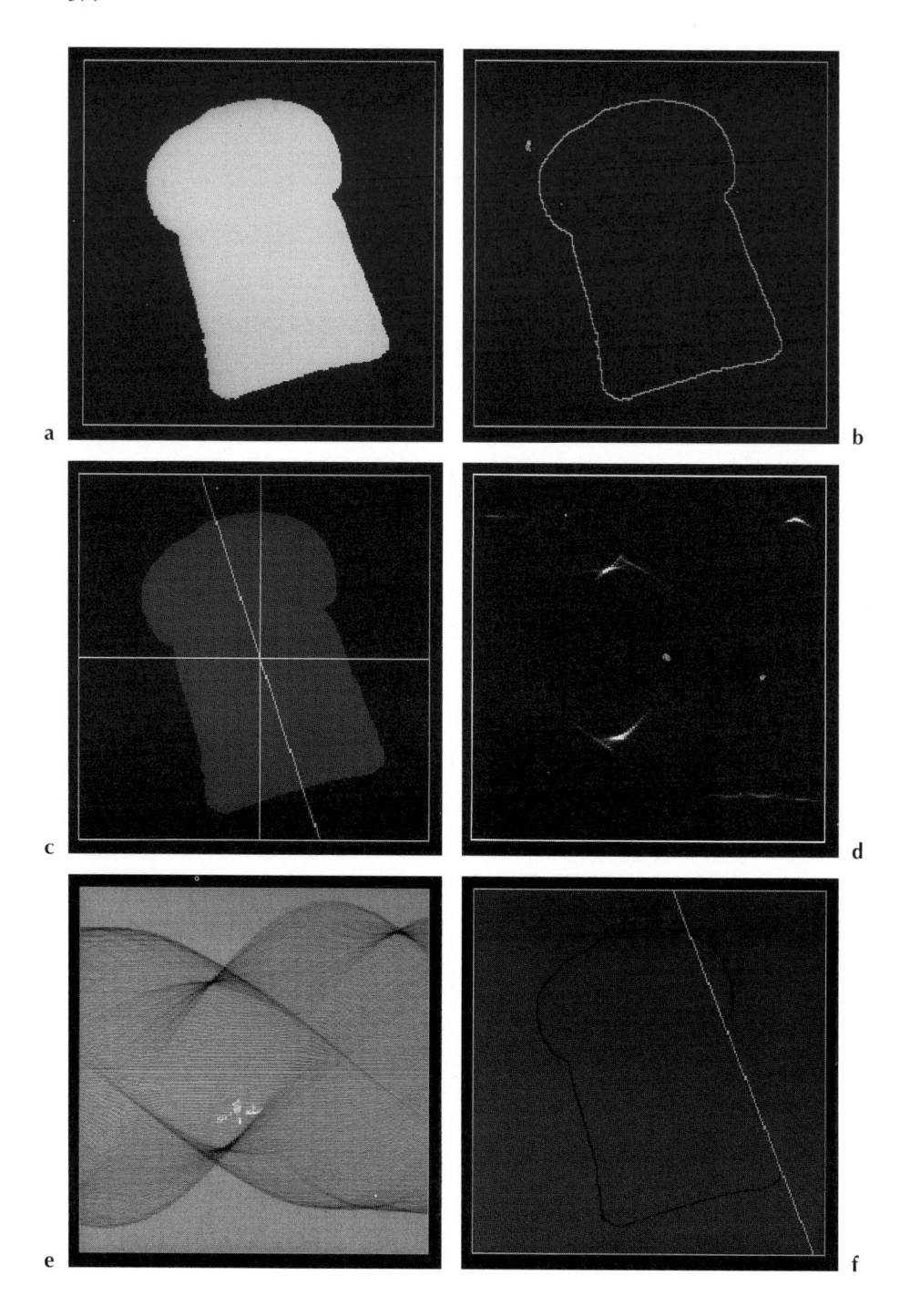

Image 7.6.2 a–f

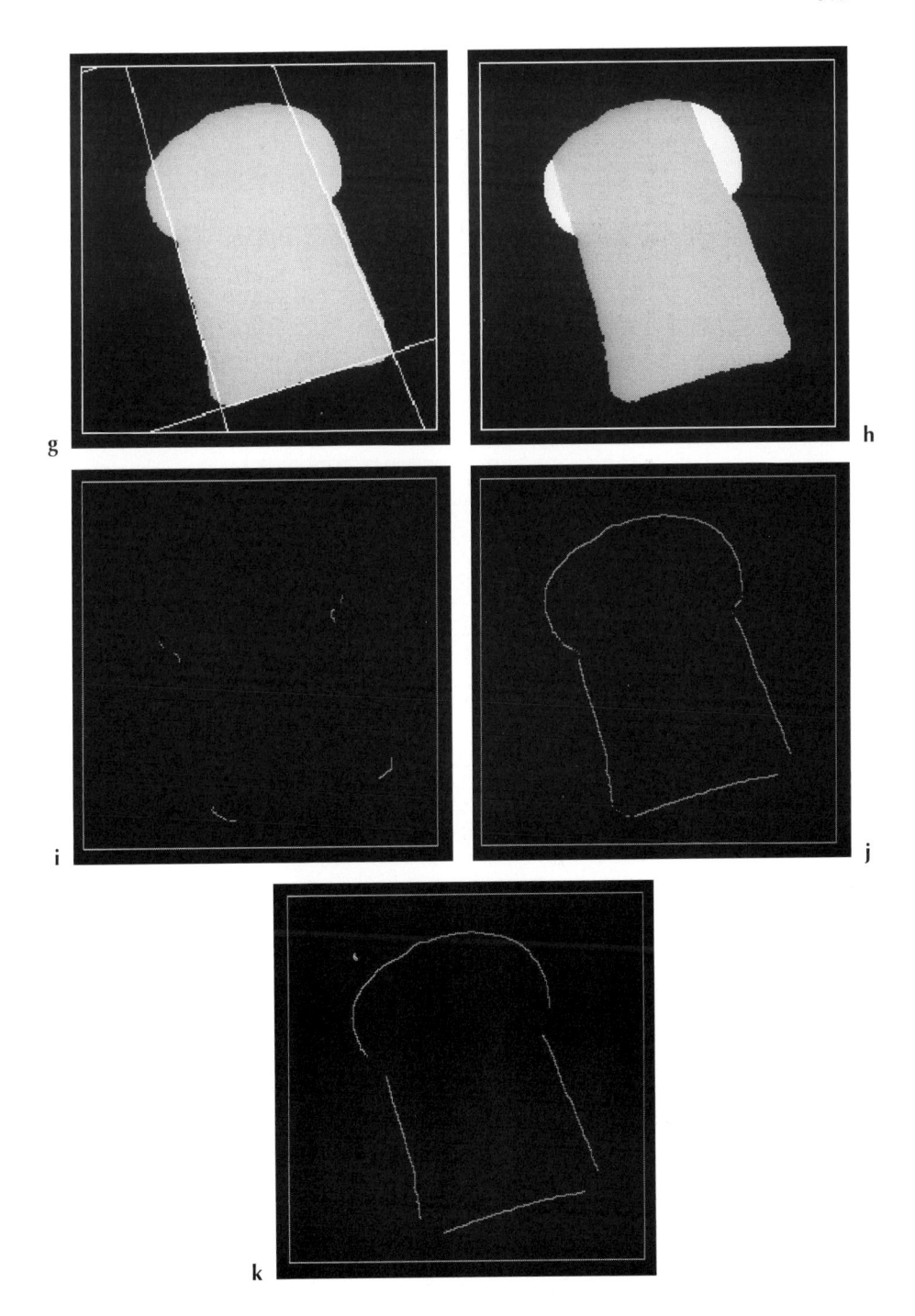

Image 7.6.2 g–k

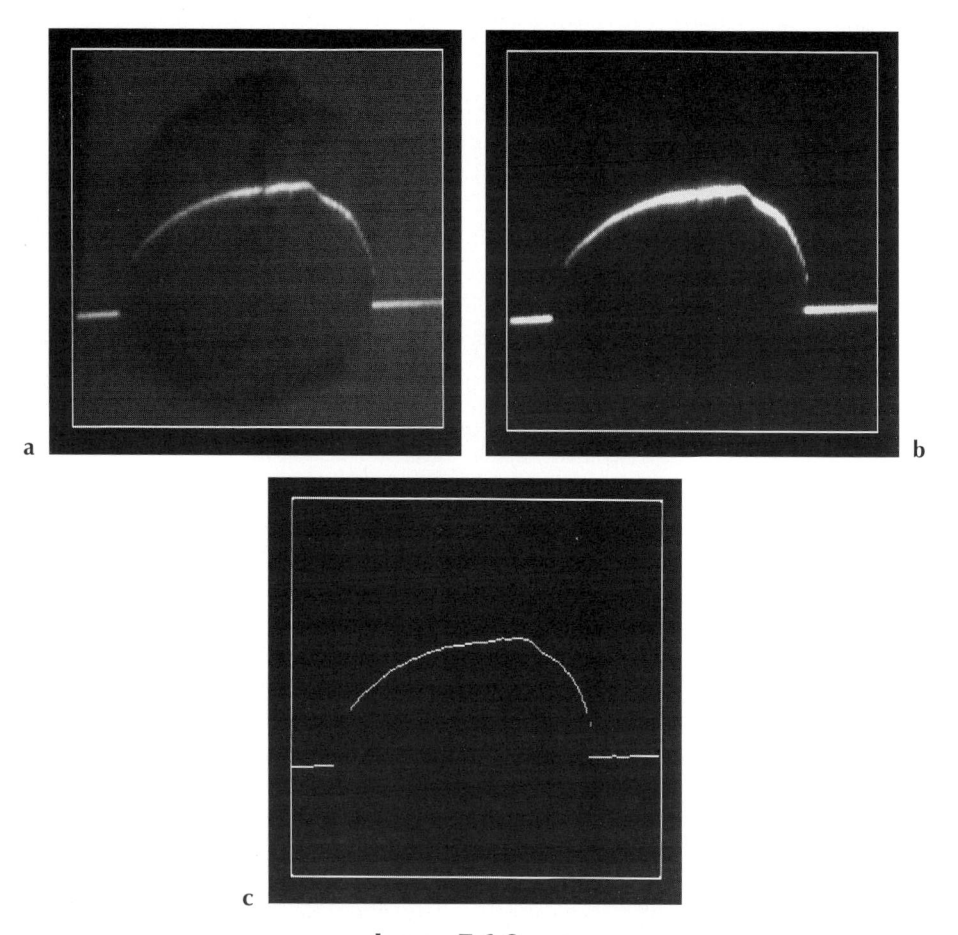

Image 7.6.3 a–c

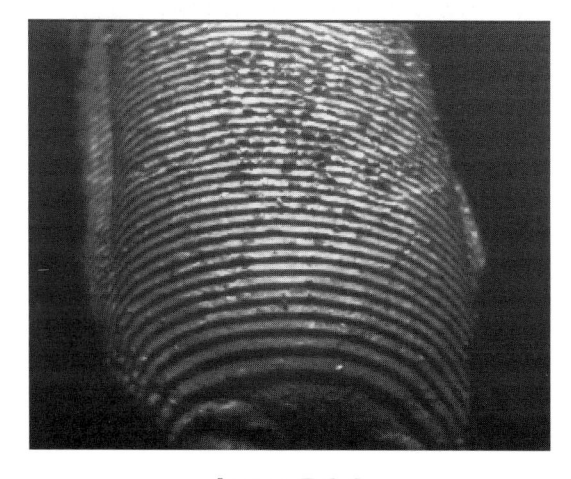

Image 7.6.4

377

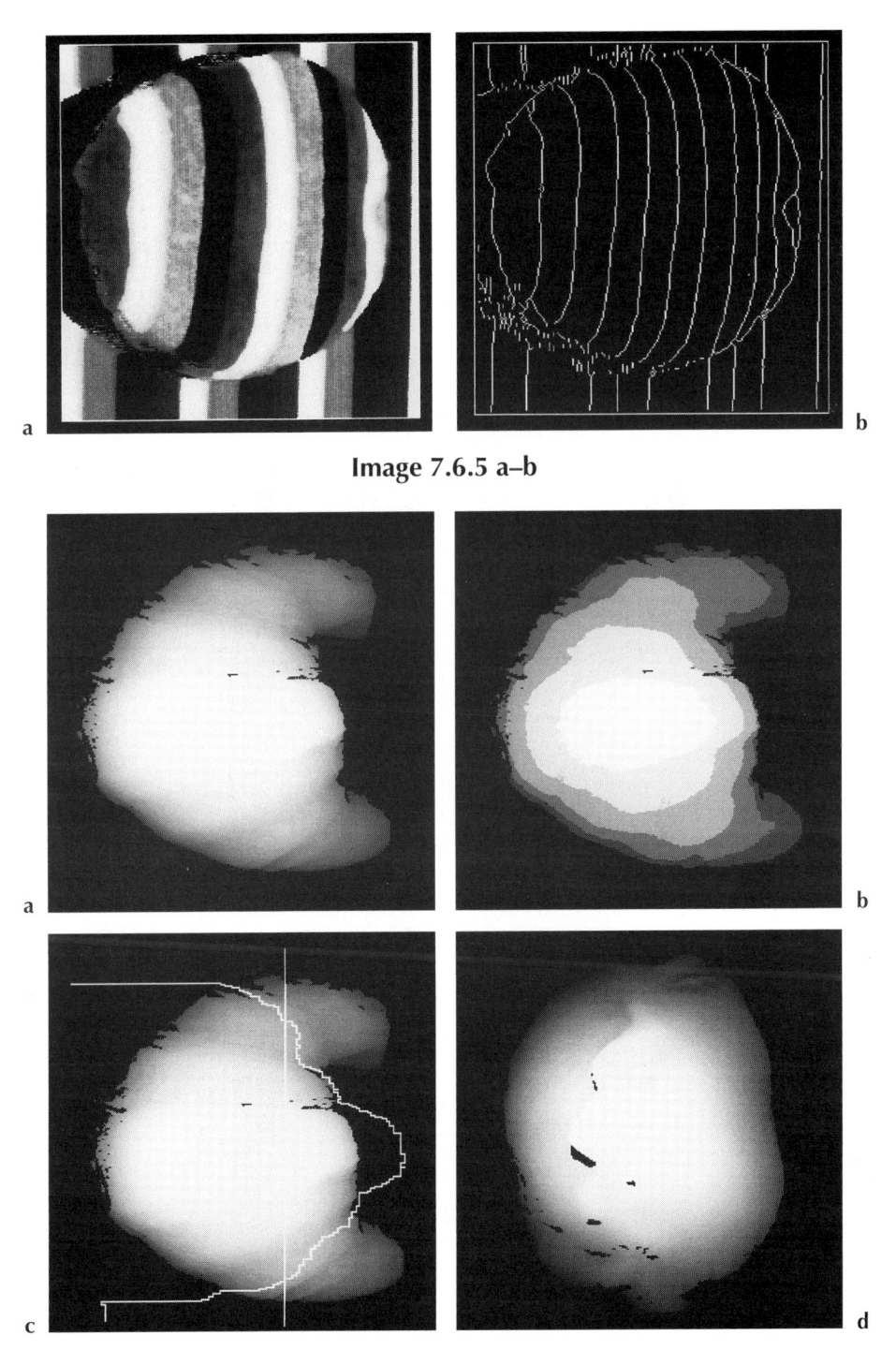

Image 7.6.5 a–b

Image 7.6.6 a–d (continued on next page)

378

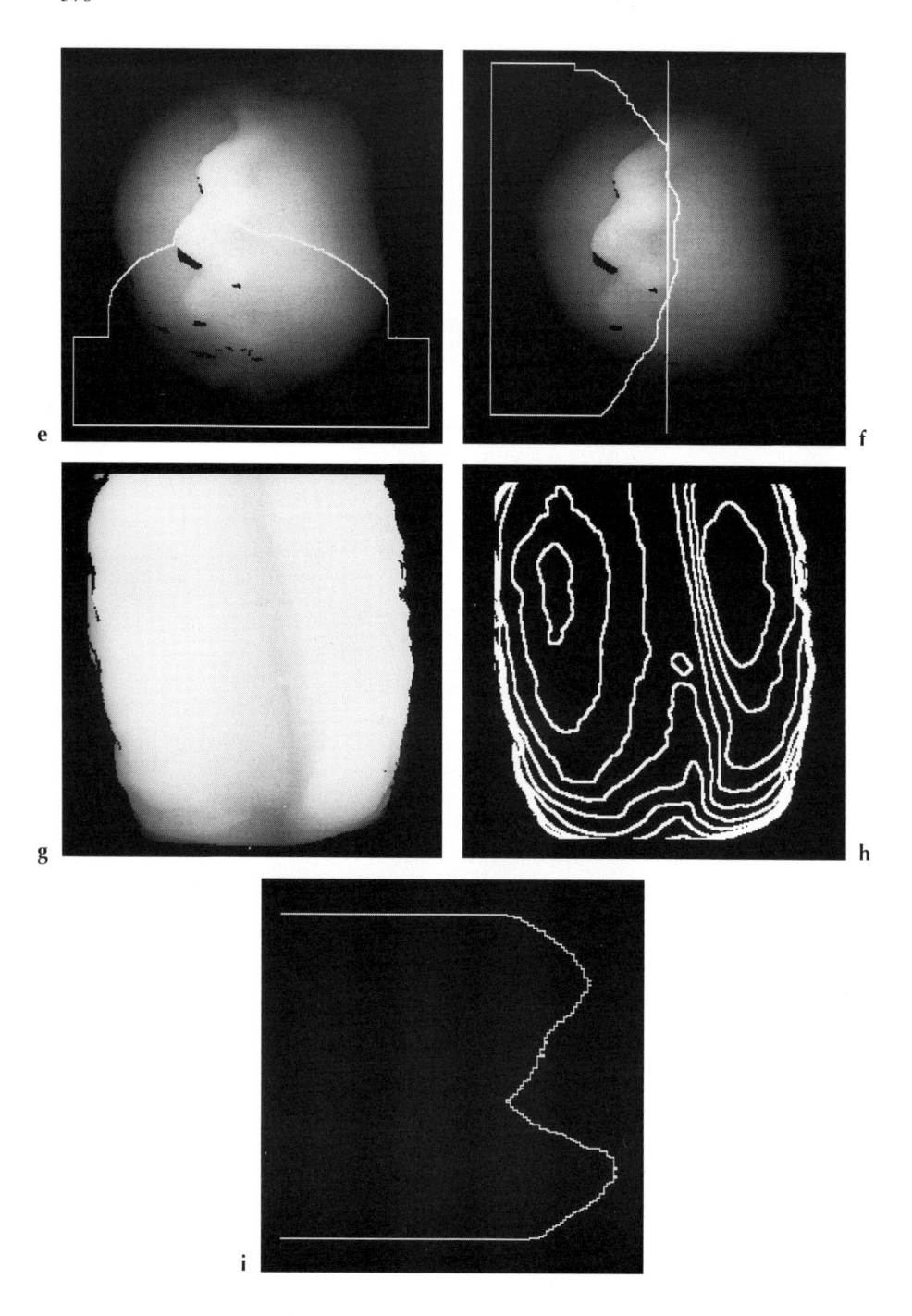

Image 7.6.6 e–i (continued)

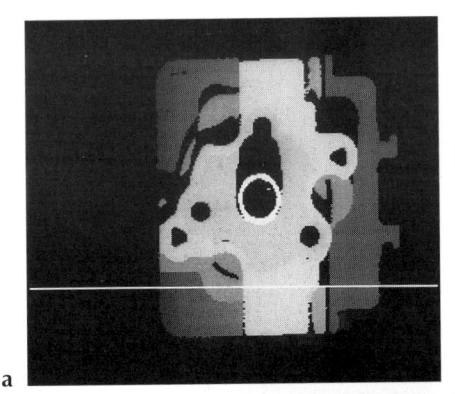

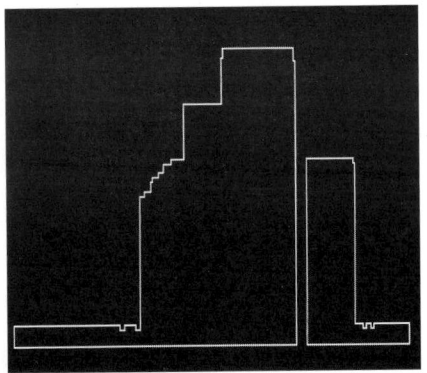

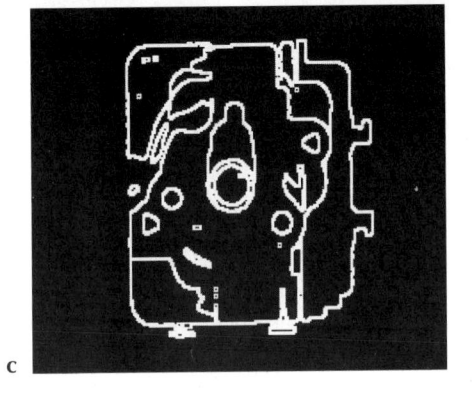

Image 7.6.7 a–c

8

Concluding Remarks

Industrial applications of machine vision are extremely varied in their nature and requirements. Despite this, it seems to be a universal truth that however fast, cheap, or smart we make them, somebody wants a machine vision system that is faster, cheaper and smarter than anything that has been made so far. Whichever computer and language we chose to use, somebody will want us to use a different one. If we have solved a problem for one industry, there is a feeling among some people that it has no relevance to another, even though the inherent nature of the two tasks are very similar. The converse is also true: if we can solve one problem for a certain industry, then, some people believe that we can also solve another, whatever its relationship in terms of application requirements. The authors hope that by now the reader will be aware that these are nothing but simple fallacies. Machine vision does not necessarily conform to our naive, uninformed expectations. Some of the ideas and methods that we have encountered in the earlier pages are counter-intuitive. In particular, introspective thought is not a viable means of designing vision systems. We cannot design a vision system by simply asking ourselves the question "How do *I* see this pattern?" A lot of people have tried this approach and *all of them* have failed. Introspection simply does not work! However, confident that the reader is that he can design a system whilst sitting at a desk, without suitable experimentation, *it is impossible*. Let us make it absolutely clear that frustration, disappointment and despair inevitably lie at the end of that particular road. How do we know that with such unshakeable certainty? The answer is simple: experience and observation. The collective experience of the authors is over thirty years and, at no time, have they ever seen the "introspective approach" to machine vision system design even come close to working effectively. Any speaker, or author of a book or paper who indicates otherwise, is simply mistaken or misrepresenting reality. To summarise, we need good design tools for machine vision. Prolog+ and its modern implementation, PIP, provide one such a facility.

To date, there have been many thousands of successful industrial applications of machine vision and there have been many unsuccessful one too! Like the ancient builders of bridges and cathedrals, many of the designers of machine vision systems were operating on *ad hoc* principles. As a result, their designs were based on weak foundations and often collapsed. After a number of failures, sometimes spectacular, the engineering community has, at last, gained a sense of realism about machine vision and designers are more often inclined now to the view that the "system" principles that we have emphasised here are of crucial importance. One of the key reasons for these failures has been (and often still is) the innate belief that most people have that they are experts on vision. In fact very few people really are experts at human vision or at *machine vision*. Accepting this is one of the great steps forward, that a person has to take, in order to become proficient at the design process. No electronics engineer would try to analyse the behaviour of a circuit without taking detailed measurements with a suitable meter. In the same way, no good vision engineer would attempt to analyse images without the use of an interactive image processor.

There is a great shortage of well trained vision systems engineers. A good honours degree in Engineering, Electronics, Physics or Computer Science is merely the minimal entry qualification for training in this subject. Both authors have taught machine vision as an option on Master's degree programmes, and we still feel that the graduates are inadequately prepared for the real needs of industry. We have also taught numerous short courses (2 - 3 days) and feel that even this route leaves engineers with scant knowledge of the real problems that they will face as designers of vision systems. Much more effective use could and, the authors believe, should be made of the available human resources by promoting active collaboration between academic and industrial staff.

Apart from the urgent need for longer and more concentrated specialist training for machine vision systems engineers, there is a requirement for improved design aids. In this book, we have discussed several of these. Interactive image processing is central to understanding how images should be processed. A facility like Prolog+ is important because it allows engineers to construct prototypes quickly and easily. Ready access to a range of tools such as the Lighting Advisor is important, to train / remind engineers of techniques which they might otherwise forget or ignore. Of course, it is impossible in a book such as this to encompass all of the relevant knowledge. All that we can hope to do is to provide the reader with a "snap-shot" of what the authors believe to be the important issues at this point in time. We have deliberately restricted our attention to *industrial* applications of machine vision. We have stated emphatically on several occasions that this book is not about computer vision. The two subjects are quite different, as we have taken great pains to point out in Appendix A. The subject of machine vision is evolving at a rapid rate and the tools that we have developed are themselves evolving. We envisage, for example, that before the end of the decade, it will be possible to choose whatever happens to be the most convenient computer language, for use at the top level in a prototyping system; it should be possible to "plug in" image processing facilities into otherwise perfectly standard languages, such as Lisp, C, SmallTalk, Basic and of course, Prolog. The Lighting Advisor will no doubt expand in the next few years and the authors hope to extend the range of advisory programs that are available. There is an on-going development process for these design tools and the authors are particularly anxious to receive suggestions, comments, feedback, etc. so that they can be made even more effective. Suggestions would also be welcome for further "Proverbs", which encapsulate the wisdom needed to design successful machine vision systems. (See Appendix A.)

What unsolved problems do we envisage will be important in the near future? The authors would suggest that the following will be among the most important ones:

- Design of an inspection system for the aesthetic appearance, of wood, marble, food products, etc.
- Learning by showing from a "Golden Sample". This will, of necessity, require the use of meta-knowledge. To suggest that a person or machine accepts that a sample should *"look like this example that I am showing you"*, presupposes that there is higher-level knowledge about what constitutes acceptable similarity.
- Declarative programming involving natural language will be refined to a much higher level.
- The user interface will continue to develop and improve, through the use of multi-media techniques. Prolog+, for example, should be provided with at least rudimentary graphics facilities, to aid feedback to the human operator.
- Multi-camera / multi-processor systems will become more common-place and networking will develop to allow really effective co-operative action between vision systems.
- Closed-loop process-control, through the use of Expert Systems with visual inputs, will become more common in manufacturing industry.
- Techniques will be devised for giving advice about which image processing method to use in a given application.

We referred earlier in this chapter to the large number of successful applications that have been studied. In Chapter 7, we have illustrated the use of the tools described earlier in this book. While such tasks as telling the time and recognising playing cards may seem to be remote from industrial applications, they are, in fact, models for "real world" industrial applications that we are not at liberty to discuss in detail. There are literally thousands of diverse applications that designers have had to face in the past. However, we do know that this number is minuscule compared to the huge quantity of potential applications. How many ways can you find to use your eyes? If we are ever to realise the full potential of this fascinating technology, we must have many more properly trained personnel and we must develop even better design tools. This book is an attempt to encourage both of these objectives. The rewards, in terms of improved methods for monitoring and controlling manufacturing processes, could be truly enormous! A famous entertainer used the catch-phrase *"You ain't seen nuthin yet"*. This could well be our watch-word, as we approach the end of the decade!

References

AHL-91	R.-J. Ahlers, "Case studies in machine vision integration", *Proc. SPIE Machine Vision Systems Integration*, vol. CR36, 1991, 56-62.
ALIS	ALIS™ 600 and Micro ALIS are products of Dolan Jenner Industries., Inc., Blueberry Hill Industrial Park, PO Box 1020, Woburn, MA 01801, USA.
AVA-85	*AVA Machine Vision Glossary*, Automated Vision Association, 1985.
BAT-74	B. G. Batchelor, *Practical Approach to Pattern Classification*, Plenum, London & New York, 1974.
BAT-79	B. G. Batchelor, Interactive Image analysis as a Prototyping Tool for Industrial Inspection, *Computers & Digital Techniques*, vol. no. 2, 1979, 61-69.
BAT-80	B. G. Batchelor, B. K. Marlow, B. D. V. Smith and M. J. Werson, "A research laboratory for automatic visual inspection", *Proc. 5th Int. Conf. Automated Inspection & Product Control*, Stuttgart, Germany, June, 1980, 13-32.
BAT-85	B. G. Batchelor, D. A. Hill & D. C. Hodgson (Ed's), *Automated Visual Inspection*, IFS Publications Ltd., Bedford, England, 1985.
BAT-85b	B. G. Batchelor & A. K. Steel, "A flexible inspection cell", *Proc. 5th Int. Conf. on Robot Vision & Sensory Control*, IFS Publications Ltd., Bedford, England, October 1985, 449-468.
BAT-89	B. G. Batchelor, "A Prolog lighting advisor", *Proc. SPIE Intelligent Robots & Computer Vision VIII: Systems and Applications*, Philadelphia, USA, vol. 1193, 1989, 295-302.
BAT-91	B.G. Batchelor, *Intelligent Image Processing in Prolog*, Springer-Verlag, 1991.
BAT-91b	B. G. Batchelor, "Tools for designing industrial vision systems", *Proc. SPIE Machine Vision Systems Integration*, vol. CR36, 1991, 138-175.
BAT-91c	B.G. Batchelor, "Interpreting the radon transform using Prolog", *Proc. SPIE Machine Vision Architectures, Integration, and Applications,* vol. 1615, 1991, 87-97.
BAT-92	B. G. Batchelor, "Design aids for visual inspection systems", *Sensor Review* 12(3), 1992, 3-4.
BAT-92b	B. G. Batchelor & F. N. Waltz, *Interactive Image Processing*, Springer-Verlag, London, 1992.

BAT-94	B.G. Batchelor and P.F. Whelan (Eds.), *Selected Papers on Industrial Machine Vision Systems*, SPIE Milestone Series MS 97, SPIE Optical Engineering Press, 1994.
BAT-94b	B.G. Batchelor, M.W. Daley, and E.C. Griffiths, "Hardware and software for prototyping industrial vision systems", *Proc. SPIE Machine Vision Applications, Architectures, and Systems Integration III*, vol. 2347, 1994, 189-197.
BAT-95	B.G. Batchelor and P.F. Whelan, "Ethical, environmental and social issues for machine vision in manufacturing industry", *Proc. SPIE Machine Vision Applications, Architectures and Systems Integration IV*, vol. 2597, 1995, 2-15.
BAT-95b	B. G. Batchelor and P. F. Whelan, "Real-time colour recognition in symbolic programming for machine vision systems", *Machine Vision and Applications 8*(6), 1995, 385-398.
BAX-94	G.A. Baxes, *Digital Image Processing: Principles and Applications*, John Wiley, 1994.
BIE-91	L. H. Bieman & J. A. Peyton, "Building an infra-structure for system integration", *Proc. SPIE Machine Vision Systems Integration*, vol. CR36, 1991, 3-19.
BOR-86	G. Borgefors, "Distance transformations in digital images", *Computer Vision, Graphics, and Image Processing* 34, 1986, 344-371.
BRA-90	I. Bratko, *Prolog: Programming for Artificial Intelligence*, 2nd Edition, Addison Wesley, 1990.
CHA-80	G. J. Chamberlain, *Colour: Its Measurement, Computation and Application*, Heyden and Son, Ltd., 1980.
CHA-93	T. Chang and C.-C. Jay Kuo, "Texture analysis and classification with tree-structured wavelet transforms", *IEEE Trans. on Image Processing* 2(4), 1993, 429-441.
CHA-95	John Chan, *Application of Machine Vision in The Food Industry*, Ph.D. Thesis, University of Wales College of Cardiff, UK, 1995.
CHI-74	Y.P. Chien and K.S. Fu, "Recognition of X-ray picture patterns", *IEEE Trans. Syst, Man and Cybern.* SMC-4, 1974, 145-156.
CHI-88	R.T. Chin, "Automated visual inspection: 1981 to 1987", *Computer Vision, Graphics and Image Processing* 41, 1988, 346-381.
CHI-82	R.T. Chin and C.A. Harlow, "Automated visual inspection: A survey", *IEEE Transactions on Pattern Analysis and Machine Intelligence* 4(6), 1982, 557-573.
CIE-31	*Commission Internationale de l'Eclairage*, the International Committee on Colour Standards.
CLO-87	W.F. Clocksin & C.S. Mellish, *Programming in Prolog*, 3rd Edition, Springer-Verlag, 1987.
COE-88	H. Coelho & J.C. Cotta, *Prolog by Example*, Springer-Verlag, Berlin, 1988.

DAU-92	R. Daum & K. Harding, "The machine vision lighting testbed", *Proc. Conf. Applied Machine Vision Conference '92*, Atlanta, SME Machine Vision Association, 1992, 92-183.
DEL-92	A. Delchambre, *Computer-Aided Assembly Planning*, Chapman & Hall, 1992.
DOU-92	E.R. Dougherty, *An Introduction to Morphological Image Processing*, Tutorial Text TT9, SPIE Press, 1992.
DOU-95	E.R. Dougherty and P.A. Laplante, *Introduction to Real-time Imaging*, Tutorial Text TT19, SPIE/IEEE Press, 1995.
DOW-85	W.B. Dowsland, "Two and three dimensional packing problems and solution methods", *New Zealand Operational Research* 13(1), 1985, 1-18.
DRE-86	H.L. Dreyfus and S.E. Dreyfus, *Mind over Machine*, The Free Press, 1986.
DUF-73	M.J.B. Duff, D.M. Waston, T.M. Fountain and G.K. Shaw, "A cellular logic array for image processing", *Pattern Recognition*, 1973.
EUR-89	EUREKA, *Robotics and Production Automation*, European Community, 1989.
FOS-84	J. Foster, P.M. Page and J. Hewit, "Development of an expert vision system for automatic industrial inspection", *Proc. 4'th Intl. Conf. on Robot Vision and Sensory Control*, London, 1984, 303-311.
FRE-88	H. Freeman (Ed.), *Machine vision: Algorithms, Architectures, and Systems*, Academic Press Inc., 1988.
FU-87	K.S. Fu, R.C. Gonzalez and C.S. Lee, *Robotics: Control, Sensing, Vision and Intelligence*, McGraw-Hill, 1987.
GAR-79	M.R. Garey and D.S. Johnson, *Computers and Intractability - A Guide to the Theory of NP-Completeness*, W.H. Freeman and Co., 1979.
GAZ-89	G. Gazdar and C. Mellish, *Natural Language Processing in Prolog*, Addison-Wesley, Wokingham, UK, 1989
GON-87	R.C. Gonzalez and P. Wintz, *Digital Image Processing*, Addison Wesley, 1987.
GON-92	R. C. Gonzalez & R. E. Woods, *Digital Image Processing*, Addison Wesley, 1992.
HAR-79	R.M. Haralick, "Statistical and structural approaches to texture", *Proceedings of the IEEE* 67(5), 1979, 786-804.
HAR-87	S.J. Harrington and P.J. Sackett, "Study of robotic assembly systems", *Assembly Automation* 7(3), 1987, 122-126.
HAR-87b	R.M. Haralick, S.R. Sternberg and X. Zhuang, "Image analysis using mathematical morphology", *IEEE Trans. Pattern Anal. Machine Intell.* 9(4), 1987, 532-550.
HAR-92	R. Haralick, "Performance characterization in computer vision", *Proceedings of the British Machine Vision Conference*, Springer-Verlag, 1992, 1-8.

HAR-92b	R.M. Haralick and L.G. Shapiro, "Mathematical Morphology", Chapter 5 of *Computer and Robot Vision: Volume 1*, Addison Wesley, 1992.
HEI-91	H.J.A.M. Heijmans "Theoretical aspects of grey-level morphology", *IEEE Trans. Pattern Anal. Machine Intell.* 13(6), 1991, 568-582.
HOC-87	J. Hochberg, "Machines should not see as people do, but must know how people see", *Computer Vision, Graphics and Image Processing* 37, 1987, 221-237.
HOL-79	S.W. Holland, L. Rossol and M.R. Ward, "CONSIGHT-I: A vision-controlled robot system for transferring parts from belt conveyors", *Computer Vision and Sensor-based Robots*, Plenum Publishing, 1979, 81-100.
HOL-84	J. Hollingum, *Machine Vision, the Eyes of Automation*, IFS Publications, 1984.
HOL-92	Holmes, Newman & Associates, *Automated Meat Piece Grading System*, Oakhampton, Devon, England, UK, Patent Pending.
HOS-90	D.R. Hoska, "Fixtureless assembly/manufacture", *Proceedings of Robots and Vision Automation 1990*, 1990, (4-17)-(4-28).
HUT-71	T. C. Hutson, *Colour Television Theory*, McGraw-Hill, London, 1971.
IMA	*NIH-Image*, National Instiutes of Health, USA. Anonymous FTP: *zippy.nimh.nih.gov*
INT	*Intelligent Camera*, Image Inspection Ltd., Unit 7, First Quarter, Blenheim Road, Kingston, Surrey, KT19 9QN, UK.
IPL	*IP-Lab*, Signal Analytics Corporation, Vienna, VA, USA.
ITI-89	*The Lighting Science Database*, Sensor Center for Improved Quality, Industrial Technology Institute, PO Box 1485, Ann Arbor, MI 48106, USA.
JON-94	A. C. Jones & B. G. Batchelor, "Software for intelligent image processing", *Proc. SPIE Intelligent Robots & Computer Vision XIII: Algorithms, Techniques Active Vision Materials Handling*, Boston, MA., Nov. 1994.
KEL-86	J. M. Keller, "Color image analysis of food", *Proc. IEEE Computer Society on Computer Inspection and Pattern Recognition*, Florida, 1986.
KID	*Optical Design Software*, Kidger Optics Ltd/, 9a High Street, Crowborough, East Sussex, TN6 2QA, England, UK.
KLI-72	J.C. Klien and J.Serra, "The texture analyzer", *J. Microscopy* 95(2), 1972, 349-356.
KRU-81	R.P. Kruger and W.B. Thompson, "A technical and economic assessment of computer vision for industrial inspection and robotic assembly", *Proceedings of the IEEE* 69(12), 1981, 1524-1538.
LEE-89	M.H. Lee, *Intelligent Robotics*, Open University Press, 1989.
LEV-88	P. Levi, "Image processing robotics applied to citrus fruit harvesting", *Proc. ROVISEC7 Conf*, Feb. 1988, Zurich, Switzerland.

LPA	Logic Programming Associates Ltd., Studio 4, Royal Victoria Patriotic Building, Trinity Road, London, SW18 3SX, UK
MAC	*MacProlog*, Logic Programming Associates Ltd. MacProlog is also sold in USA by Quintus Inc., Mountain View, CA.
MCC-92	K.M. McClannon, P.F. Whelan and C. McCorkell, "Machine vision in process control", *Procs. of the Ninth Conference of the Irish Manufacturing Committee*, University College Dublin, 1992, 637-646.
MCC-93	K.M. McClannon, P.F. Whelan and C. McCorkell, "Integrating machine vision into process control", *Procs. of FAIM'93*, University of Limerick, CRC Press, 1993, 703-714.
MCG-94	T. McGowan, "A text to speech synthesiser for the MacProlog environment", *Internal Technical Report, School of Electronic Engineering*, Dublin City University, Ireland, 1994.
MIC-86	D. Michie, *On Machine Intelligence*, Ellis Horwood Ltd, 1986.
MOL-90	J Mollon, "The tricks of colour", *Images and Understanding*, H. Barlow, C. Blakemore, & M. Weston-Smith, Cambridge University Press, Cambridge, England, 1990.
MUN	Munsell Color Co. *Munsell Book of Color*. 2441 North Calvert St., Baltimore, MD.
MVA	*Machine Vision Lens Selector*, Machine Vision Association of the Society of Manufacturing Engineers, 1992, manuf. by American Slide-Chart Corp., Wheaton, IL 60187, USA.
NEM-95	Sasha Nemecek, "Employment blues: Nothing to do with being green", *Scientific American*, June 1995, 25.
OPT	*OptiLab & Concept Vi*, Graftek France, Le Moulin de l'Image, 26270 Mirmande, France.
OWE-85	T. Owen, *Assembly with Robots*, Prentice-Hall, 1985.
PAV-92	T. Pavlidis, "Why progress in machine vision is so slow", *Pattern Recognition Letters* 13, 1992, 221-225.
PEA-84	J. Pearl, *Heuristics: Intelligent Search Strategies for Computer Problem Solving*, Addison Wesley, 1984.
PEN-84	A.P. Pentland, "Fractal-based descriptors of natural scenes", *IEEE Trans. on Pattern Analysis and Machine Intelligence* 6(6), 1984.
PEN-88	*Lighting Advisor Expert System*, © Penn Video, Inc. (a subsidiary of Ball Corporation, Inc.,) Industrial Systems Division, 929 Sweitzer Avenue, Akron, Ohio 44311, USA.
PER-91	S. J. Perry, *Colour Machine Vision*, M.Sc. dissertation, University of Wales College of Cardiff, 1991.
PET-90	P. Peters, *Camera Advisor*, M.Sc. dissertation, University of Wales College of Cardiff, UK, 1990.
PHO	*Photoshop*, Adobe Systems, Inc., 1585 Charleston Road, PO Box 7900, Mountain View, CA 94039-7900, USA.
PIT-93	I. Pitas, *Digital Image Processing Algorithms*, Prentice-Hall, 1993.

PLU-91	A. P. Plummer, "Inspecting coloured objects using grey-scale vision systems", *Proc. SPIE Machine Vision Systems Integration*, vol. CR-36, 1991, 64-77.
QUC	*QuickCam*, Connectix, 2600 Campus Drive, San Mateo, CA 94403, USA.
RED-91	A. Redford, "Guest editorial", *Int. J. Prod. Res.* 29(2), 1991, 225-227.
RJA	*Regulated Lighting Unit for Fibre Optic Illuminators*, R-J Ahlers, A-Tec, Mittlerer Kirchaldenweg,10, D-70195 Stuttgart, Germany.
ROB-89	S.L. Robinson and R.K. Miller, *Automated Inspection and Quality Assurance*, Dekker, 1989.
RUM-89	P. Rummel, "Applied robot vision: Combining workpiece recognition and inspection", *Machine Vision for Inspection and Measurement* (Ed. H. Freeman), 1989, 203-221.
SER-82	J. Serra, *Image Analysis and Mathematical Morphology Vol: 1*, Academic Press, 1982.
SER-86	J. Serra, "Introduction to Mathematical morphology", *Comput. Vision Graph. Image Process.* 35, 1986, 283-305.
SIE-88	L.H. Siew, R.M. Hodgson and E.S. Wood, "Texture measures for carpet wear assessment", *IEEE Trans. on Pattern Analysis and Machine Intelligence* 10(1), 1988.
SIL-80	E.A. Silver, R.V.V. Vidal and D. de Werra, "A tutorial on heuristic methods", *European Journal of Operational Research* 5, 1980, 153-162.
SIM-81	H.A. Simon, *The Sciences of the Artificial*, 2nd Edition, MIT Press, 1981.
SNY-92	M. A. Snyder, "Tools for designing camera configuration", *Proc. SPIE Machine Vision Architectures, Integration and Applications*, vol. 1823, 1992, 18-28.
SON-93	M. Sonka, V. Hlavac and R. Boyle, *Image Processing, Analysis and Machine Vision*, Chapman & Hall, 1993.
STE-78	S.R. Sternberg, "Parallel architecture for image processing", *Proc. IEEE Int. Computer Software and Applications Conf.*, Chicago, 1978, 712-717.
STE-86	S.R. Sternberg, "Grey scale morphology", *Computer Vision, Graphics and Image Processing* 35, 1986, 333-355.
STE-91	I. Stewart, "How to succeed in stacking", *New Scientist*, July 1991, 29-32.
SWI-89	K. Swift, "Expert system aids design for assembly", *Assembly Automation* 9(3), 1989, 132-136.
TAY-88	W.A. Taylor, *What Every Engineer Should Know About Artificial Intelligence*, MIT, 1988.
VCS	*VCS Image Processing Software*, Vision Dynamics Ltd., Suite 7a, 1 St. Albans Road, Hemel Hempstead, HP2 4XX. UK
VIN-91	L. Vincent, "Morphological transformations of binary images with arbitrary structuring elements", *Signal Processing* 22, 1991, 3-23.

VIS	*Visilog*, Neosis, Immeuble Nungesser, 13 Avenue Morane Saulnier, 78140, Velizy, France.
VOG-89	R.C. Vogt, *Automatic Generation of Morphological Set Recognition Algorithms*, Springer-Verlag, 1989.
WAL-88	A.M. Wallace, "Industrial applications of computer vision since 1982", *IEE Proceedings* 135(3), 1988, 117-136.
WAL-88b	F.M. Waltz, "Fast implementation of standard and 'fuzzy' binary morphological operations with large, arbitrary structuring elements", *Proc. SPIE Intelligent Robots and Computer Vision VII*, vol. 1002, 1988, 434-441.
WAL-94	F.M. Waltz, "Application of SKIPSM to binary morphology", *Proc. SPIE Machine Vision Applications, Architectures, and Systems Integration III*, vol. 2347, 1994, 396-407.
WES-76	J. Weska, C. Dyer and A. Rosenfeld, "A comparative study of texture measures for terrain classification", *IEEE Trans. Syst., Man and Cybern.* 6(4), 1976, 269-285.
WHE-91.	P.F. Whelan and B.G. Batchelor, "Automated packing of arbitrary shapes", *Proc. SPIE Machine Vision Architectures, Integration, and Applications*, vol. 1615, 1991,. 77-86.
WHE-93	P.F. Whelan and B.G. Batchelor, "Flexible packing of arbitrary two-dimensional shapes", *Optical Engineering* 32(12), 1993, 3278-3287.
WHE-96	P.F. Whelan and B.G. Batchelor, "Automated packing systems - A systems engineering approach", *IEEE Trans. on Systems, Man and Cybernetics - Part A: Systems and Humans*, 26(05), 1996, 533-544.
WHI-94	K. White, *Opto*Sense®*, Visual*Sense*Systems, 314 Meadow Wood Terrece, Ithaca, NY 14850, USA, 1994.
WIL-92	G. Wilson, *A Prototype Knowledge-Based System for Specifying Industrial Imaging Systems*, M.Sc. dissertation, University of Wales College of Cardiff, UK, 1992.
WIL-93	R. H. R. Williams, *Lighting Advisor - A HyperCard Version*, M.Sc. dissertation, University of Wales College of Cardiff, UK, 1993.
WWW-1	*The Lighting Advisor* , *http://www.cm.cf.ac.uk/lad/text.intro.html*
ZHU-86	X. Zhuang and R.M. Haralick, "Morphological structuring element decomposition", *Computer Vision, Graphics, and Image Processing* 35, 1986, 370-382.

Appendix A

Proverbs, Opinions and Folklore

The following is a list of observations, comments, suggestions, etc. based upon our direct and our colleagues' experiences. It is offered in a light-hearted manner but encapsulates some important lessons that we have learned but which are unfortunately not universally acknowledged. We hope it is will bring enlightenment and promote discussion among our colleagues. By its very nature, this list is dynamic and additions to it are always welcome. The current version of this list can be found at the following web site:

http://www.eeng.dcu.ie/~whelanp/proverbs/proverbs.html

General

There is more to machine vision than meets the eye.
> *A machine vision system does not see things as the human eye does.*

An eye is not a camera. A brain is not a computer.
> *Machine vision systems should not necessarily be modelled on, or intended to emulate human vision.*

Machine vision is not a scientific discipline.
> *Machine vision is not an exercise in philosophy but an engineering project.*

No vision system should be required to answer the general question "What is this?"
> *It is better for vision systems to answer more specific questions, such as "Is this widget well made?" Verification (i.e. checking that the widget is well made) is better than recognition, where few or no a priori assumptions are made.*

Intelligence ≠ Computing power.
> *Making the computer more powerful does not necessarily make the system smarter.*

Optimal solutions do not always exist.
> *If they do exist, optimal solutions may be too complex, or impossible to find. We should therefore be prepared to search for and accept satisfactory solutions, rather than optimal ones.*

Use a standard solution to a vision problem but only if it is sensible to do so.
> *Wherever possible we should provide standard solutions to industrial problems, since this helps to broaden the application base.*

Avoid the application of machine vision techniques for their own sake.
> *It is vanity on the part of the vision engineer to do so. There are plenty of other methods of solution available. Most of them are cheaper than vision.*

Defect prevention is better than cure.
> *We should consider using vision in closed loop feedback control of the manufacturing process.*

Do not rely on second-hand information about the manufacturing process and environment.
> *The vision engineer should always see the manufacturing process for himself. If the customer is unwilling to let the vision engineer into the factory, it may be necessary to abandon the application.*

Vision systems need not be fully automatic.
> *While it is more usual to use a fully automatic vision system, it can be used instead to enhance images for subsequent human analysis.*

Systems

No system should be more complicated than it need be.
> *This is a reformulation of Occam's Razor, which in its original form is "Entia non multiplicanda sunt." In its English translation, excessive complication is attributed to mere vanity. In colloquial use, this is often referred to as the KISS principle. (Keep it simple, stupid.) Simple systems are almost always the best in practice.*

All parts of a properly designed machine vision system bear an equal strain.
> *Of course, it is impossible to measure strain in any formal sense. The point is that no part of a vision system should be made more complicated because a sloppy attitude has been adopted during the design of other parts. A particularly common error is the tendency to concentrate on the image processing, to the detriment of the image acquisition (i.e. pose of the object being inspected, lighting, optics and sensor).*

If it matters that we use the Sobel edge detector rather than the Roberts operator, then there is something fundamentally wrong, probably the lighting.
> *This remark is not about the relative merits of the various edge detection operators but is a statement about the need for a broader "systems" approach. A common error is to pay much more attention to*

> the image processing process but ignore the fact that the image contrast is low because the lighting sub-system is poorly designed.

The following inequality is always true: Vision-system ≠ PC + Framegrabber + Camera + Software.

> To many people, these are the only components needed to build a vision system. However, this neglects many important issues: lighting, optics, systems integration, mechanical handling, ergonomics and standard industrial inspection practice.

Problem constraints allow the vision engineer to simplify the design.

> By taking systems issues into account, it may well be possible to design a simpler, faster, cheaper and more robust system.

Vision systems can use the same aids as people to reduce task complexity.

> For example, special optical/lighting techniques, X-rays, fluoroscopy, multi-spectral imaging, specialised sample preparation can all be used.

Documentation is an essential part of the system.

> A vision system will not survive for long without sufficient documentation.

Customer

Whatever software and hardware that a machine vision system uses, the customer will want it to be different, so don't tell them.

> Many customer companies have a policy of using certain types of computer hardware / software, which will often conflict with the vision system. It is wise to regard the vision system as a closed box.

The customer must not be allowed to tinker with the system after it is installed.

> The customer should be dissuaded from making internal adjustments to the system, since this requires rare and specialised skills (lighting, optics, camera, algorithms, systems integration).

The customer's company just does not make defective widgets; the vision system is simple intended "to improve product quality".

> Companies are often sensitive about the way that quality (or lack of it) in their products is discussed. This must be borne in mind when designing a vision system and particularly when reporting developments at conferences, in publications, etc.

Everybody (including the customer) thinks that they are an expert on vision and will tell the vision engineer how to design the machine.

> This is, regrettably, one of the great truths. As a result, everybody will feel it is their right and duty to tell the vision engineer how to do his job. In many instances, prototyping tools need to be used for the specific purpose of convincing the customer that his intuitive approach just does not work reliably.

The widgets that were provided for the feasibility study were specially cleaned and chosen by the customer for the project.
> *Beware of the pernicious habit of some customers who deliberately, or through ignorance, select good quality products to show the vision company, rather than providing a more representative sample.*

Customer education is an integral part of vision system design.
> *A well educated customer can help to reduce the project cost and may well help to reach a better system design.*

A little knowledge is a dangerous thing.
> *The customer will suggest many changes to the system design if he is ignorant of the subtleties which led to the present design. It is best to tell the customer all or nothing. For example, the vision engineer should not tell the customer that the system uses a camera costing $5000, because the latter will know of a camera that costs only $100 but will not appreciate the benefits of the more expensive device.*

Financial

The vision system must pay for itself in 6 months.
> *The vision engineer must be prepared to argue against the simple-minded attitude which attempts to judge the value of a vision system solely on financial grounds. When a company buys a vision system, it is investing in the improvement of the quality/safety of its products.*

Component cost is not the same thing as system cost.
> *By purchasing one relatively expensive component, it may be possible make the overall system cheaper, faster and more reliable.*

Only ten percent of the cost of installing a vision system is directly attributable to the requirements of image formation, acquisition and processing.
> *The remaining ninety percent of the project cost is due to making the system work properly in the factory.*

$1 spent on inspection is worth $10 in improved profits.
> *Investing a little in automated visual inspection can lead to significant gains in improved efficiency.*

System Specification

The specification of the vision system is not what the customer wants.
> *Do not try to take short cuts in the initial dialogue. The vision engineer should be prepared to spend a considerable amount of time finding what the customer really wants.*

The system specification must be agreed and fully understood by all concerned.
> *All specifications should be in writing with negotiable and non-negotiable specifications noted before the design proper begins.*

No machine vision system can solve the problem that the customer forgot to mention when placing the order.
> *Undertake a proper and complete design study for each type of product.* The specification of a vision system should indicate its functionality and performance.
> *It should not be used merely as a marketing tool.*

Beware when the customer says "By the way! We would like to be able to inspect these objects as well."
> *We repeat the point just made above: undertake a proper design study for each type of product.*

Simple accept/reject labelling is easier than classifying defects by type.
> *If the customer wants to classify defects, they should be made aware that this could have a major bearing on the cost of the inspection system. Detailed classification of defects can greatly increase the speed/cost of the vision system.*

It may not be possible to classify defects reliably.
> *The classification process may not always be clear-cut. A certain product may, for example, have a combination of faults. The vision system supplier and customer must agree beforehand what bounds are to be imposed on the classification process.*

Specify the operating environment.
> *It is relatively easy to make a system that works well in the laboratory. However, it is much more difficult to build a target system that will work reliably in a hostile factory environment.*

Defect types must be realistically prioritised.
> *The ranking of defect types in order of importance can have a major influence on the approach taken, and hence the final cost of the solution. For example, it may be the case that 90% of defect types can be detected for a cost of 90% of the total project budget, whereas detecting the remaining 10% of defect types would cost another 90%. (This is an example of the 90:90 rule.)*

Choosing Inspection System Design Samples

Maximise the number of product samples.
> *The feasibility study, the target system design process, the testing and evaluation of the target system and any demonstrations to the customer should all be based on a large number of representative sample parts. These samples should cover the full range of part variability.*

Choose design samples following proper statistical sampling techniques.
> *Their selection should be made according to a carefully planned and agreed protocol.*

If necessary, choose inspection samples manually using agreed criteria.
> *If samples are chosen manually they will need to be cross-checked to ensure that the variation found in manual inspection is minimised. It is critical that the vision engineer establishes a reliable training set.*

The customer said his widgets were made of brass. He did not think to state that they are always painted blue and oily.
> *To the vision engineer, the surface finish is more important than the underlying material. This contrasts sharply with the customer who often regards surface finish as being of mere cosmetic value.*

Classify sample defects.
> *There are many different ways in which a product can fail to meet its criteria. Any specific application knowledge that the customer can add concerning the type and origin of the fault, will be useful in the design process.*

Vision Company

A sales-person who says that their company's vision system can operate in uncontrolled lighting is lying.
> *No. We are not exaggerating. The human eye cannot. No machine can either.*

A happy vision team has (at least) seven players.
> *This consists of engineers who specialise in mechanical handling, lighting, optics, video sensor technology, electronic hardware, software, vision system integration.*

Alternative Solutions

What a person cannot see, generally cannot be detected by the machine vision system.
> *The human eye is remarkably adept and versatile. In contrast, a vision system is clumsy and unsophisticated, although it may be faster and more reliable. It is a good maxim to admit defeat sometimes as this will gain customer confidence, in the long term.*

It may be cheaper to hire a person to inspect the widgets.
> *However, a machine may be faster, more consistent and reliable. Be prepared to argue this point with the customer.*

Machines can do some things better than humans.
> *Machines can sense outside the visible spectrum (X-rays, IR, UV). Line-scan cameras and laser scanners can produce high resolution images that cannot be seen directly by the eye. Depending on the technology used, a machine vision system would be expected to achieve a substantially higher inspection efficiency, and it can theoretically do*

this for 24 hours a day, 7 days a week. Machine vision can also be useful at detecting gradual changes in continuous processes that appear over long time periods. For example, inspecting gradual colour variations in the production of web materials. Such a gradual change in colour is unlikely to be detected by a human operator.

People can do some things better than machines.

So far, no machine has been built that can reliably guide a car through busy traffic, safely and at speed. No machine can yet judge the aesthetic qualities of a person's dress or a fine painting.

Even the best human inspector is only 70% efficient.

This is one of the best arguments in favour of using machine vision. A person is easily distracted, for example by a good-looking member of the opposite sex walking past. The performance of a human inspector falls as a results of boredom, dissatisfaction with employment, distress due to a recent argument, illness, fatigue, hunger, discomfort, pain, alcohol and drug ingestion.

Machines can work in situations that people cannot tolerate.

Machines can work in radioactive, chemical and biological hazards, where there are high levels of noise, IR, UV, X-ray and microwave radiation, or it is very hot. Machines can tolerate flashing lights, which would induce epileptic fits and migraine attacks in people. A camera can operate under very high, very low, or suddenly changing pressure, and can also be used safely where there is a danger of explosion, or brittle materials are likely to shatter suddenly. A camera can be placed close to a laser cutter, which would be dangerous to a human being. A person cannot inspect the inside of a working jet engine, nor even a drain pipe.

Human inspection often comes free.

Packing and assembly operators can inspect objects without adding (significantly) to the overall cost of the manufacturing process.

Neither a human inspector, nor a fully automated vision system, will always get the best results.

It is sometimes better to employ a person working in symbiosis with a machine vision system.

Mechanical Handling

However deformed the widgets are, they must all pass through the inspection system without jamming.

If the full range of defective widgets cannot be fed properly through the inspection system, then it is of no use whatsoever. It is an irony that one of the main aims of automated visual inspection is to preventing jamming of a mechanical sub-system, such as an assembly machine.

If the parts feed mechanism of the inspection system can go wrong, it most certainly will and the camera will be crushed.
> *Be prepared to sacrifice the camera, lighting and/or optical sub-systems, in the event of a failure of the feed mechanism. Design the system accordingly.*

Lighting and Optics

Many hands make light work.
> *... but not very well. However, some people do apply proper engineering principles to the design of the optical sub-system and inevitably obtain better results.*

The lighting is not constant.
> *Lighting is never constant in either time or in space.*

Never use software to compensate for a poor lighting system.
> *It is not cost effective and will result in a poor system design.*

It is cheaper to add a light-proof shroud to keep sun-light away from the object under inspection than to modify the software.
> *Another universal truth which is often forgotten.*

Nothing exceeds the speed of light.
> *Any processing that can be done optically will save a lot of computer processing later.*

It is all done by mirrors.
> *Wishful thinking, in view of the previous remark.*

Image Resolution

Any feature whose diameter is equal to 0.1% of the width of the camera's field of view, requires an image resolution better than 2000x2000.
> *Nyquist's Sampling Theorem places a fundamental limit on the image resolution. This is often forgotten / ignored by advertisers on broadcast television, who frequently place disclaimer notices about their products on the screen, using printing that cannot be read properly because it is too small The same principal applies to machine vision.*

A (100x100) picture is worth 10000 words.
> *The ancients were very astute when they realised that a digital image requires the storage and processing of a lot of data.*

One high-quality image is better than 5 fuzzy pictures.
> *Few people would dispute this point.*

Five fuzzy pictures are better than one high-quality image.
> *No! This does not conflict with the previous proverb. It may be cheaper and easier to obtain the required information from a small set of low-resolution images than to process one very high resolution image. For*

example, it may be necessary to see several small features within a large scene. In such a case, it might be appropriate, say to use 5 low resolution images (e.g. 256*256), rather than one image of much higher resolution (e.g. 2000*2000).

Related Disciplines

Machine Vision ≠ Computer Vision.
Machine vision is concerned with Systems Engineering and the solution of practical problems, such as guiding industrial robots, inspection and process monitoring. On the other hand, Computer Vision concentrates on the concepts and scientific basis of vision . The latter is concerned with generic issues and takes inspiration from and is often used to model human and animal vision.

Machine vision research is not a part-time activity for workers in Image Processing, Pattern Recognition, or Artificial Intelligence.
Some people think it is, unfortunately. The solutions they offer to industrial inspection problems are, at best, unreliable and over-complicated, because they are unaware of the broader "systems issues", such as image acquisition, QA practices, industrial engineering etc..

Environmental Protection

Protect the machine from the work place.
A factory is a hostile place, with lots of dirt, tampering fingers, etc.
Protect the work place from the machine.
Protect eyes from flashing lights, lasers, etc. Make sure that the inspection machine does not shed bits, such as nuts, bolts, etc. to contaminate food products, pharmaceuticals, etc.
It is cheaper to pay for a shroud to enclose strobed light than to pay compensation for causing epileptic fits.
Flashing lights can trigger epileptic fits and migraine attacks.
The lens may not fit the workman's camera at home, but he thinks it will.
Be aware of light fingered workers causing damage by removing pieces of equipment.
"He is a good worker and likes to keep things clean - he washes down all of the equipment using a hose-pipe, every afternoon".
This is quotation from one factory manager about a dedicated, but uninformed worker who did not realise the potential damage and danger his actions could cause. It is imperative therefore that the vision equipment be made safe and robust.
Adjustment of the camera is achieved using a 1kg hammer.
Vision engineers will be horrified at this prospect but it may happen.

Factories are dirty places.
> *The electrical power supply is noisy. The air supply, for pneumatic equipment, also carries dirt, moisture and oil. Dirt, dust, moisture, fumes, spray, etc. all abound in the local atmosphere.*

Proving and Working with the System in the Factory

Do not assume that the factory workers are computer literate.
> *Software should be designed in such a way that it can be used with minimal computer skills.*

The people who will make sure that the machine will not work are standing beside it.
> *So, the vision engineer should try to persuade them that it is actually in their best interests (as well as his) to work in co-operation with the treasured vision system, not against it.*

A picture is worth ten thousand words.
> *Give the workers a television program to watch. A visual display, showing performance statistics of the vision system and explaining its operation is well worth having, even though it may not seem to be essential.*

People "understand" pictures.
> *A visual display is a useful way of building the confidence of factory personnel. It is also a valuable diagnostic tool: a person can easily recognise whether a sequence of images, showing the operation of the vision system is being repeated properly.*

The service schedule of the vision system should be compatible with the production line.
> *If it is not, the vision system will not fit into the factory environment properly.*

For every hour you spend on inspecting products, your smarter competitor spends 10 hours improving their process.
> *Automated inspection is not always the best way to get the desired results.*

Document all experiments to validate the system.
> *All laboratory and on-site trials in the customer's premises should be fully documented. This should include details about the hardware and software used, parameter settings, optical and lighting set-ups, lens distance, aperture settings and mechanical handling features, how the products were selected.*

Quantify the system performance.
> *The ability of the system to perform to the agreed specification should be demonstrated and quantified. Accuracy, repeatability, robustness, feature delectability and tolerance of product variation should all be*

measured and recorded. All demonstrations should be attended by the vision application engineer(s) who are ultimately responsible for the system design and implementation.

Results may not be reproducible.

Wherever possible, the results of all system performance tests should be reproducible and statistically characterised as to repeatability. In certain applications, for example the inspection of natural products, the variation in product characteristics make it difficult to implement this approach.

Align, calibrate and then test the system before it is used.

A badly aligned system, or one which has not been calibrated, is likely to produce erroneous but seemingly reasonable results.

Appendix B

Factors to be Considered when Designing a Vision System

Mechanical Handling

Motion: Continuous linear; indexed; continuous rotating; sliding; free fall; direction; velocity.

Presentation

Known or random position; known or random orientation; arbitrary or limited number of attitudes possible when dropped on table; touching; separate; overlapping; on table; held in machine chuck; hanging (on chain); stacked; palletised; jumbled heap; jumbled in a bin.

Will faulty parts feed without jamming? Number of components/minute; separation between components; vibration; physical space available for illumination; optics and camera.

Illumination

Spectral characteristics: Visible waveband; colour; infra-red; ultra-violet.
Intensity: Variation with time; lamp ageing; variation when new lamps are fitted; power supply variations.
Spatial distribution: Uniform; patterned (structured); filament shadow; dark spots to uneven packing and broken fibres in fibre-optic bundles.
Temporal variation (short term): Constant; strobed; computer controlled; feedback to compensate for falling light output of lamps as they age.
Polarisation: None; linear; circular.
Coherence: Incoherent; coherent.
Illumination optics (also see below): Mirrors; lenses; fibre optics; filters; filters and mirrors for heat removal.
Servicing and maintenance: Lamp life; lamp replacement procedure; cleaning optics.

Environmental protection: Heat; water; dust; dirt; fumes; splashes etc.; tampering; malevolence; theft; ambient light; protecting people from dangerous light (lasers & stroboscopes); ionising radiation.

Optics

Lenses: Custom or standard; magnification; aperture; focal length; depth of focus; resolution (line pairs/mm); aberrations; anamorphic; materials; glass; quartz; plastic.
Filters: Long pass; short pass; band pass; notch; infra-red; ultra-violet; effects of heat and moisture.
Beamsplitters: Pellicle or cube type; vibration.
Polarisers: Linear; circular; spectral performance.
Fibre optics: Fibre material; ambient light.

Image Sensor

Type: CRT; solid state; laser scanner.
Camera characteristics: Spatial resolution; sensitivity, dynamic range; gamma/linearity; geometric precision; intensity scale fidelity; lag; image burn-in; blooming; comet tail effect; noise level; monochrome or colour; weight; radiation damage.
Physical characteristics: Weight; size; lens mounting; magnetic susceptibility; damage by ionising radiation; operating voltages.
Protection of camera: Heat; infra-red; moisture; vibration; accidental knocks; fibre optics.

Image Processing

Hardware: Architecture/technology; processor; bus; analog pre-processing; analogue to digital converter (ADC); digital pre-processing; image analysis and measurement.
Image coding and representation methods: Array representation of an image; run length code; sparse array code.
Software: Operating system; language.
Algorithm "intelligence": Smart; dumb.

System Level

Engineering: Robustness; Reliability; equipment protection; safety of equipment.

Economic: Direct cost of installation; indirect cost of installation; running costs; pay-back period.

Speed: Throughput rate; delay.

Human interface: Ease of use; level of skill needed by operator; ease of reprogramming for new task; user education; machine driven operating dialogue.

Output type: Qualitative; quantitative; image.

Performance: Collection of statistics on reject rates; definition of "gold standard" for inspection.

Co-ordination with other machines: Synchronisation; immediate feedback to manufacturing plant.

System test: Calibration; standard test samples; self test; test images in backing store files.

Servicing and maintenance procedures.

Appendix C

General Reference Material

Machine Vision

I. Alexander, *Artificial Vision for Robots*, Kogan Page Ltd.(1983).
AVA Machine Vision Glossary, Automated Vision Association (1985).
H. Bassman and P.W. Besslich. *AdOculos - Digital Image Processing*, International Thomson Publishing (1995).
B.G. Batchelor, *Pattern Recognition, Ideas in Practice*, Plenum (London) (1978).
B.G. Batchelor, *Intelligent Image Processing in Prolog*, Springer-Verlag (1991).
B.G. Batchelor, D.A. Hill and D.C. Hodgson, *Automated Visual Inspection*, IFS Ltd/North Holland (1984).
B.G. Batchelor and F.M. Waltz, *Interactive Image Processing*, Springer-Verlag (1993).
B.G. Batchelor and P.F. Whelan (Eds.), *Selected Papers on Industrial Machine Vision Systems*, SPIE Milestone Series MS 97, SPIE Optical Engineering Press, (1994).
A. Browne and L. Norton-Wayne, *Vision and Information Processing for Automation*, Plenum Press, New York, (1986).
E.R. Davis, *Machine Vision: Theory, Algorithms, Practicalities*, Academic Press (1990).
G. Dodd and L. Rossol, *Computer Vision and Sensor Based Robots*, Plenum press (1979).
E.R. Dougherty and P.A. Laplante, *Introduction to Real-time Imaging*, SPIE/IEEE Press, SPIE Vol. TT19 (1995).
M. Eijiri, *Machine Vision A Practical Technology for Advanced Image Processing*, Gordon and Breach Science Publishers (1989).
H. Freeman, *Machine Vision: Algorithms, Architectures, and Systems*, Academic Press (1987).
H. Freeman (Ed.), *Machine Vision for Inspection and Measurement*, Academic Press (1989).
H. Freeman, *Machine Vision for Three Dimensional Scenes*, Academic Press (1990).

K.S. Fu, R.C. Gonzalez and C.S. Lee, *Robotics, Control, Sensing, Vision and Intelligence*, McGraw-Hill (1987).

L.J. Galbiati, *Machine Vision and Digital Image Processing Fundamentals*, Prentice-Hall (1990).

R.C. Gonzalez and P. Wintz, *Digital Image Processing*, Addison Wesley (1987).

R.C. Gonzalez and R. E. Woods, *Digital Image Processing*, Reading, Mass, (1992).

R.M. Haralick and L.G. Shapiro, *Computer and Robot Vision: Volumes I and II*, Addison Wesley (1992).

J. Hollingum, *Machine Vision: The Eyes of Automation*, IFS Ltd. (1984).

B.K.P. Horn, *Robot Vision*, MIT press (1986).

R. Jain, R. Kasturi and B.G. Schunck, *Machine Vision*, McGraw-Hill (1995).

M.D. Levine, *Vision in Man and Machine*, McGraw-Hill (1985).

H.R. Myler and A.R. Weeks, *Computer Imaging Recipes in C*, Prentice Hall (1993).

H.R. Myler and A.R. Weeks, *The Pocket Handbook of Image Processing Algorithms in C*, Prentice Hall (1993).

A. Pugh, *Robot Vision*, IFS/Springer-Verlag (1983).

S.L. Robinson and R.K. Miller *Automated Inspection and Quality Insurance*, Marcel Dekker (1989).

A. Rosenfeld and A. C. Kak, *Digital Picture Processing (2nd edition)*, Academic Press, New York, (1982).

R.J. Schalkoff, *Digital Image Processing and Computer Vision*, Wiley (1989).

M. Sonka, V. Hlavac and R. Boyle, *Image Processing, Analysis and Machine Vision*, Chapman & Hall (1993).

C. Torras (Ed.), *Computer Vision, Theory and Industrial Applications*, Springer-Verlag, Berlin (1992).

D. Vernon, *Machine Vision*, Prentice-Hall (1991).

N.J. Zimmerman and A. Oosterlinck (Eds.), *Industrial Applications of Image Analysis*, D.E.B. Publishers (1983).

Computer Vision

D.H. Ballard and C.M. Brown, *Computer Vision*, Prentice-Hall (1982).

H. Barlow, C. Blakemore and M. Weston-Smith (Eds), *Images and Understanding*, Cambridge University Press (1990).

A. Basu and X. Li, *Computer Vision: Systems, Theory and Applications*, World Scientific (1993).

G.A. Baxes, *Digital Image Processing:- A Practical Primer*, Prentice-Hall/Cascade Press (1984).

G.A. Baxes, *Digital Image Processing: Principles and Applications*, John Wiley (1994).

R.D. Boyle and R.C. Thomas, *Computer Vision: A First Course*, Blackwell (1988).

M. Brady and H.G. Barrow, *Computer Vision*, North-Holland (1981).
K.R. Castleman, *Digital Image Processing*, Prentice-Hall (1996).
R. Chellappa and A.A. Sawchuk, *Digital Image Processing and Analysis: Volume 1: Digital Image Processing*, IEEE Computer Society (1985).
R. Chellappa and A.A. Sawchuk, *Digital Image Processing and Analysis: Volume 2: Digital Image Analysis*, IEEE Computer Society (1985).
C.H. Chen, L.F. Pau and P.S.P Wang, *Handbook of Pattern Recognition and Computer Vision*, World Scientific (1993).
E.R. Dougherty, *An Introduction to Morphological Image Processing*, Tutorial Text Vol. TT9, SPIE press (1992).
R.O. Duda and P.E Hart, *Pattern Classification and Scene Analysis*, John Wiley (1973).
M.A. Fischler, *Readings in Computer Vision: Issues, Problems, Principles and Paradigms*, M. Kaufmann Publishers (1987).
K.S. Fu (Ed.), *Syntactic Pattern Recognition and Applications*, Springer-Verlag (1981).
C.R. Giardina and E.R. Dougherty, *Morphological Methods in Image and Signal Processing*, Prentice Hall (1988).
W.B. Green, *Digital Image Processing - A Systems Approach*, Van Nostrand Reinhold (1983).
E.L. Hall, *Computer Image Processing and Recognition*, Academic Press (1979).
T.S. Huang, *Image Sequence Analysis*, Springer-Verlag (1981).
M. James, *Pattern Recognition*, BSP (1987).
A.K. Jain, *Fundamentals of Digital Image Processing*, Prentice-Hall (1989).
C.A. Lindley, *Practical Image Processing in C*, Wiley (1991).
A. Low, *Introductory Computer Vision and Image Processing*, McGraw Hill (1991).
D. Marr, *Vision*, W.H Freeman and Company (1982).
A.D. Marshall and R.R. Martin, *Computer Vision, Models and Inspection,* World Scientific (1992).
R. Nevatia, *Machine Perception*, Prentice-Hall (1982).
L.F. Pau, *Computer Vision for Electronic Manufacturing*, Plenum Press (1990).
A.P. Pentland, *From Pixels to Predicates: Recent Advances in Computational and Robotic Vision*, Ablex Publishing Corp. (1986).
I. Pitas, *Digital Image Processing Algorithms*, Prentice-Hall (1993).
W.K. Pratt, *Digital Image Processing*, Wiley (1978).
J. C. Russ, *The Image Processing Handbook (2nd edition)*, CRC Press, Boca Raton, (1995).
J. Serra, *Image Analysis and Mathematical Morphology Vol. 1*, Academic Press (1982).
J. Serra, *Image Analysis and Mathematical Morphology Vol: 2 Theoretical Advances*, Academic Press (1988).
L. Uhr, *Parallel Computer Vision*, Academic Press (1987).

Related Material

M. Brady, L.A. Gerhardt and H.F. Davidson (Eds.), *Robotics and Artificial Intelligence*, Springer-Verlag (1984).
J.P. Chan and B.G. Batchelor, "Machine vision for the food industry", *Food Process Monitoring Systems*, Blackie and Sons, Glasgow (1992).
H.L. Dreyfuss and S.E. Dreyfuss, *Mind over Machine*, Free Press (1986).
E.A. Feigenbaum, *The Fifth Generation: Artificial Intelligence and Japans Computer Challenge to the World*, Pan, London, (1984).
M.H. Lee, *Intelligent Robotics*, Open University Press (1989).
D. Michie, *On Machine Intelligence* (2nd Edition), Ellis Horwood Ltd (1986).
T. Owen, *Assembly with Robots*, Prentice-Hall (1985).
S.K. Rogers and M. Kabrisky, *An Introduction to Biological and Artificial Neural Networks for Pattern Recognition*, Vol. TT4, SPIE Press (1991).
H.A. Simon, *The Sciences of the Artificial*, MIT Press, Cambridge MA (1969).
W.A. Taylor, *What Every Engineer Should Know About Artificial Intelligence*, MIT (1988).

Special Issues

Special section on Shape Analysis, *IEEE Transactions on Pattern Analysis and Machine Intelligence* 8(1) (1986).
Special Issue on Industrial Machine Vision and Computer Vision Technology - Part I, *IEEE Transactions on Pattern Analysis and Machine Intelligence* 10(1) (1988).
Special Issue on Industrial Machine Vision and Computer Vision Technology - Part II, *IEEE Transactions on Pattern Analysis and Machine Intelligence* 10(3) (1988).
Special section on Computer Architectures, *IEEE Transactions on Pattern Analysis and Machine Intelligence* 11(3) (1989).
Special Issue on Shape Analysis in Image Processing, *Pattern Recognition* 13 (1981).
Special Issue on Digital Picture Processing, *Proceedings of the IEEE* 60(7) (1972).
Special Issue on Designing Autonomous Agents, *Robotics and Autonomous Systems* 6(1,2) (1990).
Special Issue on Image Analysis and Processing, *Signal Processing* 3 (1981).

Survey/Review Papers

R.T. Chin, "Automated visual inspection: 1981 to 1987", *Computer Vision, Graphics and Image Processing* 41, 346-381 (1988).

R.T. Chin and C.A. Harlow, "Automated visual inspection: A survey", *IEEE Transactions on Pattern Analysis and Machine Intelligence* 4(6), 557-573 (1982).

R.C. Gonzalez, "Syntactic pattern recognition - Introduction and survey", *Proc. Natl. Elec. Conf.* 27, 27-31 (1972).

R.P. Kruger and W.B. Thompson, "A technical and economic assessment of computer vision for industrial inspection and robotic assembly", *Proceedings of the IEEE* 69(12), 1524-1538 (1981).

T. Pavlidis, "A review of algorithms for shape analysis", *Comput. Graph. Image Processing* 25, 68-88 (1984).

A. Rosenfeld, "Image analysis: problems, progress and prospects", *Pattern Recognition* 17(1), 3-12 (1984).

P.K. Sahoo, "A survey of thresholding techniques", *Computer Vision, Graphics and Image Processing* 41, 233-260 (1988).

A.M. Wallace, "Industrial applications of computer vision since 1982", *IEE Proceedings* 135(3), 117-136 (1988).

J. Weska, "A survey of threshold selection techniques", *Comput. Graph. Image Processing* 7, 259-265 (1978).

Periodicals/Journals/Magazines

Advanced Imaging. PTM Publishing.
Artificial Intelligence - An International Journal, Elsevier Science Publishers.
Communications, Speech and Vision. IEE-The Institution of Electrical Engineers.
Computers and Digital Techniques. IEE.
Computer Graphics and Image Processing.
Computer Vision, Graphics and Image Processing.
IEE Proceedings on Vision, Image and Signal Processing, IEE.
IEEE Transactions on Acoustics, Speech and Signal Processing. IEEE.
IEEE Transactions on Pattern Analysis and Machine Intelligence. IEEE.
IEEE Transactions on Systems, Man and Cybernetics. IEEE.
IEEE Transactions on Robotics and Automation. IEEE.
Image and Vision Computing. Butterworths.
Image Processing. Reed Publishing.
Imaging Systems and Technology. John Wiley and Sons.
Intelligent Systems Engineering, IEE.
International Journal of Computer Vision. Kluwer Academic Publishers.
Journal of Visual Communications and Image Representation. Academic Press.
Journal of Systems Engineering, Springer International.
Machine Vision and Applications. Springer International.
Optical Engineering, SPIE-The International Society for Optical Engineering.
Pattern Recognition. Pergamon.
Pattern Recognition Letters. IARP, North Holland.
Photonics: Spectra. Laurin Publishing.

Real-time Imaging, Academic Press.
Robotica. Cambridge University Press.
Sensor Review.
The Photonics Design and Applications Handbook. Laurin Publishing.
The Photonics Dictionary. Laurin Publishing.
Vision. SME (Machine Vision Association).

Conference Proceedings

Applications of Digital Image Processing, SPIE.
Applied Machine Vision Conference, SME.
British Machine Vision Conference, Springer-Verlag.
European Conference on Computer Vision - ECCV, Springer-Verlag.
IEEE Conference on Robotics and Automation, IEEE.
Intelligent Robots and Computer Vision, SPIE.
International Conference on Assembly Automation, IFS/North-Holland.
International Conference on Automated Inspection and Product Control, FS/North-Holland.
International Conference on Robot Vision and Sensory Controls, IFS Publishers.
International Robots and Vision Automation Conference.
Machine Vision Applications, Architectures and Systems Integration, SPIE.
Optics, Illumination and Image Sensing for Machine Vision, SPIE.
Topical Meeting on Machine Vision, Optical Society of America.
Vision Systems in Robotic and Industrial Control, IEE Computing & Control.

Internet Resources[1]

Newsgroups

```
alt.3d
     Three-dimensional Imaging
comp.ai.vision
     Computer Vision
comp.robotics
     Robotics and Robot Vision
sci.image.processing
     Scientific Image Processing
```

---

[1] Although correct at the time the book went to print, these links may change.

Mailing Lists

```
vision-list@ads.com
```
Computer Vision
```
pixel-request@essex.ac.uk
```
'The Pixel' Digest
```
morpho@cwi.nl
```
Morphological Digest

FTP Sites

```
mom.spie.org
```
International Society for Optical Engineering (SPIE).
```
peipa.essex.ac.uk
```
Pilot European Image Processing Archive. Also see `peipa/info/IP-tools.review` *for a review of image processing tools.*
```
ftp://ftp.wmin.ac.uk/pub/itrg/coloureq.txt
```
Colour spaces and colour transforms

World Wide Web (URL)

```
http://www.cs.cmu.edu/~cil/txtvision.html
```
Computer Vision Home Page. This site contains a comprehensive list of computer vision research groups on the World Wide Web. It also includes topics related to computer vision, conference and symposia notifications, frequently asked questions, a list of news groups and archives, publications, test images and source code. A very useful source of information.
```
http://www.eeng.dcu.ie/~whelanp/vsg/vsghp.html
```
Vision Systems Group (DCU)
```
http://www.vision.auc.dk/LIA/NORVIC/index.html
```
NORVIC: Nordic Research Network in Computer Vision

```
http://afrodite.lira.dist.unige.it/fullservice.html
```
ECVNet
```
http://www.vision1.com/links.html
```
Vision and Imaging Resource Links
```
http://www.epm.ornl.gov/~batsell/imaging.html
```
Imaging on the Internet: Scientific/Industrial Resources
```
http://www.sme.org/memb/mva.html
```
Machine Vision Association of SME (MVA/SME)
```
http://piglet.cs.umass.edu:4321/robotics.html
```
Robotics Internet Resources

```
http://www.wiley.com/wileychi/electronic/hipr/
```
HIPR - Hypermedia Image Processing Reference. (Available on CD-ROM from John Wiley & Sons Ltd.)
```
http://arachnid.cs.cf.ac.uk/Lad/text.intro.html
```
The Lighting Advisor

    http://www.cm.cf.ac.uk/Dave/Vision_lecture/Vision_lecture_
    caller.html
        Vision Systems Courseware
    http://www.cogs.susx.ac.uk/users/davidy/teachvision/
    vision0.html
        Sussex Computer Vision Teach Files
    http://www.ph.tn.tudelft.nl/Software/TimWin/timwww2.html
        TIMWIN: A program for scientific image processing
    http://www.khoros.unm.edu/khoros/
        The Khoros Page

    http://pasture.ecn.purdue.edu/~precetti/
        Colour Classification Tutorial
    http://www.isc.tamu.edu/~astro/color.html
        Colour Science
    http://www.cis.rit.edu/mcsl/
        Munsell Color Science Laboratory

    http://wwwwhite.media.mit.edu/vismod/imagery/VisionTexture
    /vistex.html
        VisTex Vision Texture Database
    http://moralforce.cc.gatech.edu/
        ARPA Image Database Browser
    http://www.cwi.nl/projects/morphology/
        The Morphology Digest

    http://www.cs.washington.edu/research/vision/pamitc.html
        IEEE - PAMI TC Home Page
    http://www.elsevier.nl:80/section/computer/416/525443/
    menu.htm
        Image and Vision Computing
    http://scorpions.ifqsc.sc.usp.br/ifsc/ffi/grupos/instrum/
    visao/meetings/rti.htm
        Real-Time Imaging

    http://iris.usc.edu/Information/Iris-Conferences.html
        Computer Vision Conferences
    http://www.rpd.net/Info/conferences/index/
    Machine_Vision.html
        WWW Virtual Library on Conferences: Machine Vision

Design Aids

Lighting

ALIS 600 is a sophisticated multi-function lighting system, which provides a variety of illumination devices, mounted inside a light-proof cabinet. The lights are operated from regulated power supplies and can be switched by a computer.

Micro-ALIS [ALIS] is an experimental tool-kit consisting of a set of useful illumination, optical and fibre-optic devices. In addition, there is a set of versatile mechanical fixtures for holding lamps, optical devices and samples.

Optics

Sigma 2100 [KID] is a program for designing optical systems from simple objectives to complex multi-configuration systems, including zoom lenses, switchers, multi-channel lenses, multi-waveband lenses and scanners.

KDP is a general-purpose optical design and analysis program. It has provisions for modelling ray-tracing and optimising a wide variety of types of optical systems. It has an extensive optical analysis capability which is enhanced by a semi-compiled macro programming language. KDP is free and runs on a PC. (*KDP, Optical Design Software*, Engineering Calculations, 1377 East Windsor Road, #317 Glendale, CA 91205, USA. Available via WWW: *www.kdpoptics.com.*)

TracePro is a ray-tracing program for optical analysis. It accounts for optical flux as it propagates through a solid model, defined in terms of geometric objects, such as spheres, elliptical and conical cylinders and cones, blocks and tori. *TracePro* can calculate absorption, specular reflection, refraction, scattering and aperture diffraction effects. (*TracePro*, Optical Systems Analysis Program, Lambda Research Corp., PO Box 1400, Littelton, MA 01460-4400, USA.)

Optica has a large collection of data relating to lenses, mirrors, prisms, gratings. It provides a full range of geometric ray-tracing functions for designing optical systems and components. It is a based on Mathematica. (*Optica, Optical Design Software*, Wolfram Research Inc., 100 Trade Centre Drive, Champaign, IL 61820-7237, USA.)

Zemax is an optical design program.

OPTICAD provides optical layout and analysis software. (*OptiCAD* and *Zemax, Optical Design Software*, Focus Software, PO Box 18228,Tucson, AZ 85731-8228, USA.)

Lens Selector Program. Optimum Vision Ltd., Unit 3a, Penns Road, Petersfield, GU32 2EW, UK.

LensVIEW is a compilation of lens design data on CD-ROM.

Machine Vision Lens Selector [MVA] is a slide rule and performs basic lens design calculations.

Camera Calculator [SNY-92] is a Macintosh desk-accessory. It solves the standard lens formulae, given any sufficient sub-set of variables. It allows the user to specify virtually any sub-set of known values for such features as object size, object distance, image size, image distance, magnification, depth of field, f number, image resolution, and then calculates the unknown values.

Lighting-Viewing Subsystem

HyperCard Lighting Advisor provides a catalogue of over 150 different lighting and viewing techniques. For each lighting-viewing technique, there are three cards: one provides notes in a fixed-format frame; another shows a sketch of the optical layout and the third card provides a sample image obtained using that lighting-viewing method. The Lighting Advisor is available on a shareware basis.

Email :	*Bruce.Batchelor@cs.cf.ac.uk*
WWW:	*http://www.cs.cf.ac.uk/User/Bruce.Batchelor/*
	and *http://www.cs.cf.ac.uk/lad/text.intro.htm*
FTP:	*http://bruce.cs.cf.ac.uk/FTP/Light.sit.hqx*

Lighting Science Database [ITI-89], Prolog Lighting Advisor [BAT-89] and Lighting Advisor Expert System [PEN-88] all provide a broadly similar function to the HyperCard Lighting Advisor.

Equipment / Software Suppliers

Opto*Sense® [WHI-94] is a comprehensive database of machine vision vendors' names and addresses. It runs on a PC.

Training Courses

"Success with Vision" is a set of six video tapes, describing the basic principles of machine vision system design. (Visual*Sense*Systems, 314 Meadow Wood Terrace, Ithaca, NY 14850, USA.)

On-line Training Course in Machine Vision, Automated Vision Systems, 1550 La Pradera Drive, Campbell, CA 95008-1547, USA.
    WWW: *http://www.autovis.com/autovis/*

Appendix D

PIP - Software Implementation of Prolog+

This appendix was written in conjunction with **Andrew Jones** and **Ralf Hack** (University of Wales College of Cardiff). They, together with Stephen Palmer and BGB, are the joint authors of PIP.

D.1 Availability of the PIP Software

Copies of the PIP software described below may be obtained by contacting:
Bruce.Batchelor@cs.cf.ac.uk or *Andrew.C.Jones@cs.cf.ac.uk*.
Up-to date information about the status and availability of PIP is available on the World Wide Web (*http://bruce.cs.cf.ac.uk/bruce/index.html*).

D.2 Introduction

In this appendix, we describe PIP (mnemonic for *Prolog Image Processing*), a software system for interactive image processing and which provides the ability to write Prolog+ programs. PIP runs on an Apple Macintosh computer but differs from Prolog+ in two important respects. The first is that PIP performs image processing, implemented in software, using the C programming language, whereas Prolog+ originally relied on the availability of dedicated image processing hardware. The second difference is that, although the present version of PIP supports almost all of the Prolog+ commands mentioned elsewhere in this book, these are implemented in terms of lower-level Prolog predicates, which enable a more flexible approach to image manipulation to be taken. In principle, other operating paradigms, such as processing colour and multi-spectral images, or maintaining a history of past results using an image stack, are possible in PIP.

We shall discuss the impact of the Apple Macintosh operating system upon the implementation of the image processing functions, and the interface between these and the Prolog sub-system. We also explain how the Prolog+ commands have been implemented. We will outline the principles upon which the PIP system is built, explaining in detail why this particular software-based approach is attractive. We shall then describe the infra-structure that has been implemented for image processing in which Prolog operates as a "top-level" controller. It is anticipated, however, that the system will not normally be used at this (i.e. infrastructure) level. One way in which a more accessible command set may be realised is to implement Prolog+ commands above this infra-structure. We discuss how this has been achieved. While we do not explain in detail how the other operating paradigms just mentioned may be implemented, it is fairly obvious how this can be achieved.

D.3 Software for Image Processing

Using specialised hardware for image processing has the obvious advantage over a software implementation that the hardware is tailored to image processing and will often give substantially better performance. If a software implementation is capable of providing *adequate* performance for a particular application, then such an implementation offers a number of benefits, including the following:

- Apart from initial image capture, no investment in specialised hardware is required. Indeed, a complete image processing system may be assembled by merely purchasing a standard CCIR/RS320, or RS170, video camera and installing the PIP software on a Macintosh computer fitted with a standard "AV" (Audio-Video) card. Alternatively, a low cost camera (QuickCam[2]) may be used without any other hardware. (This device is interfaced to the computer via the serial port.) A third option is to use scanned images.
- As a user upgrades his computer, he will obtain a corresponding improvement in image processing performance, without incurring the additional cost of investing in new hardware.
- The software can be extended indefinitely, whereas image processing hardware is typically packaged in a closed "black box", providing a predetermined range of functions.

It should be clearly understood that Prolog is not an appropriate language for implementing "low-level" image processing operations, such as image addition, thresholding, filtering, skeletonisation, convex hull, etc. (These are often described colloquially as "pixel pushing" operations.) A procedural high-level language, or of course, assembly language, is much better suited to rapid, iterative processing of large arrays of data. Thus, an essential feature of the PIP system is the interface between Prolog and the image processing software.

D.4 Choice of Hardware and Software Platforms

We have chosen the Apple Macintosh computer, LPA MacProlog32 and Symantec Think C for system development. The software has been tested on

---

[2] QuickCam, Connectix Corporation, San Mateo, CA, USA

several Apple computers, including those based on Motorola 680X0 processors and the PowerPC family. At the time of writing (October 1996), there is no "native code" version of the MacProlog32 software for the PowerPC family, so the software runs in emulation mode on these machines. The stand-alone version runs successfully under the *Macintosh Application Environment 2.0* (Apple Computer, Inc.) on a Sun or Hewlett-Packard workstation. Unfortunately, the PIP software will not run under the WINDOWS 95 or MS-DOS operating systems. However, the promised enhancements to the *Executor 2* software[3], which emulates a Macintosh computer, should make this possible soon.

We chose to use the Apple Macintosh for a number of reasons. Historically, our previous work on Prolog+ has been carried out on Apple Macintosh computers, due to the availability of a good implementation of Prolog. The LPA MacProlog environment used by the authors provides a full implementation of Prolog, user-interface development facilities, the ability to call functions written in C or Pascal, and the ability to act upon low-level events, such as activation of a window, in a user-defined manner.

We are using THINK C because this is one of the languages supported by LPA MacProlog. The former offers an integrated programming environment, which has proved useful in developing and testing image processing functions, before attempting to integrate them into the PIP system.

D.5 Why not Implement Prolog+ Commands Directly?

Prolog+ is centred mostly around just two images: the *current* and *alternate* images. In previous implementations of Prolog+, both of these images were continually visible on a video (i.e. not the computer) monitor, whereas in PIP, they appear in windows on the Macintosh computer screen.

We elected not to implement C routines which perform Prolog+ functions directly. Instead, a new image is created by each image-to-image mapping operator (e.g. *neg, add, lpf, chu,* etc.) and we have provided separate routines for such tasks as, disposal of images which are no longer required, creation of windows to display images and association of a new image with a window.

The reasons for this approach include the following:

- It is fairly easy to implement a 2-image (i.e. Prolog+) operating paradigm on top of this, by writing appropriate Prolog code.
- The idea of leaving the source images unchanged is more in keeping with the spirit of the Prolog language.

---

[3] The *Executor 2* software is available from Ardi Software, Inc., Suite 4-101, 1650 University Boulevard, Albuquerque, NM 87102, USA. Also consult the following WWW site: *http://www.ardi.com*

- It will not always be desirable to have a continuous display of the images when the system is working. (The user may, for example, wish to hide intermediate results from a customer.)
- There is freedom to implement and explore other image processing models, if desired. (For example, we may wish to defined operations such as *add, subtract, multiply*, etc on 3-component colour images.)

D.6 Infra-structure for Image Processing Using Prolog

LPA MacProlog allows the programmer to call functions written in C, or Pascal. On the Apple Macintosh family of computers, files have two separate parts: the *data* and *resource* forks. MacProlog requires that a new code resource be created containing the compiled foreign code. Having opened the file containing the resource, the *call_c* (or *call_pascal*) predicate is used to invoke the required function. A collection of 'glue' routines must be linked into the foreign code resource, which allow the programmer to access arguments of the *call_c* routine and manipulate the data structures supported by Prolog, such as lists.

In order to obtain a system which successfully coexists with the Macintosh Finder (the Graphical User Interface) and other applications, it was necessary to build our own application within the framework provided by Apple Computer, Inc. In particular, there is a wide range of Toolbox routines, for managing entities such as windows and menus. *QuickDraw* and offscreen graphics worlds are among the facilities provided for creating, manipulating and displaying graphical data. Using these features, in a way consistent with the Apple Computer Company's recommendations, should ensure the future portability of the PIP system.

In the following, we shall first consider how images are stored, displayed and manipulated in our system, and then consider how the interface between Prolog and the C routines is built.

D.7 Storing, Displaying and Manipulating Images

It is generally best to avoid accessing the screen display directly on the Apple Macintosh computer. Instead, drawing is carried out using *QuickDraw* routines, via a *graphics port*, which is normally a window. The operating system ensures that only visible parts of the window are drawn, and generates *update* events when part of a window needs to be redrawn, perhaps as a result of another overlapping window being moved. If a pixel map must be manipulated directly, then an *offscreen graphics world (GWorld)* may be used. One creates an offscreen GWorld and draws into it using QuickDraw or accessing the GWorld's pixels directly. The result may be copied to the appropriate window, using the QuickDraw *copyBits* routine. In our system, we use an offscreen GWorld to represent each image currently in use. Not all of these offscreen GWorlds necessarily have a corresponding window, but each image display window in PIP does have a corresponding offscreen GWorld.

Using offscreen GWorlds offers a number of benefits to the programmer:

- An offscreen GWorld can be associated with a window for future redrawing, as necessary.
- Whatever the pixel depth of the display, it is possible to make an offscreen GWorld of appropriate depth for the *image*. (So far, the images we have dealt with have had a depth of 8 bits per pixel, and our program assumes this pixel depth when it creates a new offscreen GWorld.) If a display mode is selected in which not all the image colours are available, the *copyBits* routine will select the nearest possible colour from the current palette.
- It is possible to associate a colour lookup table (CLUT) with an offscreen GWorld which is different from the default. This is useful because the default Macintosh CLUTs assign white to a pixel value of 0, whereas our grey-scale image-processing operations assume a grey-scale gradient, in which 0 signifies black. So, in the present PIP system, which deals with 8-bit grey-scale images, we set the CLUT for a new off-screen GWorld to a gradient of 256 grey levels, in which 0 signifies black and 255 signifies white. It is not necessary to change the screen CLUT (which would corrupt the colours of other items on the screen), since mapping between CLUTs is performed automatically by the QuickDraw routines.
- Since the Macintosh operating system has its own memory management routines, memory occupied by an offscreen GWorld may be released as soon as it is no longer needed.

Tables D.1 and D.2 illustrate how the above functionality is implemented in our C code. Table D.1 contains annotated extracts from the *negate_image* routine, which indicates how a new offscreen GWorld is created and accessed. Table D.2 contains extracts from *update_window*, which indicates how an offscreen GWorld is associated with a window and how the window is updated.

D.8 Prolog-C Interface

Information concerning the current images and windows is stored by the Prolog program. In this section we explain how parameters are passed between Prolog and the C routines, and then discuss the implementation of the predicates which call the C routines and provide the infra-structure for the PIP system. Finally, we discuss how user events are handled.

Code	Comments
externOSErr negate_image(GWorldPtr iml, GworldPtr *im2) { ...	iml:inputi image; im2:output image.
GetGWorld(&origPort, &origDev);	Store current GWorld for later restoration.
sourcePM = GetGWorldPixMap(iml);	Get input image pixel map (NB This contains a *reference* to the memory where the pixels themselves are located, and other information.)
good=LockPixels(sourcePM);	Prevent it from moving.
boundRect=(*iml).portRect;	Get boundaries of pixel map.
ctable = GetCTable(129);	Obtain the greyscale CLUT.
errNo = NetGWorld(im2, 8, &boundRect, cTable, nil, 0);	Create new offscreen GWorld of depth 8, with our special CLUT.
DisposeCTab(cTable);	Free memory
SetGWorld(*im2, nil);	Drawing to occur in this new GWorld.
destPM=GetGWorldPixMap(*im2);	Obtain the new pixel map.
good=LockPixels(destPM);	
srcAddr=(unsigned char*)GetPixBaseAddr(sourcePM); srcRowBytes=(**sourcePM).rowBytes & 0x3fff; destAddr=(unsigned char*)GetPixBaseAddr(destPM); destRowBytes=(**destPM).rowBytes & 0x3fff; width = boundRect.right - boundRect.left; height = boundRect.bottom - boundRect.top;	Calculate where pixels are stored and prepare to copy the pixels,
for (row=0; row<height; row++)	Copy pixels negating. NB we assume a greyscale image with pixel values between 0 and 255 inclusive.
{ srcAddrl=srcAddr; destAddrl=destAddr; for (column=0; column<width; column++) *destAddrl++ = 255-(*srcAddrl++); srcAddr=srcAddr+srcRowBytes; destAddr=destAddr+destRowBytes; }	
UnlockPixels(destPM); UnlockPixels(sourcePM);	Allow pixel maps to move again.
SetGWorld(origPort, origDev); ... }	Restore original GWorld (screen).

Table D.1 The *negate_image* routine.

Code	Comments
extern OSErr update_window(WindowPtr theWindow) { ... GetGWorld(&origPort, &origDev); theImage=(GWorldPtr) GetWRefCon(theWindow); SetPort(theWindow); BeginUpdate(theWindow); display_in_window(theImage, theWindow); EndUpdate(theWindow); SetGWorld(origPort, origDev); ...}	 Store current GWorld for later restoration. Retrieve pointer to offscreen Gworld associated with the window Drawing to occur in this window. Indicate to OS that an update event is being processed; call the image drawing routine; indicate update is complete. Restore original GWorld (screen).

Table D.2 The *update_window* routine.

Passing Parameters

The routines such as *negate_image*, described earlier, cannot be called directly from Prolog: some pre- and post-processing is required, in order to retrieve and set the parameters passed between Prolog and the C routine. A major reason why this extra processing is not bundled with each routine is that, in their present form, it is fairly easy to write a stand-alone C program that allows the functions to be tested and debugged separately, from within the Symantec programming environment. A C routine is invoked from LPA MacProlog using a call of the form:

```
call_c(<parameter list>, <resource type>, <resource id>)
```

In our case we have a single code resource of type 'MINE' and resource id 0. We always use a parameter list of the form:

```
[<input param. list>, <output param. list>, <function no.>, <err. code>]
```

As an example of the implementation of the image processing predicates, consider the following extract from our Prolog code:

```
neg_im(Im1, Im2):-call_c([[Im1],Var,6,Err],'MINE',0), Err=0,
Var=[Im2], recd_new_im(Im2).
```

The *neg_im* predicate only succeeds if the error number were zero; the output list should contain a single element (i.e. the value of a pointer to the new image) and the Prolog system records that this new image has been created. This kind of data is recorded using the *properties* feature of MacProlog. An example of how this feature is used is given below, when considering the implementation of Prolog+ commands.

User Events

We need to be able to handle the following two kinds of event:

- *update* events, generated when a window needs to be redrawn; and
- *mouse_down* events, when they occur in the close box of the window.

MacProlog provides a way of trapping these events and acting upon them, provided that the windows' *windowKind* is greater than 32. (All windows created by our C routines have *windowKind 33*.) When an update event is received, MacProlog calls the user-defined *x_update* predicate (if any); a mouse-down event calls the *x_mousedown* predicate, etc. As an example, *x_update* is defined thus in our system:

```
x_update(Win)   :- call_c([[Win], _, 2, _], 'MINE', 0).
```

When the Prolog system invokes this predicate, the parameter Win is bound to the value of a pointer to the window that must be updated. Our *x_update* code calls the C routine which we have written to process update events for the specified window.

D.9 Using Infra-structure Facilities Directly

To use the infra-structure facilities directly is a somewhat laborious procedure. No images are automatically disposed of and an image is only displayed in response to an explicit command to do so. The following example illustrates how these facilities may be used; the values of pointers to the original and final images are returned in variables X and Y respectively:

```
example(Im,Im4) :-
     new_im(Im),              % Read new image from disk
     new_win_for_im_disp(Im,Win,"Original Image",1,50)
                              % Create new window for image titled
                              % "Original Image", with top left hand
                              % corner at (1,50)
     inop3_im([2,3,2,3,5,3,2,3,2],Im, Im1),
                              % Local operator - blur
     new_win_for_im_disp(Im1,Win1, "New Image", 320,50),
                              % New window for this image
     linop3_im([2,3,2,3,5,3,2,3,2],Im1,Im2),
                              % Local operator - blur
     new_win_im_disp(Im2,Win1),
                              % Display this new image
     kill_im(Im1),            % Dispose of previous image
     sobel_im(Im2,Im3),       % Sobel edge detector
```

In the case of the image negation function, the call is composed thus:

```
call_c([[Iml], Var, 6, Err], 'MINE', 0).
```

A single parameter, the value of the source image pointer, is passed to the C routine; the routine binds *Var* to a single-element list which contains the destination image pointer's value; the negation routine is routine number 6, and *Err* will be bound to the error value (which is 0, unless an error occurred).

Table D.3 contains extracts from the main C routine which chooses the appropriate image processing function based on the function number it receives, and passes parameters between the Prolog environment and that function. Table D.4 lists predicates forming the image processing infra-structure.

Code	Comments
bool main(long argc, void *link()) { ... inListTag=get_arg(1); outListTag=get_arg(2); fNoTag=get_arg(3); errNoTag=get_arg(4);	Get 4 arguments passed by call_c
switch (get_int_val(fNoTag)) { ... case 6; errNo = do_negate_image(inListTag, outListTag); break; ... }	Recover the value of the function number; call routine which handles the specified function number.
put_int_val(errNoTag, errNo); return SUCCESS;	Store error number. Return value indicating predicate succeeded
OSErr do_negate_image(cellpo inListTag, cellpo outListTag) { ... lml = (GWorldPtr)get_int_val (get_list_head(inListTag)); errNo = negate_image(iml, &im2); outListTag = put_list(outListTag);	Get 1st (and only) input parameter (head of the input list). Call the negate routine. Create a list to hold output parameter(s).
put_int_val(get_list_head(outListTag), (long) im2); put_nil(get_list_tail(outListTag));	Tail of output list is empty, i.e. the list has only one element
return errNo; ... }	Return error number to main routine.

Table D.3 The main C routine, interfacing with Prolog.

D.10 Predicates Forming the Image Processing Infra-structure

It would be undesirable to use the *call_c* predicate directly in Prolog programs for the following reasons:

- The semantics we have imposed upon the *call_c* predicate are very opaque.
- When an image processing operation is called, additional processing at the Prolog level is required, such as keeping a record of any new images and windows that have been created.

The following program segment explains how this can be achieved:

```
new_win_im_disp(Im3,Win1),
kill_im(Im2),
thresh_im(Im3,Im4,15,255),    % Pixels with intensities 15 255 are
                              % set to white. All others are black
new_win_im_disp(Im4,Win1),
kill_im(Im3).
```

Predicate Format	Description
new_im(Im)	Select a new file (in PICT format) from a dialog box, and draw that file's contents into a new image. "Im" points to the new image on return from this predicate.
kill_im (Im)	Dispose of the image, freeing the memory it occupied.
new_win_for_im (Im, Win, Name, OffH, OffV)	Create a new window for the image. New window has title Name, & its top left hand corner is located at (OffH, OffV). *Win* points to the new window on return from this
new_win_for_im_disp (Im, Win, Name, OffH, OffV)	Perform the same function as *new_win_for_im*, but display the window's contents immediately (rather than waiting for an update event).
new_win_im (Im, Win)	Associate the existing window *win* with new image "Im".
new_win_im_disp(Im, Win)	As *new_win_im*, but display the window's contents immediately.
close_win (Win)	Close the window, freeing the memory it occupied.
copy_im (Im1, Im2)	Copy image "Im1" to new image. "Im2" points to this new image on return.
kill_wins_and_ims	Dispose of all windows and images currently in use, freeing the memory they occupy.

Table D.4 Predicates forming the image processing infra-structure discussed in the text. *Note*: Variables *Im, Im1, Im2* and *Im3* denote integers which are pointers to images; *Win* denotes an integer which is a pointer to a window.

D.11 Implementing Prolog+ Commands

Prolog+ is built essentially around two images: the *current* and *alternate* images. When a Prolog+ image processing operation occurs, the current image will be replaced with the result of the operation, and the alternate image will be replaced with the previously current image. These images are displayed continuously on a monitor.

Our predicates implementing these commands use the lower-level predicates previously described, and perform additional 'housekeeping', such as disposal of old images and maintaining information about the current and alternate images. As an example, the following is an implementation of the *neg* operator:

```
neg:-
        recall(curr_im,CurrIm),          % Retrieve present current & …
        recall(alt_im,AltIm),            % … alternate images & windows
        recall(curr_win,CurrWin),        % … stored as properties
        recall(alt_win,AltWin),
        neg_im(CurrIm,NewIm),            % Negate the image
        new_win_im_disp(NewIm,CurrWin),
                                         % Display new image in current
                                         % window
        new_win_im_disp(CurrIm,AltWin),
                                         % Display previous current
                                         % image in the alternate window
        kill_im(AltIm),                  % Dispose of previous alternate
                                         % image
        remember(curr_im,NewIm),         % Store references to new
                                         % current …
        remember(alt_im,CurrIm).         % … and alternate images
```

Other operators are implemented in a broadly similar manner. A complete list of Prolog+ operators currently supported by PIP is given in Appendix E.

Appendix E

Prolog+ and PIP Commands

The predicates listed below are all available on the PIP or Prolog+ systems. PIP has a rather larger range of commands than Prolog+ and is being developed actively at the University of Wales Cardiff. A few Prolog+ commands are not yet available in PIP but are included here, since they will be added very soon.

Image Processing Primitives		
Mnemonic	**Arity**	**Description**
aad	3	Aspect adjust
abs	0	"Absolute value" of intensity
acn	1	Add constant to each intensity value
add	0	Add current and alternate images
and	0	Logical AND of corresponding pixels in current & alternate images
ang	6	Orientation and length of line
avg	1	Average intensity
bay	0	Bays (indentations) of a blob
bbt	6	Is biggest bay above second biggest bay?
bed	1	Edge detector for binary images
bic	1	Clear A-th bit of intensity of each pixel in current image
bif	1	Flip the A-th bit of the intensity of each pixel in the current image
big	1	Find A-th biggest blob
bis	1	Set A-th bit of intensity of each pixel in current image to 1
blb	0	Fill holes (lakes) in a binary image
blk	0	Set every pixel in the image to black (level 0)
blo	1	Expand the central intensities.
blp	1	Find blob parameters
box	5	Set (hollow) rectangle to defined grey level
bpt	2	Find co-ordinates of centre of bottom-most chord
bsk	0	Copy image at bottom of stack into current image
bve	5	Find all points where given vector intersects edge of object
cal	1	Copy each pixel with intensity > A. All other pixels are black
cbl	1	Count blobs

\multicolumn{3}{c}{Image Processing Primitives}		
Mnemonic	Arity	Description
ccc	0	Draw circumcircle of a blob placed at the blob centroid
cct	1	Concavity tree
cgr	0	Co-ordinates of geometric centroid of all white pixels
chf	0	Flip horizontal axis if longest vertical white chord is to left of image centre
chf	1	Flip hor. axis if longest vertical section is left of image centre
chu	0	Draw the convex hull around a blob-like object
cin	0	Column integrate
cir	5	Set an ellipse to a defined grey level
clc	0	Column run length coding
cnw	0	Count number of white 8-neighbours in each 3*3 neighbourhood
com	1	Count number of points with different intensities
con	9	General purpose linear convolution operator based on 3*3 window
cox	0	Column maximum
cpy	0	Copy current image into alternative image
crk	1	Crack detector
crp	4	Crop image
csh	0	Copy intensity horizontally from RHS
csk	0	Clear image stack
ctp	0	Cartesian to Polar axis transformation
cur	4	Cursor
cvd	0	Convex deficiency
cwp	1	Count the white points in the current image.
dab	2	Draw some defined feature (e.g. centroid, principal axis) for each blob in image
dbn	0	Direction of the brightest neighbour
dcg	0	Draw geometric centroid and print its co-ordinates
dci	7	Draw centre of the image
dcl	3	Draw a pair of cross lines through a given point
dcn	1	Divide each pixel intensity by a constant
dgw	4	Get image size
dif	0	Subtract alternate and current images ignoring the sign.
dil	1	Dilate image along given direction
dim	4	Extreme X and Y values for all white pixels
din	0	Double all intensities
div	0	Divide current image by alternate image
dlp	2	Difference of low-pass filters
doc	0	Suspend Prolog and enter HyperCard HELP facility
dpa	1	Draw principal axis
dsl	3	Draw straight line given one point on it and its slope
eab	1	Analyse each blob in turn
ect	0	Threshold mid-way between minimum & maximum intensity
edc	3	Euclidean distance between two vectors, specified as lists
edd	1	Non-linear edge detector
edg	2	Set the border of width W to grey level G

Image Processing Primitives

Mnemonic	Arity	Description
egr	1	Grow the ends of an arc in a binary image
enc	0	Enhance contrast
ero	1	Erode image along given direction
eul	3	Euler number
exp	0	Exponential intensity transformation
exw	0	Expand white
fac	0	Flip the image about its centroid
fbr	0	Find and remove all blobs touching the border
fcb	9	Fit circle to three points on a blob
fcd	4	Fit circle to three points
fil	5	Set (solid) rectangle to defined grey level
fld	4	Fit straight line to data (2 points)
gft	0	Grass fire transform
gli	2	Get limits of intensity
gob	1	Get one blob and delete it from stored image
gra	0	Gradient, a simple edge detector
gry	1	Set every pixel in the image to defined level
hfl	8	Synonymous with *blb*
hgc	0	Cumulative histogram
hge	0	Histogram equalisation
hgi	1	Intensity histogram
hgr	0	Horizontal gradient
hid	0	Horizontal intensity difference operator
hil	3	Highlight intensities in given range
him	1	Hide a displayed image
hin	0	Halve all intensities
hmx	2	Histogram maximum
hol	0	Obtain the holes (lakes) of a blob in given binary image
hpf	0	High pass filter (3*3 window)
hpi	0	Plot intensity histogram of current image
huf	0	Hough transform
iht	2	Inverse Hough transform of given point
ior	0	OR corresponding pixels in current and alternate images
itv	0	Enter interactive mode
jnt	0	Joints (of a skeleton)
kgr	0	Keep blobs with area greater than defined limit.
ksm	0	Keep blobs with area smaller than defined limit
lak	0	Obtain the holes (lakes) of a blob in given binary image
lat	1	Local averaging with threshold
lav	1	Local averaging (blurring) filter.
lgr	0	Largest gradient of each 3x3 neighbourhood
lgt	2	Transfer intensities along a line into a Prolog list
lin	1	Normalise orientation so that longest straight side is horizontal
lme	0	Limb ends (of a skeleton)
lmi	4	Geometric centroid & orient. of axis of min. second moment
lnb	0	Largest neighbour

\multicolumn{3}{c}{**Image Processing Primitives**}		
Mnemonic	**Arity**	**Description**
log	0	Logarithmic intensity transformation
lpc	1	Laplacian operator (4 or 8-neighbour)
lpf	0	Low Pass (Blurring) Filter (3* window)
lpt	2	Find co-ordinates of centre of left-most chord
lrt	0	Left-to-right transform
lut	1	Apply one of the standard look-up tables to the current image
mar	0	Draw the minimum-area rectangle
max	0	Maximum of current and alternate images (pixel by pixel)
mbc	0	Draw minimum bounding circle around a blob
mcn	1	Multiply each pixel intensity by a constant
mdf	1	Median filter
mdl	0	Skeleton. Synonym for *ske*
min	0	Minimum of current and alternate images (pixel by pixel)
mma	2	Find lengths of blob projected onto principal axis and axis normal to it
mul	0	Multiply current image by alternate image
ndo	0	Numerate (shade) distinct objects in a binary image
neg	0	Negate image
nlk	0	Normalise position and orientation using largest and second largest lakes
nlk	5	Normalise position and orientation based on lakes A & B
nmr	0	Normalise position; put middle of Min. rect. at centre of image
nnc	0	Nearest Neighbour classifier
not	0	Logical negation of all pixels in a binary image
npo	3	Normalise position and orientation of a blob in a binary image
nxy	2	Normalise [X,Y] position; put centroid at centre of the image
per	1	Perimeter
pex	2	Picture expand (increase image size)
pfx	3	Set the pixel whose address is (X, Y) to level G
pic	8	Save/load named image
pis	0	Push an image onto the stack
plt	1	Plot the intensity profile along a specified column
pop	0	Remove image from top of stack and put it into current image
psh	2	Picture shift
psk	0	Push an image onto the stack
psq	2	Picture squeeze (reduce image size)
psw	2	Picture shift with wrap around
ptc	0	Polar to Cartesian axis transformation
pth	2	Percentage threshold
raf	1	Repeated averaging filter
rbi	0	Recover both current and alternate images from image stack
rea	1	Read image
red	0	Roberts edge detector
rim	1	Read image from RAM disc
rlc	0	Row run length coding
roa	0	Rotate image counterclockwise by 90°

\multicolumn{3}{c	}{Image Processing Primitives}
Mnemonic	Arity	Description
roc	0	Rotate image clockwise by 90°
rpt	2	Find co-ordinates of centre of right-most chord
rsh	0	Copy intensity vertically from bottom
sbi	0	Save both current and alternate images on image stack
sca	1	Reduce number of bits in each intensity value to A
sco	0	Circular wedge
sed	0	Sobel edge detector
set	0	Set every pixel in the image to white (level 255)
shf	1	Shape factor
shp	1	Sharpen image
sim	2	Generate a new image display window
skw	0	Shrink white regions
snb	0	Smallest neighbour
sqr	0	Square all intensities
sqt	0	Square-root of all intensities
ssk	0	View (see) the images on the stack
sub	0	Subtract images
swi	0	Swap current and alternative image
tbt	0	Flip the vertical axis of the image
thr	2	Threshold
tpt	2	Find co-ordinates of centre of top-most chord
tsk	0	Copy image at top of stack into current image
tur	1	Rotate an image by A degrees about its centre point
usm	1	Unsharp masking (High pass filter)
vgr	0	Vertical gradient
vgt	1	Store grey-levels along RHS of image in a Prolog list
vid	0	Vertical intensity difference operator
vpl	5	Draw a digital straight line
vpt	1	Set intensities along RHS of image to values in Prolog list
vsk	1	Transfer the A-th image on the stack in the current image
vsm	0	Vertical smoothing
wdg	1	Draw an intensity wedge
wgx	0	Draw an intensity wedge
wim	1	Save image in RAM disc
wri	1	Write image in RAM
wrm	0	Remove isolated white pixels
xor	0	Exclusive OR of current and alternate images
yxt	0	Transpose the image axes
zer	0	Make image black

\multicolumn{3}{c}{Synonyms used in this Book}
Predicate	Arity	Description
angle	6	Synonym for *ang*
bays	0	Synonym for *bays*
biggest	0	Synonym for *big*
big_blobs	1	Synonym for *kgr*
count	2	Similar in operation to *cbl* but can also counts other features e.g. bays, lakes
crack	0	Synonym for *crk*
draw_disc	3	Draw a solid white disc of given radius in a given position
draw_one_disc	3	Synonym for *draw_disc*
fetch	1	Similar in operation to *rea(saved_im)*
fit_circle	9	Fit a circle to 3 points
keep	1	Similar in operation to *wri(saved_im)*
label blobs	0	Synonym for *ndo*
speak	1	Synonym for *utter*
normalise	0	Synonym for *npo*

\multicolumn{3}{c}{Colour Image Processing, Partial Listing}
Predicate	Arity	Description
colour_scattergram	0	Calculate the colour scattergram
colour_similarity	2	Program the PCF to measure similarity to a defined colour
create_filter	0	Program the colour filter from the current image
draw_triangle		Draw outline of the colour triangle in the current image
generalise_colour	0	Colour generalisation
grab_3_images	0	Digitise the RGB colour separations as 3 distinct images
hue	0	Program the PCF for the *hue* filter
initialise_pcf_lut	0	Set all PCF LUT values to zero (black) initially
normal_pcf	0	Reset PCF to normal operation for monochrome image processing
pseudo_colour	1	Pseudo-colour on/off
redness	0	Program the PCF to measure *redness*
saturation	0	Program the PCF to measure *saturation*
video	1	Select R, G, B or monochrome channel for future *grb* operations

\multicolumn{3}{c}{**Gauge Predicates** [Bat-91]}
Predicate	**Arity**	**Description**
balloon	5	Get co-ordinates of closest white pixel to a given point. For each radius, scan anti-clockwise, starting at 3 o'clock position.
circle	9	Calculate centre & radius of circle passing through 3 given points.
compass	6	Get co-ordinates of first white pixel on circumference of circle, given its centre and radius. Start searching at 3 o'clock position.
edge	7	Find position of largest absolute value of gradient along given line.
fan	7	Get co-ordinates of closest white pixel found in fan-shaped search area, given fan position, orientation and spread parameter
gap	8	Get position of minimum & maximum intensity gradient along the line joining two given points
lmn	6	Find position of smallest value of intensity along line defined by its two end points
lmx	6	Find position of largest value of intensity along line defined by its two end points
mid_point	6	Calculate the mid-point of line joining 2 given points.
perdendicular_bisector	8	Given 4 points (A, B, C, & D), check that line [C,D] is perpendicular to line [A,B] and *vice versa*.
protractor	6	Get co-ordinates of the first white pixel encountered along a line, given it starting points and orientation.
triangle	9	Calculate perimeter, perpendicular height and area of triangle defined by co-ordinates of its vertices

Properties and Relationships Between Objects in Images

Predicate	Description
about_same	Are two numbers about the same. Tolerance is specified by user.
about_same_horizontal	Test whether two blobs are at about the same horizontal position
about_same_vertical	Test whether two blobs are at about the same vertical position
above	Test whether one blob is above another
adjacent	Are two named objects adjacent to one another?
below	Test whether one blob is below another
bigger	Test whether one blob has larger area than another
brighter	Test whether one point is brighter than another
circular	Is given blob approximately circular
concentric	Are centroids of two named objects at same position?
connected	Are two given points parts of the same blob? Are they 8-connected?
contains	Test whether one blob is inside another
convex	Is object convex
darker	Test whether one point is darker than another
encloses	Synonym for *contains*
inside	Test whether one blob is inside another
left	Test whether one blob is to the left of another
parallel	Are two lines defined by their end points parallel?
right	Test whether one blob is to the right of another
right_angle	Are two lines specified by their end points at right angles?
smaller	Test whether one named blob is smaller than another
straight_line	Is arc with specified end points a straight line?
top_of	Is one named object in top part of another named object?

Operators and Control Predicates

Operator	Description
&	AND operator (infix) - can be used in lieu of ',' in compound goals
->	Conditional evaluation of a goal
case	Conditional evaluation of list of goals
for	FOR i = N1 STEP N2 UNTIL N3 DO GOAL
if	Use in lieu of ':-' in defining Prolog clauses (infix operator)
if_then	Conditional evaluation of a goal. Synonymous with '→' operator.
if_then_else	IF P THEN Q ELSE R
or	OR, use *in lieu* of ';' in definitions of compound goals. (Infix operator)
•	Repeat defined goal a given number of times. (e.g. 6•lpf)
¶ (prefix)	Device control operator. Used to operate MMB interfacing unit
△(A,B)	Send given Prolog goal to remote computer.

Miscellaneous	
Predicate	**Description**
cut	Use "!,cut" in lieu of "!" when using speech synthesiser to follow program flow
do_it	Performs an operation on behalf of HyperCard. Normally used only in conjunction with HyperCard
fails	Use lieu of *fail* when using speech synthesiser to follow program flow
gob_init	Initialise *gob*
gob_modified	Similar to *gob* but can be used inside recursive loop
help	Switch to PIP manual (Bring HyperCard to the front)
recursive_eab	Similar to *eab* but can be used inside recursive loop
repeats	Use *in lieu* of *repeat* when using speech synthesiser to follow program flow
utter	Use speech synthesiser to say phrase or list of phrases

Controlling External Devices	
Predicate	**Description**
all_lights	Switch all lights on/off
aperture	Set aperture of selected camera to defined value
calibrate_axes	Calculate mapping function parameters between (X,Y,Theta)-table and camera co-ordinates
camera_state	Find camera state-vector
convert_axes	Convert between (X,Y,Theta)-table and camera co-ordinate axes.
focus	Set focus of selected camera to defined value
grasp	Operate FIC gripper (Suction on)
home	Send the (X,Y,theta)-table to its home position.
in	Put pick-and-place arm in the IN position
input_port	Find state of a given input port
laser	Switch the laser light stripe generator on/off
light	Set given lamp to defined brightness level.
move_to	Move (X,Y,Theta)-table to given point and orientation
nudge	Move (X,Y,Theta)-table by a defined amount.
out	Put pick-and-place arm in OUT position
output_port	Set given parallel output port to defined bit pattern
pan(A)	Adjust pan of selected camera.
pick	Pick up object from the (X,Y,Theta)-table using pick-and-place arm
place	Place object on the (X,Y,Theta)-table
projector(A)	Switch the slide projector on/off
release	Release FIC gripper
select_camera	Choose camera
table_at	Where is the (X,Y,Theta)-table?
tilt(A)	Set tilt of selected camera.
up	Put pick-and-place arm in the UP position
utter(A)	Utter the phrase or list of phrases defined by "input" parameter.
zoom(A)	Set zoom of selected camera

Glossary of Terms

Algorithm, A well-defined set of rules for performing a particular mathematical calculation. (c.f. heuristic)

Aliasing, Phenomenon which gives rise to spurious results if an image is sampled at a spatial frequency below the Nyquist limit.

Analogue to digital converter (ADC), An electronic hardware device which converts an analogue voltage into a digital representation of the value of that voltage. An ADC is characterised by its resolution (i.e. the number of bits used to represent the voltage) and its conversion time.

Anamorphic mirror. A mirror that produces different magnifications along different directions in the image plane.

Aperture, The aperture controls the amount of light passing through a lens. The *F number* scale (1.4, 2, 2.8, 4, 5.6, 8, 11, 16) for a standard photographic lens reduces the amount of light passing through the lens by half with each step increase in the scale.

Astigmatism, Optical aberration in which, instead of a point image being formed of a point object, two short line images are produced at right angles to each other.

Autocollimation. A procedure for collimating an optical instrument with variable objective lens and cross hairs. The instrument is directed towards a plane mirror and the cross hairs and lens are adjusted so that the cross hairs coincide with their reflected image.

Autocorrelatlon function. See Section 2.7.

Automatic gain control. Attribute of a circuit (e.g. video amplifier) whose gain is automatically adjusted to allow for changes (in ambient light).

Auto iris lens. Lens whose aperture automatically adjusts itself to allow for variations in the ambient light.

Back focal length. The distance from the rear surface of a lens to its focal plane.

Beam expander. An optical system for increasing the width of a light beam.

Beam splitter. A partially silvered or aluminised mirror which splits an incident beam of light into a reflected beam and a perpendicular transmitted beam. Other forms of beam splitter are also available.

Binary image. An image in which each pixel can be represented by only the two binary digits 0 or 1, (i.e. black or white). (See Chapter 2.)

Blooming. An effect by which a highly illuminated point image on an image sensor spreads out to form a disc; caused by the high intensity of the incident beam saturating the image sensor at that point.

Borescope. A telescope in the form of a straight tube containing a mirror or prism used for inspecting cylindrical cavities. Also called endoscope or intrascope.

Buried channel CCD. Type of CCD with a buried layer of doping material which together with the electrodes causes the charge packets to move below the surface; giving a high-charge transfer efficiency.

Byte. Unit of information or memory size equal to eight bits; memory size is normally measured in kilobytes (1024 bytes) or megabytes.

"C" mount. This is a 1" diameter threaded standard lens mount. Preferred mount in industrial applications due to its size and weight.

Carbon dioxide laser. A powerful, continuous, infrared laser that can emit several hundred watts at a wavelength of 10.6µm. Used for welding and cutting applications.

Chain code (Digital). Code used for describing a curve such as the periphery of an object. Each discrete point on the curve is represented by an integer from 0 to 7, representing the direction of the next point as an angle ranging from 0° to 315° in 45° steps. (Section 2.3.1)

Charge coupled photodiode array (CCPD). An image sensor which combines the best properties of CCDs (low noise, high clock rate) and photodiode arrays (good spectral response, resistance to blooming), i.e. the image sensors are photodiodes but the scanning function follows the principles of operation of a CCD.

Charged coupled image sensor. CCD in which each element generates a charge proportional to the light intensity falling on it. Associated circuitry moves these charges bodily through an analogue shift register on the same chip to form a serial representation of the incident image at the output.

Charge injection device (CID). A charge transfer device used as an image sensor in which the image points are accessed by reference to their Cartesian co-ordinates. CIDs have low dark current, are resistant to blooming but are relatively noisy.

Charge transfer efficiency. Efficiency with which charges are transferred between neighbouring locations in a CCD.

Chromatic aberration. An optical aberration in which the component colours in a point source of white light are brought to a focus at different points; caused by the variation of the refractive index of the glass with wavelength.

Classification. An object is classified as belonging to some group or class on the basis of the features extracted. (Section 7.2.4)

Closing is a mathematical morphology operator and consists of a combination of erosion and dilation operations. It has the effect of filling in holes and blocking narrow valleys in the image set, when a structuring element (of similar size to the holes and valleys) is applied. It of is the dual morphological operation of opening. (Section 2.4)

Coherent illumination. Monochromatic light with a definite phase relation between different points in space. Applies particularly to laser light.

Collimator. Optical device for producing a parallel beam of light.

Colour is the general name for all sensations arising from the activity of the retina of the eye and its associated nervous system. Colours vary in three different ways: hue, saturation and intensity. (Chapter 6)

Colour cube refers to a 3-dimensional space in which we plot the RGB colour information. (Section 6.5.4)

Coloured is used to signify an object or scene that has some discernible colour, other than *neutral*, which has low values of saturation.

Colour scattergram is a graphical representation of the colour variation to be found in a scene and consists of a set of points superimposed onto the colour triangle. It is convenient, in practice, to plot the colour scattergram as an image, so that it can be filtered, thresholded and processed using the usual image processing software. (See Section 6.5.7.)

Colour temperature. That temperature of a black body which radiates energy with the same spectral distribution as that from a given surface.

Colour triangle is an abstract geometric figure that is useful when analysing and discussing colours. Each point in the triangle corresponds to a different colour in the scene being viewed. Points that are close together usually have similar colours. (See Section 6.5.4.)

Coma. Optical aberration of an optical system which gives a point object a pear-shaped image.

Connectivity. Topological property of a binary image relating to the number of 'holes', or 'lakes', it contains. (Section 2.5)

Contrast. The difference in light intensity between two adjacent image points, normalised by dividing by the sum of those intensities.

Convex deficiency. The set of points within a convex hull that are not in the object. It includes *lakes* (regions totally enclosed by an object), and *bays* (regions lying between the convex hull perimeter and the object). (Section 2.3)

Convex hull. Given an arbitrary two-dimensional shape, the perimeter of its convex hull could be obtained by stretching a rubber band around the shape.

Correlation, two-dimensional. An image-processing operation used to search for a particular image pattern within a picture, i.e. a template matching operation.

Cross talk. A process by which an unwanted signal is induced in a circuit because of its close proximity to a neighbouring circuit; can be applied to adjacent image sensors in a solid-state array.

Dark current. Current that flows in the output circuit of an image sensor even in the absence of illumination.

Dark-field illumination. A method of microscope illumination in which the illuminating beam is a hollow cone of light formed by an opaque stop at the centre of the condenser large enough to prevent direct light from entering the camera's objective lens; the specimen is placed at the apex of the cone and is seen only with light scattered, diffracted or refracted by the specimen.

Depth of field. The range of object distances over which a camera gives a sufficiently sharp image.

Depth of focus. The range of image distances over which the image remains sharp for a given object distance.

Descriptive syntactic process. A pattern recognition technique which models an object by a set of features and by the spatial relationships between these features. (Section 3.5.2)

Diffraction. Wave phenomenon of light whereby the intensity distribution in a wave is spatially redistributed after encountering an obstacle. It accounts for the ability of waves to go round corners.

Diffuser. Translucent material, e.g. polypropylene, used to produce diffuse illumination.

Digital to analogue converter (DAC). A piece of electronic hardware, typically a single chip, used to convert a binary number into an analogue voltage.

Digitisation. The conversion of an analogue or continuous waveform (typically video) into a sequence of numbers suitable for digital processing. The conversion process is accomplished by an analogue to digital converter (ADC).

Dilation. A mathematical morphological operation (also referred to as filling and growing) which is concerned with the expansion of an image set by a structuring element. (Section 2.4)

Distortion. Defect of an optical system in which magnification varies with angular distance from the axis, causing straight lines to appear curved.

Dual. The duality relationship that exists between certain morphological operators, such as erosion and dilation, means that the equivalent of such an operation can be performed by its dual on the complement image and by taking the complement of the result.

Dye laser. A type of tuneable laser in which the active medium is a dye such as acridine red, with very large molecules.

Dye penetrant. A liquid dye used for detecting cracks or surface defects in non-magnetic materials.

Dynamic aperture. The effective transverse aperture of a linear image sensor which is being mechanically scanned in the transverse direction.

Dynamic range. A characteristic property of any measuring instrument. It is equal to the ratio of the maximum to minimum measurable values of the physical quantity which the instrument was designed to measure.

Edge detection operator. An image-processing operator whose effect is to highlight the edges of an image, e.g. Sobel or Roberts edge detection operators. (Section 2.2.5)

Endoscope. A rigid arrangement of optical fibres with an objective at one end and an eyepiece at the other. Unlike a fibrescope it cannot be bent and is used for direct in-line viewing.

Erosion. This is the dual morphological operation of dilation and is equivalent to the shrinking (or reduction) of the image set by a structuring element. (Section 2.4)

Euler number. Topological property of a binary image equal to the number of distinct 'blobs' minus the number of 'holes'.

Extension tubes Hollow, cylindrical "C" mount devices that can be used to increase the distance between the sensor and the lens, thereby altering its working distance.

Fast Fourier transform (FFT). A particularly fast algorithm for computing the Discrete Fourier transform of a digitised signal. The signal can be a function of distance or time.

Feature extraction is the extraction of image features which are characteristic of the object and which will be used in the classification process.

Fibrescope. An arrangement of optical fibres with an objective at one end and an eyepiece at the other; unlike the endoscope the instrument can be bent as required, to view inaccessible objects.

Field. A complete scan of a TV picture using either odd (or even) numbered lines, i.e. a complete frame consists of two interlaced fields.

Field curvature. Aberration of an optical system causing the surface of sharpest focus to be curved.

Field of view. Described by an angle of arc from side-to-side and top-to-bottom of the scene viewed through a lens of a specific format. It is directly related to the lens focal length and the camera image format.

Filters (Optical) can be used to absorb light of a given wavelength before it reaches the image sensor. Correct use of filters can help to simplify the image processing. The main types of filters used are *colour* filters, but *polarising* filters are frequently used to reduce specular reflection. (Section 6.4.1)

Fluorescence. A process in which a material absorbs electromagnetic radiation at one wavelength and emits it at another, longer wavelength.

Fluorescence microscope. A type of compound microscope in which the specimen on being illuminated by ultraviolet or blue radiation fluoresces and is then viewed in the normal way.

Fluorescent lamp. An example of a cheap non-directional light source. Diffuse, therefore minimising the amount of shadow.

Flying spot laser scanner. A device used for optical inspection where very fine detail is required. It consists of a laser beam which is made to scan the object by mechanical, or electromechanical means, the reflected light being collected by a suitable photodetector.

Focal length. Distance from focal point to principal point of a lens or mirror.

Focal point, focus. Point at which rays parallel to the axis of an optical system converge or from which they appear to diverge. Most optical systems have two principal foci produced by rays incident from the left and from the right.

Frame. One complete TV picture usually representing a snapshot of a moving scene with an effective exposure time of 1/25'th (Europe) or 1/30'th (USA) of a second.

Frame-store. An electronic memory used for storing one or more digital representations of an image. The storage process must be fast enough to occur in real-time.

Frame transfer. A term applied to a particular type of CCD image sensor which has special scanning circuitry to minimise image smear.

Fresnel lens. A thin lens constructed in the form of a series of stepped concentric segments thus giving the optical properties of a thick lens.

Gallium arsenide laser. A laser that emits infrared radiation ($\lambda = 900$ nm) at right angles to a junction region in gallium arsenide. Can be modulated directly at microwave frequencies. Cryogenic cooling is required.

Gamma correction. The photo-electrical response I of some TV cameras is a non-linear function of the incident light intensity E of the form: $I = const.E^\gamma$. Any attempt to correct for this non-linearity using either hardware or software is called gamma correction.

Geometric transform. Type of image processing operator in which the transformed image is essentially recognisable but is in some way rotated, distorted or warped. (Section 2.6)

Grey scale. A numerical representation of intensity in which black is usually represented by 0, white by some fixed maximum number (e.g. 255) and shades of grey by intermediate numbers.

Helium-neon laser. Low-power laser in which the lasing medium is a mixture of helium and neon.

Heuristic. A heuristic method is a *"procedure for the solving of well defined mathematical problems by an intuitive approach in which the structure of the problem can be interpreted and exploited intelligently to obtain a reasonable solution"* [SIL-80]. (c.f. algorithm)

HSI Hue, Saturation and Intensity. This is a convenient method of describing colour and is an alternative to the RGB representation, from which the HSI parameters can be calculated.

Hue is the component in the description of colour, which, in effect, defines the name of a colour. Although the terms yellow, red, violet, etc. are defined subjectively, they can be related to the measurement of hue.

Image acquisition is concerned with the generation of a two dimensional array of integer values representing the brightness function of the actual scene at discrete spatial intervals. A frame-store is used for capturing video image.

Image analysis is concerned with the extraction of explicit information regarding the contents of the image.

Image format. Describes the diameter of the light sensitive area of an imager. Possible formats: 1/2", 2/3" and 1".

Image interpretation is concerned with making some decision, based on the information gleaned from image analysis.

Image processing, The principal objective of image processing is to process a given image, so that it is more suitable for a specific application. The term is also used in a generic sense, to include Image Analysis and Image Enhancement.

Image transform. An image processing operator in which each pixel is replaced by a function of, many or, all of the pixels in the original image, e.g. autocorrelation.

Incandescent lamps (Common bulb). Simple cheap light source. Relies on hot filament. Gives directional illumination which causes shadows. Emits a lot of infra-red. Not commonly used in industrial applications.

Incoherent illumination. Light in which the phase relation between different points in space varies randomly.

Infrared (IR). Term applied to that part of the electromagnetic spectrum containing wavelengths which are longer than those for visible red light but shorter than microwaves.

Integration. Solid state sensors are examples of integrating detectors, i.e. each photosensor will accumulate light during a time interval specified by electronic timing signals. For an array device this is generally done for the whole frame time. The resultant signal is proportional to the light level and the exposure time. Therefore, if we expose the sensor quickly, then a higher light level needs to be supplied.

Intensity. This quantity measures the amount of light reflected from a surface. Intensity is not sensitive to colour. (The term is also used to signify the value associated with a given pixel in a digital image. The value of the intensity at a given point may be measured, by a camera, or computed.)

Interferometer. An instrument in which a beam of coherent or partially coherent light is split into two separate beams which travel different paths before being reunited to form an interference pattern. Used for very accurate measurement of distance.

Interlaced scanning. A method for scanning TV images whereby a complete frame is scanned by first scanning the odd-numbered lines followed by the interlaced even-numbered lines. Also see *field* and *frame*.

Interline transfer. A term applied to a particular type of CCD image sensor which has special scanning circuitry to minimise image smear.

Laser. Acronym for *"Light Amplification by Stimulated Emission of Radiation"*, a device which produces a highly coherent, parallel beam of monochromatic light.

Light Emitting Diodes. LED's are often used for illumination in machine vision applications. They provide long life, a fairly even beam of low intensity light and high efficiency.

Linear array. Solid-state array in which the photosensitive elements lie along a line.

Line pairs per mm. Unit of spatial frequency; often used to describe the resolving power of a lens. For example, a good lens can resolve 100 line pairs per mm.

Local operator. An image-processing operator in which each pixel is replaced by a function of its neighbouring pixels. (Section 2.2.3)

Look-up table (LUT). A table of numbers stored in a digital memory used for quick reference. Often used to speed up computer software.

Machine vision The use of devices for optical, non-contact sensing to automatically receive and interpret an image of a real scene in order to obtain information and/or control machines or processes [AVA-85].

Mathematical morphology, involves the probing of an image with a template shape, which is called a structuring element, to quantify the manner in which the structuring element fits (or does not fit) within a given image. (Sections 2.4 and 2.5)

Matrix-matrix mapping. General class of image-processing operations in which the matrix representing the result of the operation is a function of the matrix representing the original image. (Section 2.2)

Microprocessor. The central processing unit of a microcomputer, normally fabricated as a large scale integrated circuit.

Modulation transfer function. Modulus of optical transfer function.

Moiré fringes or patterns. Pattern of lines which appears when two patterns of closely spaced lines are superimposed at an appropriate angle.

Monadic point-by-point operator. Image processing operator which acts on only one image. Each pixel in the transformed image is obtained from operations on only the corresponding pixel in the original image.

Monochromatic is used in this book to refer to either light, or the scene being viewed. Monochromatic light contains electro-magnetic radiation of a single wavelength. The term is also used to describe light having a narrow range of wavelengths. For example, that pair of narrow spectral bands giving light from a sodium lamp its characteristic yellow colour would be referred to as being monochromatic. Surfaces that reflect/emit light with a narrower spectrum would also be referred to as being monochromatic.

Monochrome is used to refer to images. (The term *monochromatic* refers to light, or the scene being viewed.) A monochrome image contains no information about colour. The signal from an RGB colour camera generates three separate images, each of which is referred to as being monochrome. In effect, these measure the amount of red, green and blue light in a coloured scene. A colour image can be represented, for the purposes of display, printing, or digital processing, by three monochrome images, called its *colour separations.*

Neutral is used to signify an object or scene that has is composed only of grey, white or black regions. It does not have any colours such as yellow, red, etc. The page of this book appears to be neutral when viewed in natural (i.e. white) light.

N-tuple operator. An image processing operator in which each pixel is replaced by a function of only a selected few (N) of its neighbouring pixels. (Section 2.2.5)

Nyquist limit. Spatial frequency equal to half the sampling frequency. If an image falls on a solid-state array of element spacing d, the image is said to be sampled at a spatial frequency of $1/d$. The Nyquist limit is $1/(2d)$ and using this array it would be impossible to discern spatial detail having a frequency greater than $1/(2d)$.

Opening is a combination of erosion and dilation operations that have the effect of removing isolated points in the image set smaller than the structuring element and those sections of the image set narrower than the structuring element. (Section 2.4.1)

Optical aberration. Any deviation from perfect image formation by an optical system.

Optical character recognition. A branch of technology concerned with the automatic optical scanning of printed words and their subsequent recognition by machine.

Optical transfer function. A complex function of spatial frequency characterising an optical system. It gives a numerical measure, in amplitude and phase, of the extent to which the contrast of object details of a given spatial frequency is degraded in the process of forming the image.

Parallel processor. Computer which has a large number of identical processors each of which can operate on data (such as a digital image) at the same time.

Phosphor. A luminescent material, normally used in cathode ray tubes, which partially converts the energy of an incident electron beam into light energy.

Photodiode. A semiconductor diode in which the reverse or leakage current varies with light intensity.

Photodiode array. Solid-state array in which the photosensitive elements consist of photodiodes.

Photomultiplier. A very sensitive light detector, capable of detecting a single photon. It consists of a light-sensitive cathode, together with a series of dynodes and an anode, in an evacuated glass envelope.

Pixel (Pel). Picture element; the smallest addressable section of a digital image. Within the pixel boundaries the grey level or colour of the image can be considered constant.

Polarised light. Light beam with an electric field vector vibrating in one plane only.

Polar vector representation. Code used for describing a curve such as the periphery of an object. Each "break point" on a piece-wise linear curve is represented by the polar co-ordinates of the next point with the current point as origin.

Polychromatic A polychromatic scene contains a number of monochromatic regions, each one being clearly distinct from all of the others. This is a specialised use in this book. (Section 6.7)

Programmable Colour Filter (PCF) is the term used in this book to refer to a real-time video filtering device, consisting of a look-up table, implemented in a random access memory, RAM, whose inputs are the digitised RGB signals. The contents of the look-up table are generated from a monochrome image. The output of a PCF is a monochrome video image, but this may often be displayed to good effect using pseudo-colour. (Section 6.6)

Pseudo colour-triangle is a computer generated image, superimposed onto the colour triangle. A series of very useful programmable colour filters can be generated by creating pseudo colour-triangles (using an image processor, or graphics program) and then applying the Prolog+ program *create_filter*. (Section 6.6.10)

Quartz halogen lamp. An intense source of white light produced by an electrically heated tungsten filament enclosed in a quartz envelope which contains a halogen vapour, e.g. iodine at low pressure.

Raster scan. The simplest way of scanning an image in which the scanning electron beam starts at the top of the image and proceeds sequentially scanning one line at a time until it reaches the bottom.

Real-time. A process which in some way models a real, live event is said to take place in real-time, if it occurs at the same rate as the real process. For example a frame-store works in real-time if it is able to store digits representing an image at the rate at which they are supplied by an ADC.

Refraction. Change of direction of a ray of light when passing from one transparent medium to another.

Resolution (of a digital image). The number of rows and columns.

Resolution (of an optical system). The smallest angular separation of two object points which produces two distinguishable image points.

Retrofocus lens. Compound lens system consisting of a diverging lens followed by a converging lens (reverse of telephoto principle); this gives a back focal length which is greater than the true focal length, facilitating convenient camera design.

Retrorelective tape. Adhesive tape with a special coating which returns an incident beam of light along its path of incidence.

RGB (Red, Green and Blue). Both photoconductive and solid-state (CCD) colour video cameras use three sets of photo-detectors, behind red, green and blue optical filters. Hence, both types of camera generate RGB signals naturally.

Roberts' edge detection operator. See Section 2.2.

Robot. The formal definition that is generally accepted as defining the functionality of a robot was drawn up by the Robot Institute of America in 1979 and states that a robot is "*a reprogrammable multi-functional manipulator designed to move material, parts, tools or specialised devices through variable programmed motions for the performance of a variety of tasks*".

Run code. A mathematical representation of image, in which each segment of a line scan along, which the grey-level or colour does not change, is represented by the number of pixels in that segment, together with its shade of grey or colour.

Saturation. Colourfulness, or strength of colour. A highly saturated red means that light from only the red part of the spectrum is present. On the other hand, pink is non-saturated, having a considerable amount of white light mixed with red. (Section 6.5.3)

Segmentation is a process that divides an image into its constituent parts or objects. It is a grouping process which identifies regions in the image as being similar, with respect to some defined criterion..

Serial processor. Computer with one processor which performs all operations sequentially. Most current computers are of this type.

Sobel edge detection operator. See Section 2.2.

Solid-state array. Type of image sensor fabricated normally in the form of a linear, or rectangular, array of photosensitive elements, constituting a single integrated circuit.

Spatial frequency. Optical term used as a measure of the amount of detail in an object or image; usually measured as a number of lines per mm.

Speckle. A phenomenon in which the scattering of coherent light by a rough surface, or inhomogeneous medium, generates an interference pattern of random intensity distribution, giving the surface or medium a granular appearance.

Specular reflection. Reflection of light in which the angle of incidence is equal to the angle of reflection. This gives rise to glinting.

Spherical aberration. Optical aberration produced by lenses or mirrors with spherical surfaces. Rays of light parallel to the optic axis, but at different distances from it, are brought to a focus at different points.

Surface channel CCD. Type of CCD in which the potential distribution used to confine the charge packets is created by the electrode voltages only. This gives a poor transfer efficiency.

Synchronisation (synch) pulse. Synchronisation pulses accompany a video signal, to trigger certain crucial events such as the start of a line scan.

Telecentric. A telescopic system whose aperture stop is located at one of the foci of the objective. Such a system is made to accept only collimated light.

Telephoto lens. A lens for imaging distant objects. It is designed to be compact so that the distance from the front of the lens to the image plane is less than the focal length of the lens.

Template. Ideal representation of an object to be found in an image.

Template Matching. Technique for shape matching, which involves the translation of the template to every possible position in the image and finding a measure of the match between the prototype (template) and the image at that position. If the match is within a specified range then the object is assumed to be present.

Thresholding. An image-processing operation which converts a monochrome image into a binary image by setting all those pixels above a certain threshold to 1 (white) and all those pixels below that threshold to 0 (black).

Translucent. Permitting partial passage of light.

Trichromacity. Idea that any observed colour can be created by mixing three different "primary" colours. These may be derived by mixing paint or coloured light.

Tungsten filament lamp Non-uniform "point" source of light which gives low intensity illumination.

Two-dimensional scattergrams. Derived by plotting one colour component against another. For example, we might plot the amount of red light against the amount of blue light, on a point-by-point basis, for a polychromatic scene.

Two-dimensional array. Solid-state array in which the photosensitive elements are arranged in a rectangular array.

Ultraviolet. Term applied to that part of the electromagnetic spectrum containing radiation in the wavelength range of approximately 400 nm to 5 nm, i.e. between visible blue light and X-rays.

Video. Pertaining to visual information. Normally used to describe the output signal of any kind of TV camera.

Vidicon. Generic term for a family of photoconductive camera tubes using a transparent signal plate and a low-velocity scanning electron beam. Advantages include small size and simplicity.

Vignetting. Dark shadows around the corners of the image "seen" by a video camera, due to insufficient coverage by the lens, e.g. a 2/3" lens used on a 1" format sensor.

Weight matrix. A matrix of constant coefficients multiplying the pixel intensities in the definition of a local or N-tuple operator. (Section 2.2.4)

Woods glass. A type of glass that transmits ultraviolet radiation well but is relatively opaque to visible radiation.

Xenon lamp. High intensity arc discharge bulb. Light from it resembles daylight. Fast response - suitable for strobe lights. Although commonly used, these lamps have a number of dangers associated with them. These include: ultra-violet emissions, high flash rates (which can induce photo-sensitive epilepsy), and they require high wattage power supplies (e.g. 14.4W at a 10Hz flash rate).

YAG laser. Yttrium-aluminium-garnet laser. Infrared laser in which the active material consists of neodymium ions in an yttrium-aluminium-garnet crystal; it can provide a continuous power output of several watts.

Additional Glossary Material

The Photonics Dictionary, Photonics Spectra.
    Covers high-technology optics, lasers, fibre optics, electro-optics, imaging and optical computing.

Digital Image Processing [BAX-94]
    Includes a glossary of image processing terms.

Precision Digital Images Corporation glossary of terms.
    Hardware orientated Online glossary
    http://www.precisionimages.com/gloss.htm

Index of Predicates, Operators and Grammar Rules

'<ERROR>'/2, 154
'<INTERRUPT>'/1, 158
'<LOAD>'/1, 106
-> operator (infix), 105
• operator (infix), 103
△ operator (prefix), 152
'S'/1, 128
'U'/1, 110, 113

about_same/3, 89
about_same_vertical/2, 92
above/2, 258
adjacent/2, 92
age/2, 77
alpha_numeric_features/8, 300
analyse_binary_image1/0, 308
analyse_binary_image2/0, 308
analyse_node/1, 330
ancestor/2, 77
anglir/0, 103
apple_logo/0, 254, 256, 257
approximate_colour_scattergram/0, 272
ardal_gwyn/1, 103
area/3, 258

bakewell_tart/0, 81
banana/0, 240
bay_rotate_options/4, 315
below/2, 88, 89
big_changes/1, 79, 81
biggest_bay_top/1, 301
born/2, 76
build_face_image/0, 348
build_menus/0, 113

camera_sees/1, 86
case/2, 104
casual_time/3, 345
check_areas/6, 258
cherry/0, 82

child/2, 76
chirality/2, 336, 340, 341, 342
chirality_database/4, 336
circle/9, 359
circular/0, 83, 353, 354
circularity_tolerance/4, 354
colour_separation/0, 241
colour_similarity1/2, 249
colour_similarity2/2, 249
component/3, 325
concavity_tree/0, 330
concentric/2, 92
connected/4, 92
contours/0, 362
count/2, 254
count_coloured_objects/2, 253
count_limb_ends/1, 87
cover_image/1, 272
crack/1, 104
crude_color_reconition/0, 259
ct_node/3, 332
cull/3, 89

disc_parameters/1, 273
draw_discs/1, 273
draw_polygon/1, 119
draw_sucker/0, 310

eab_modified/2, 330
encloses/2, 91
equal_sets/2, 88
euclidean_distance/4, 255, 309

face/1, 297
fill_occlusions/1, 362
find_object_list/2, 87
find_smile/1, 297
find_wife/1, 74
finished/0, 87
fit_circle/9, 359
for/4, 104

generalise_colour/0, 251
get_colour/1, 265
get_data/2, 184
get_image/1, 258
get_parameters/1, 270
gob_modified/1, 330, 334
grab_and_threshold/0, 79

handedness/1, 324
hue/0, 244

icing/0, 83
if operator (infix), 105
if_then/2, 105
if_then_else/3, 105
inside/2, 92
interactive_hough_analysis/0, 119
interactive_mode/0, 108
interpret/0, 133
interpret1/1, 133
isolate//1, 253
isolate_blob/2, 119
isophotes/0, 104

lamp/2, 184
learn/0, 308
learn_with_masking/0, 244
learning_colour/0, 264
learning_colour1/0, 265
learning_colour2/0, 265
learning_coloured objects/0, 260, 262
left/2, 88, 89
list_all_objects/1, 87
loaf_left_side/1, 358
loaf_right_side/1, 358
loaf_top/3, 359
locate_and_pick/0, 197
lower_case/1, 83

master_program/0, 157
measurements/1, 305
menu_builder/2, 127
menu_item/3, 112
min_blob_size/1, 330
mmb_response/1, 184
morph_pack/0, 318
must_be/2, 74
mwyaf/0, 103

naive_colour_learning/0, 264
ne/0, 103
neg/0, 424
negate/0, 102
next_blob/0, 86
nnc/3, 263, 309
normalise/0, 309
normalise_card/0, 268
normalise_loaf_orientation1/0, 357
normalise_loaf_orientation2/0, 357
number_to_words/2, 346

object_data/2, 306
object_is/1, 86
older//2, 77
one_row_height_data/1, 361
outer_edge/0, 82

packbay/0, 314
pack_bay_1/0, 314
pack_bay_main/0, 315
parameters/1, 157
pcf_with_noise_cleanup/0, 244
pick_up/0, 309
picture/0, 90, 106
playing_card/0, 295
plotit/1, 362
polypack/0, 318
poly_pack_1/0, 318
poly_pack_main/0, 318
preprocess/0, 308
process/0, 81
process_image_sequence/0, 80
pseudo_colour_value/2, 266
parent/2, 76
print_descendents/1, 77

rank_order/3, 16
rebuild_colour_scattergram/0, 273
recognise/0, 308
recognise_alpha_numeric/1, 299
recognise_upper_case/2, 84
rectangle/4, 95
rectangular_biscuit/0, 352
reduce/2, 272
remote_abolish/1, 155
remote_asserta/1, 155
remote_assertz/1, 155

remote_clear_and_assert/1, 155
remote_goal/0, 153
remote_machine_running/0, 154
remote_reset/1, 152
resistor/4, 274
right/2, 75
right_angle/4, 93
rounded_rectangular_biscuits/0, 355
run/2, 153

safe_grasp/0, 198
saturation/0, 248
save_parser_output/1, 131
scan_3_lines/6, 359
select_list_element/3, 105
select_one_blob/0, 308
shape_factor/1, 297
slave_program/0, 157
speech_terms/0, 127
stop/0, 156
stored_features/9, 300
straight_line/2, 93
subtle_colour/1, 249
subtle_colour_changes/0, 250
suit/2, 296

table_place_setting/0, 88
telling_time/3, 345
template_match/(0,1,3), 267, 268
test_face/0, 297

thr/(0,1), 103
threshold/(0,1,2), 103
top_of/2, 92
transparent/0, 153

upper_case/1, 83

value1/2, 274
value3/2, 275
vertical_scan_count/3, 300
visa_card/0, 269, 271

words_to_digits/2, 132

Grammar Rules

amount, 131
article, 130
courtesy, 130
dimension, 131
direction, 130
motion, 130
numeral, 131
preposition, 131
sentence, 130
table_command, 130
table1, 130
table2, 130
teens, 132
units, 132

Index

\# operator, 96
& operator, 105, 432
-> operator, 105, 432
△ operator, 152
@ operator, 98
¶ operator, 180, 432
• operator, 103, 104, 432
11x11 window, 28
1-dimensional histograms, 224
2-dimensional scattergrams, 224
3D structure, 360
3x3 neighbourhood, 22
3x3 processing window, 25
4-adjacency, 42
4-connected, 22
5x5 operator, 27
7x7 operator, 27
8-adjacency, 42
8-connected, 22
8-neighbours, 22

abstract concepts, 90
accumulator array, 52
adaptive visual feedback, 14
add, 24
agricultural products, 349
agri-food processing industry, 349
AI languages, 17
algorithms, 15
ALIS 600, 176, 411
alphanumeric characters, 298
alternate image, 67, 68
alternative solutions, 395
analogue shape measurements, 302
analogue, 162
AND, 35
antilogarithm, 23
AppleEvent, 122, 150
Appletalk, 147
applications, 295
approximating an image, 271
arguments, 103
arity, 103

artificial intelligence (AI), 17, 69
assignment operator, 21
auto-correlation, 56
automated material handling, 321
automatic self-calibration, 185
automation, 2, 6-7
auto-start, 106
Autoview, 66, 72

back-projection, 231
back-tracking, 75
band-pass filters, 213
barrel distortion, 51
bays, 41, 298, 326
binary image, 35, 41
binary mathematical morphology, 43
biscuits, 352
blackboard, 150, 154
blob packing, 314
blurring, 28
Boolean, 35
BS 950, 218

calibration, 185, 186, 190, 191, 193, 303
cameras, 163, 179, 215
canonical form, 334, 335
capacitor colour codes, 273
Cartesian to Polar, 51
cathode-ray tube, 215
CCD, 175
CCIR, 162
centroid, 42
chain code, 42
chirality, 323, 336, 339
chromatic adaptation, 212
chrominance amplitude (I), 220
chrominance angle (Q), 220
chrominance, 219, 220
circular food products, 354
circularity, 43
classification, 56
closed-loop process-control, 382
closing, 46, 50

coarse colour discrimination, 207
coding colour, 219
colour, 20, 206–207, 209
   atlas, 218
   differences, 207
   discrimination, 207
   filter, 234
   generalisation, 250
   histogram, 247
   image processing, 207, 430
   machine vision systems, 209
   manufacturing industry, 206
   mis-registration, 12
   object orientation, 269
   perception, 211
   proportions, 260, 262
   recognition, 252
   representation, 220
   reproduction, 213
   saturation, 248
   scattergram, 224, 226, 229, 231, 241, 242
   science, 411
   sensors, 207
   separations, 224
   similarity, 248
   spaces, 410
   standards, 218
   triangle, 221, 242
command keys, 113, 114
command-line, 70
commercial devices, 61
commercial systems, 162
commissioning, 4
complement, 45
composite video, 215
computer bus, 162
computer vision, 405, 409, 411
concavity trees, 323, 326, 329, 336, 339
conference proceedings, 409
conferences, 411
connectivity detector, 36
CONSIGHT, 352
context, 9
contours, 362
controlling external devices, 160
convex deficiency, 41, 328, 355
convex hull, 41, 328
Co-occurrence matrix, 57

corner detection, 40
count ratio, 319
count white neighbours, 36
counting coloured objects, 252
courseware, 411
crack detector, 29
create_filter, 231
cumulative histogram, 34
current image, 67, 68
customer, 392

dark adaptation, 212
decision surfaces, 259
declarative programming, 73, 213, 277
decomposition, 47
de-instantiation, 75
depletion, 311
depth map, 360, 362
describing a simple package, 90
design aids, 143, 411
design software, 411, 412
designing vision systems, 134, 401
determining orientation, 356
device control, 183
Dichroism, 212
digital image, 19
dilation decomposition, 47
dilation, 44, 49
direction codes, 31
Discrete Fourier Transform, 53
discrimination tasks, 296
distortion, 51
dual, 45
dyadic, 24

edge density, 57
edge detector, 28-31, 35
edge effects, 32
edge smoothing, 40
edge, 26
electro-magnetic spectrum, 209, 210
energy, 58
entropy, 58
environmental protection, 398
equalisation, 34
equipment suppliers, 413
erosion, 45, 50
Euclidean distance, 256, 258, 307
Euler number, 36

Exclusive OR, 35
expand white areas, 35
expert systems, 382
exponential, 23
extendible menus, 108, 109
external devices, 160, 433

faces, 297
factory, 399
FIC, 115, 122, 135, 160, 171, 177, 183, 303
filling holes, 37, 44
filter, 38, 49, 50, 213, 234
financial, 393
finite state machines, 61
FLEX, 159
Flexible Inspection Cell (FIC), 115, 122, 135, 160, 171, 177, 183, 303
flexible manufacturing systems, 6
folklore, 390
food processing industry, 350
food products, 208, 349, 254
Fourier spectral analysis, 56
Fourier transform, 54
frame-stores, 62
Freeman code, 42
Frei and Chen edge detector, 31
FTP sites, 410

Gantry robot, 187
gauge predicates, 431
general purpose interface unit, 169
generalise_colour, 251
generating concavity trees, 329
geometric figures, 93
geometric packer, 311, 312
geometric transforms, 51
global image transform, 51
global memory, 98
golden sample, 382
grammar rule, 129, 134
granularity, 60
graphical display, 115
grass-fire transform, 39
grey scale, 19
   closing, 50
   dilation., 49
   erosion, 50
   filter, 38

morphology, 49
opening, 50
gripper footprint, 321
gripper, 198, 199, 201, 321
growing, 44

handedness, 323
help, 117, 118
heuristic packer, 311, 313
heuristics, 15
highlight, 24
high-pass filtering, 26
histogram, 33-34, 56, 224, 247
   equalisation, 34
   features, 57
   intensity, 33
Hough transform, 51, 119, 317, 347, 355, 356
HSI, 221
Hue PCF, 244, 246
hue, 219, 221, 246
human analogies, 1
human colour perception, 1, 211
human vision, 1
Hypercard, 115, 122, 135-136, 183
hypermedia, 135

idempotency, 46
identifying table cutlery, 84
if operator, 105
illumination, 401
image acquisition, 352
image processing, 21, 66, 243, 402, 409, 415, 417, 423, 425
image recognition, 206
image resolution, 397
imperative languages, 74
index card, 137, 139
industrial applications, 380
industrial examples, 1, 10
inertia, 59
inspection, 2
instability in concavity trees, 339
instability, 338
instantiation, 75
integrate intensities, 51
Intelligent Camera, 96, 148, 173
intelligent systems, 13, 101, 145
intelligent vision, 13, 295

intensity, 221
   histogram, 33
   maximum, 34
   mean, 34
   minimum, 34,
   multiply, 23
   normalisation, 22
   shift, 22
interactive image processing, 66, 70, 108
interfacing, 162
interference filters, 214
internet resources, 409
interpreting colour codes, 273
inverse, 35
IQ plane, 220
isophotes, 362

joints, 298
journals, 408

Khoros, 411

lakes, 41, 298, 326
largest intensity neighbourhood, 28
learning, 254, 260, 262, 303, 306
lens selection, 143, 144
lenses, 163, 179
lidded tin loaf, 355
light absorption, 209
light sources, 216
light stripe sectioning, 360
Lighting Advisor, 135, 381, 410
lighting, 135. 164, 173, 177, 344, 397, 411
lighting-viewing subsystem, 413
light-stripe generator, 360
limb-ends, 298
limitations, 4
linear local operator, 25, 32
loaf, 355, 360
local area histogram equalisation, 34
local colour changes, 249
local memory, 98
local operators, 25
locating straight sides, 357
logarithm, 23
logical shape measurements, 302
low-pass filtering, 26
luminance, 219

machine intelligence, 9
machine vision, 1, 3, 19, 161, 209, 380, 404
Macintosh, 71
MacProlog, 72, 108
macros, 69
magazines, 408
magnetic gripper, 198
mailing lists, 410
manipulation of planar objects, 303
manipulation, 194
map card, 139
MAR, 354
master-slave system, 147, 150, 155
MAT, 40
mathematical morphology, 43
maximum intensity, 25, 34
Maximum Similarity Classifier, 307
Maxwell triangle, 221
mean intensity, 34
measurements, 41
measuring overspill, 358
mechanical actuators, 164
mechanical handling, 178, 396, 401
Medial Axis Transform (MAT), 40
median filter, 29
meta-concavities, 328
Metamerism, 212
meta-meta-concavities, 328
method card, 137
Micro-ALIS, 176, 412
minimum area rectangle (MAR), 354, 355
minimum bounding circle, 355
minimum intensity, 25, 34
mirror-image components, 323
mirrors (as lenses), 402
MMB, 169, 172, 179-180, 182
monadic, 22
monochromator, 218, 219
monochrome image, 19, 20
morphological filter, 49, 50
morphological imaging techniques, 312
morphological packing, 312
morphological texture analysis, 59
morphology, 49
moving-needle meters, 343
multi-camera, 144, 145, 382
multi-finger gripper, 199, 201, 321
multi-joint robot, 187
multiple colours, 236

multiple exemplar approach, 258
multiplexed-video, 144
multiply, 24
multi-processor, 382
multispectral image, 20
Munsell Book of Colour, 218

naming of a colour, 210
natural language understanding, 124, 127
natural textures, 56
Nearest Neighbour Classifier, 307
negate, 23
networked vision systems, 147
newsgroups, 409
noise effects, 235
non-lidded tin loaf, 355
non-linear N-tuple operators, 32
normalisation, 22
NPL Crawford method, 217
N-tuple operators, 32

object orientation, 269
occlusion, 360
OCR, 295
onion-peeling, 39
on-line help, 117
opening, 46, 50
operating the PCF, 239
opinions, 390
Opponent Process Representation, 220
optical analysis, 412
optical character recognition (OCR), 295
optical design, 412
optical filters, 213
optics, 397, 402, 412
Opto*Sense®, 143, 413
OR operator, 35, 105
orientation, 356
overlapping discs, 271

packing, 311
   defective regions, 322
   density, 319
   Heuristic, 311, 313
   Polygon, 316
   strategy, 311
parallel decomposition, 47
parallel digital bus, 162
Parlog, 159

parser, 129, 130, 132
passing parameters, 420
PCF implementation, 228
PCF, 227–229, 237–239, 242–243, 246, 263
pel, 19
perception, 211
performance index, 319
performance measures, 319
performance of vision systems, 5
periodicals, 408
pick-and-place arm, 122
picture element (pel), 19
pin-cushion distortion, 51
PIP, 380, 414, 425
pixel, 19
planar objects, 303
plastic bottle top inspection, 11
playing cards, 295
PLC, 164, 166, 167
plot_scattergram, 243
plug-in boards, 62, 63
point-by-point operator, 22, 24
Polar-cartesian axis transformation, 347
polychromatic logo, 254, 256, 257
polygon packing, 316
predicate names, 102
presentation, 401
Prewitt edge detector, 30
primary colours, 210
primitives, 425
printed patterns, 295
process control, 7, 8, 382
processing, 21, 66, 70, 108, 243, 402, 409
product shape, 351
program library, 106
Programmable Colour Filter (PCF), 227–229, 238–239
Programmable Logic Controller (PLC), 164, 166
Programming the PCF, 229, 242, 243
programming, 73, 213, 277
Prolog, 70, 73, 418, 422
Prolog+ Programs, 252
Prolog+, 71, 79, 96, 102, 183, 239, 252, 380, 414, 423, 425
Prolog++, 159
properties of concavity trees, 336
protocols and signals, 161

prototyping, 67
proverbs, 390
pseudo-colour display, 237
pull-down menus, 70, 108, 134, 239

Ra8 index, 218
radius of curvature, 358
randomly placed object, 194, 195
rank filters, 31
ray-tracing, 144, 412
recognition, 206, 252, 295, 303, 306
    bakewell tarts, 81
    digits, 299, 301
    multiple colours, 236
    printed letters, 83
    single colour, 235
    table place setting, 87
recorded speech, 116
rectangles, 95
recursion, 75, 326
reference material, 404
region labelling, 37
related disciplines, 398
remote process, 153
remote queries, 152
remove isolated white points, 35
representations of images, 19
resistor colour codes, 273
resolution, 397
review papers, 407
RGB, 215, 220, 221, 224
Roberts edge detector, 29
robot, 7, 176, 187, 410
    gripper, 321
    guidance, 201
    languages, 176
    vision, 2, 409
    work cell, 115
row maximum, 51
RS170, 162

saturation, 219, 221, 248
SCARA robot, 187
scattergram, 224, 226, 229, 231, 241, 242
script generation, 120
Segmented PCF, 248
self-adaptive learning, 254
self-calibration, 185
self-contained systems, 63

sensor, 163, 207, 402
separable operator, 27
serial decomposition, 47
serial digital data, 161
set theory, 43
shading blobs, 38
shape descriptors, 43
shape measurements, 302, 304
shape ratio, 355
shrink white areas, 35
similarity, 307
single colour, 235
skeleton, 39
SKIPSM, 61
slaves, 147
slices of bread, 355
small spots, 37
Sobel edge detector, 29
software, 64, 412–414
solid-state colour camera, 215, 216
space usage, 319
spatial dependency statistics, 57
spatial relationships, 91
spatial resolution, 20
special issues, 407
spectral analysis, 56
spectral transmission, 213
spectrum, 209, 210
speech input, 101, 124
speech manager, 116
speech recognition, 124, 126, 133
speech synthesis, 116
speech, 116
spoken instructions, 124
squaring, 24
standards, 218
stars, 296
stock cutting, 322
streak, 26
structured lighting, 360, 361
structuring element 44, 47
subtract, 24, 344, 348
suction gripper, 198
SuperVision, 72
survey papers, 407
SUSIE, 66, 72
System 77, 66
system, 391
    design samples, 394

engineering, 2, 9
implementation, 61
level, 403
specification, 393
synchronisation, 161, 167

teaching by showing, 263
teaching the PCF, 238, 263
television, 20
telling the time, 343
template matching, 266, 267
texels, 59
texture, 56
   analysis, 34, 56, 59
   attributes, 58
   classification, 56
   measures, 58
three-dimensional imaging, 409
threshold, 23
Thyristor, 174
tiling, 47
tone-texture concept, 56
training the PCF, 238
training course, 413
transform, 39, 50, 51, 119, 317, 347, 355, 356
transpose, 45
triangulation, 360

Trichromacity, 210, 211
turn-key systems, 64

union decomposition, 47
unsolved problems, 382

valley, 34
VCS, 66, 72
Very Simple Prolog+ (VSP), 98
video, 162, 215
Video-Multiplexed (V-M), 144
Vidicon, 215
vision company, 395
vision system co-ordinates, 188
Vision Systems Group (DCU), 410
V-M, 144
VN, 124, 133
Voice Navigator (VN), 124, 133
VSP, 98

warping, 51
weight matrix, 25, 27
World Wide Web, 410
WWW, 410

X,Y,θ-table, 115, 122, 123, 129, 178, 188, 303

YIQ, 220